ELEMENTARY
ANALYSIS

FURTHER
ELEMENTARY ANALYSIS

BY R. I. PORTER

Fourth Edition (Metricated)

This continuation of *Elementary Analysis* completes a two-year Sixth Form course in Pure Mathematics for all but the mathematical specialist.

FURTHER MATHEMATICS

BY R. I. PORTER

Second Edition (Metricated)

This third volume completes the course for mathematical specialists and extends it to S level of the G.C.E. examination.

Details from

BELL & HYMAN LIMITED

Denmark House
37/39 Queen Elizabeth Street
London SE1 2QB

ELEMENTARY ANALYSIS

BY

A. DAKIN, O.B.E., M.A., B.Sc.

and

R. I. PORTER, M.B.E., M.A.

BELL & HYMAN

London

This edition
first published in 1980 by
BELL & HYMAN LIMITED
Denmark House
37–39 Queen Elizabeth Street
London SE1 2QB

Fourth edition: amended and enlarged
Reprinted 1981, 1982

ISBN 0 7135 1848 0

Typeset at The Universities Press (Belfast) Ltd.
Printed and bound in Great Britain
at The Pitman Press, Bath

PREFACE TO AMENDED AND ENLARGED EDITION

The object of this book is to introduce pupils, who have completed the elementary course in Secondary Mathematics, as quickly as possible to fresh Mathematics fields and to make them acquainted with new Mathematical ideas. We hope that it will provide a suitable year's work for students undertaking an A level Mathematics course and a complete course for those students undertaking Additional Mathematics at O level. It will be found that the topics covered are fundamental and are included in the syllabuses for Additional and A level Mathematics of virtually all the examining Boards.

The book contains a large number of examples, most of which are original but we are indebted to the Joint Matriculation Board and to the University of London for permission to include certain examples which are marked JMB and LU respectively. The examples are divided into two kinds—those marked (A) which are easy questions of a routine type and those marked (B) which are questions of greater length and difficulty.

Several of the topics are independent of previous content and can be dealt with at any stage of the course. For example, many teachers will want to give their students some facility in handling surds and simple inequalities at an early stage and these may not have been covered by the O level or Certificate Syllabus.

I am especially grateful to my daughter Mrs Elizabeth Lucas, B.Sc., herself a teacher of Mathematics, for undertaking the task of reviewing the new material in this enlarged edition, offering constructive criticism on it and checking its content as well as the answers to the additional exercises.

The book is fully metricated and SI units are used throughout.

R.I.P.

CONTENTS

CHAPTER 1

Co-ordinates of a point. Equation of a curve. Distance between two points. Midpoint of line joining two points. Area of a triangle and of any polygon. Implicit and explicit functions.

CHAPTER 2

Gradient of a straight line. Interpretation of the equations, $y = mx$ and $y = mx + c$. The equation of a straight line having a given gradient and passing through a given point. The equation of the straight line passing through two given points. Parallel and perpendicular lines. The straight line law.

CHAPTER 3

Numerical method of determining the gradient of a curve at a given point. Gradient of a curve at any point. The δ notation. The differential coefficient as a gradient. Differentiation from first principles.

CHAPTER 4

The differential coefficient of cx^n, where n is an integer and c is a constant. Differentiation of sums and differences. Tangents and normals. Subtangent and subnormal.

CHAPTER 5

Velocity and acceleration of a point moving in a straight line. Rates of change. Approximations.

CONTENTS

of exponential and logarithmic functions. Integration of functions of the forms e^{ax+b} and $\dfrac{1}{ax+b}$.

CHAPTER 23

FURTHER ALGEBRAIC METHODS 325

Remainder and factor theorems. Manipulation of surds. Finite series and the Σ notation. Mathematical induction. Algebraic inequalities. Graphical representation of a number of simultaneous linear inequalities. Asymptotes parallel to the axes. Rational algebraic functions.

CHAPTER 1

CO-ORDINATES OF A POINT
EQUATION OF A CURVE

Rectangular Co-ordinates

The position of a point in a plane is determined if its distances from two fixed lines intersecting at right angles are known.

The fixed lines OX, OY are called *rectangular axes*.

The usual sign convention is used, as indicated in the diagram.

The distance of a point from the y axis is called the x *co-ordinate* of the point.

The distance of a point from the x axis is called the y *co-ordinate* of the point.

NOTATION. The point P with x co-ordinate α and y co-ordinate β, is represented as the point (α, β).

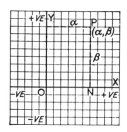

FIG. 1

The x co-ordinate of P is sometimes called the *abscissa* of P and the y co-ordinate is then called the *ordinate* of P. In Fig. 1, ON is the abscissa and NP the ordinate of P.

Ex. 1. *Take a set of axes with suitable scales and indicate the positions of the following points:*

A(1, 3), B(−2, 2), C(3, 0), D(−3, −4).

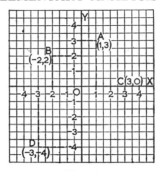

FIG. 2

The distance between two points whose co-ordinates are given

Ex. 2. *Find the distances between the points*

    (*a*) $A(1, 3)$ *and* $B(3, 5)$;   (*b*) $A(-1, 2)$ *and* $B(2, -3)$.

(*a*) Draw the ordinates AM and BN and draw AC parallel to OX to meet BN in C (Fig. 3). Then

$$AC = MN = ON - OM,$$

    = difference between abscissae
        of B and A,

    $= 3 - 1 = 2.$

$$BC = BN - CN = BN - AM,$$

    = difference between ordinates
        of B and A,

    $= 5 - 3 = 2.$

But

$$AB^2 = AC^2 + BC^2 = 2^2 + 2^2 = 8.$$

$$\therefore AB = 2\sqrt{2} \text{ units.}$$

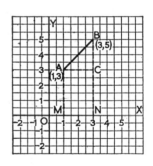

FIG. 3

(*b*) From the diagram (Fig. 4)

$$AC = AM + MC$$

$$= 2 + 3 = 5,$$

and $CB = MO + ON$

$$= 1 + 2 = 3.$$

$$\therefore AB^2 = 5^2 + 3^2 = 34.$$

$$AB = \sqrt{34} \text{ units.}$$

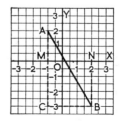

FIG. 4

GENERAL CASE. *To find the distance between the points* (x_1, y_1), (x_2, y_2).
We have,

$$AC = x_2 - x_1; \quad BC = y_2 - y_1.$$
$$\therefore AB = \sqrt{\{(x_2 - x_1)^2 + (y_2 - y_1)^2\}}.$$

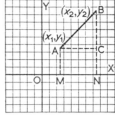

Ex. 3. *Distance between the points* $(-1, 3)$, $(2, -4)$ *is*

$$\sqrt{\{(-1-2)^2 + (3+4)^2\}} = \sqrt{9 + 49}$$
$$= \sqrt{58}.$$

FIG. 5

Mid-point of the line joining two points

Consider the line joining the points $A(x_1, y_1)$, $B(x_2, y_2)$; we require the co-ordinates $(\bar{x}, \bar{y})$ of P, the mid-point of AB.

From Fig. 6, as $AP = PB$,

clearly $AQ = \frac{1}{2}AC$ and $PQ = \frac{1}{2}BC$.

$$\therefore \bar{x} - x_1 = \frac{1}{2}(x_2 - x_1); \quad \bar{y} - y_1 = \frac{1}{2}(y_2 - y_1),$$
$$\text{i.e.} \quad \bar{x} = \frac{1}{2}(x_1 + x_2); \quad \bar{y} = \frac{1}{2}(y_1 + y_2).$$

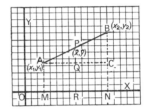

FIG. 6

Area of a plane polygon

The method of finding the area of a polygon whose vertices are given is illustrated by the following worked examples.

Ex. 4. *Find the area of the triangle* ABC *where*

    A *is the point* (1, 2),
    B   ”   ”   (0, 4),
    C   ”   ”   (3, 3).

From the diagram (Fig. 7),

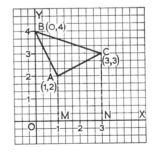

FIG. 7

area of $\triangle ABC$ = area of trapezium BCNO − trapezium AMNC − trapezium OMAB.

Trapezium $BCNO = \frac{1}{2}(OB + NC) . ON = \frac{1}{2} . 7 . 3 = \frac{21}{2}$ sq. units.

Trapezium $AMNC = \frac{1}{2} . 5 . 2 = 5$ sq. units.

Trapezium $\mathbf{OMAB} = \frac{1}{2} . 6 . 1 = 3$ sq. units.

$\therefore$ Area of $\triangle \mathbf{ABC} = \frac{21}{2} - 5 - 3 = 2\frac{1}{2}$ sq. units.

Ex. 5. Find the area of the quadrilateral $\mathbf{A}(2, -4)$, $\mathbf{B}(-1, 0)$, $\mathbf{C}(4, 1)$, $\mathbf{D}(3, -2)$.

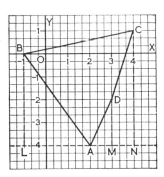

FIG. 8

From the diagram Fig. 8,

area quad. $\mathbf{ABCD}$ = area trapezium $\mathbf{LBCN} - \triangle \mathbf{LBA}$

$- \triangle \mathbf{DMA} -$ trapezium $\mathbf{DCNM}$,

$= \frac{1}{2}(4+5) . 5 - \frac{1}{2} . 4 . 3 - \frac{1}{2} . 2 . 1 - \frac{1}{2}(2+5) . 1$,

$= 22 \cdot 5 - 6 - 1 - 3 \cdot 5$,

$= 12$ sq. units.

(A) EXAMPLES 1a

1. On the same diagram, plot the following pairs of points and obtain the distances between them:

    (i) $(3, 2)$, $(5, 4)$;   (ii) $(0, 0)$, $(1, 3)$;

    (iii) $(-1, 2), (2, -1)$;   (iv) $(3, -1), (-3, -4)$;

              (v) $(0, -2), (-3, -1)$.

2. Calculate the lengths of the sides of the $\triangle \mathbf{ABC}$ and determine whether or not the triangle is right-angled when

    (i) $\mathbf{A}(0, 0)$, $\mathbf{B}(2, 3)$, $\mathbf{C}(1, 4)$;

    (ii) $\mathbf{A}(3, -1), \mathbf{B}(1, 7)$, $\mathbf{C}(-7, 5)$;

    (iii) $\mathbf{A}(2, -1), \mathbf{B}(-1, 1)$, $\mathbf{C}(1, 5)$.

3. Find the areas of the triangles in Question 2.

4. Plot the points $\mathbf{A}(-3, 2)$, $\mathbf{B}(3, 1)$, $\mathbf{C}(4, -1)$, $\mathbf{D}(0, -2)$. Calculate the lengths of the sides and the area of the quadrilateral $\mathbf{ABCD}$.

5. Find the area of the pentagon with vertices $(0, 0)$, $(6, 1)$, $(4, 3)$, $(2, 7)$, $(-1, 1)$.

6. Prove that the point $(4, 4)$ is equidistant from the points $(1, 0)$ and $(-1, 4)$.

7. What is the distance between the points (α, β) and $(2, 3)$?

8. If the point (α, β) is equidistant from the points $(0, 0)$ and $(3, 4)$, find the equation connecting α and β.

9. The distance between the points $(\alpha, 0)$ and $(0, \alpha)$ is equal to the distance between the points $(1, 2)$ and $(-1, 3)$. Find α.

10. The distance of the point (a, b) from the origin is twice its distance from the point $(-1, -2)$. Find the relation between a and b.

11. A rod **AB** (Fig. 9) of length $10\,\text{cm}$ slides with its ends on each of two fixed rectangular axes **OX**, **OY**. If $\mathbf{OA} = \frac{1}{2}\mathbf{OB}$, find the co-ordinates of **A** and **B**. Find also the co-ordinates of **P**, the mid-point of **AB**, and verify that **OP** is $\frac{1}{2}\mathbf{AB}$.

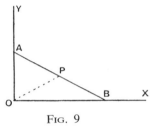

12. Prove that the triangle whose vertices are the points $(1, 2)$, $(3, 4)$ and $(-1, 6)$ is isosceles.

Fig. 9

13. The triangle **ABC** has vertices **A**$(1, 1)$, **B**$(3, 2)$, **C**$(2, 5)$. **D** is the middle point of **AB**. Write down the co-ordinates of **D**. Find the lengths **CD**, **AD**, **AC**, **BC**.
Verify the result $\mathbf{AC}^2 + \mathbf{CB}^2 = 2\mathbf{CD}^2 + 2\mathbf{AD}^2$.

The Equation of a Curve

The diagram shows the graph of the expression $y = 2x^2 + 1$, obtained in the usual way by giving x particular values and finding the corresponding values of y, and then plotting the points obtained. Suppose any point **P** is taken on this curve. Let the co-ordinates of **P** be (α, β). Then as **P** lies on the curve its y and x co-ordinates are connected by the equation $y = 2x^2 + 1$, i.e. $\beta = 2\alpha^2 + 1$. This relation between the co-ordinates holds for every point on the curve, and is called the *equation of the curve.*

i.e. The equation of the given curve is $y = 2x^2 + 1$.

Generally, the equation of a curve is the connection between the y and x co-ordinates of any point on the curve.

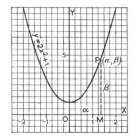

Fig. 10

Ex. 6. *Find the y co-ordinate of the point on the curve, equation* $3y = 4x^3 - 2x + 1$, *whose x co-ordinate is* 2.

We have $3y = 4(2)^3 - 2(2) + 1 = 29.$

$\therefore y = 9\frac{2}{3}.$

Ex. 7. *Find the co-ordinates of the points where the curve whose equation is* $y = (x - 2)(2x + 1)$, *cuts the x axis.*

For all points on the x axis $y = 0$, and so we require the points on the curve whose y co-ordinates are zero.

Putting $y = 0$ in the equation of the curve, we get

$(x - 2)(2x + 1) = 0,$ *i.e.* $x = 2$ and $x = -\frac{1}{2}.$
$\therefore$ the points are $(2, 0)$ and $(-\frac{1}{2}, 0).$

Intersections of Two Curves whose equations are known

The method of finding the co-ordinates of the points of intersection of two given curves is illustrated by the following example.

Ex. 8. *Find the points of intersection of the curve and straight line whose equations are* $y = x^2 (1)$ *and* $y = x + 2 (2).$

At a point of intersection, the co-ordinates of the point must satisfy equations (1) and (2). So we require the values of x and y which satisfy the equations (1) and (2), *i.e.* the solution of a pair of simultaneous equations.

We get $x^2 = x + 2.$

i.e. $x^2 - x - 2 = 0.$

$\therefore x = 2$ and $-1,$

and $y = 4$ and $1.$

$\therefore$ the points of intersection are $(2, 4)$ and $(-1, 1).$

(A) EXAMPLES 1b

1. On the curve whose equation is $y = 3x^2 + 2x - 1$ find the co-ordinates of

(i) the point whose x co-ordinate is -2;
(ii) the points of intersection of the curve with the x axis;
(iii) the point of intersection of the curve with the y axis.

2. The diagram (Fig. 11) represents the curve $y = x^2$. The abscissae of P, Q are 2 and -1. Find the ordinates of P and Q and the length of the chord PQ.

3. Find the distance between the two points on the curve $y = 2x^2 - 1$ whose abscissae are -1 and 2.

4. Find the length of the chord joining the points on the curve $y^2 = 16x$ whose x co-ordinates are each equal to 2.

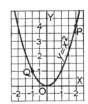

FIG. 11

5. Find the co-ordinates of the point of intersection of the straight lines whose equations are $3y - x = 1$ and $y - 2x = 4$. Does the line $2x - y = 1$ pass through this point?

6. Prove that the straight lines whose equations are $y - x = 1$, $y + 3x = 5$ and $y = 4x - 2$ are concurrent, and find the point of concurrence.

7. Prove that the line $y = 2x$ meets the curve $y = 3x^2 - 1$ in two points, and find their co-ordinates.

8. Find the length of the chord cut off on the line $y = x + 1$ by the curve $y = 2x^2 - x + 1$.

9. Find the co-ordinates of the points of intersection of the curves $y = 2x^2$ and $y^2 = 4x$. Deduce the length of the common chord.

10. The sides of a triangle consist of the following lines, $y + x = 2$, $2y - 3x = 1$, $4y = x - 6$. Find the co-ordinates of the vertices of the triangle and the lengths of the sides.

11. P and Q are neighbouring points on the curve $y = x^2$, the abscissa of P being 1.

Find PM, QN, QR, PR, $\dfrac{QR}{PR}$ when the abscissae of Q are (a) 1·1, (b) 1·01.

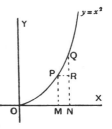

12. The diagram (Fig. 13) represents a circle, radius 4 cm. If the abscissa of P is 3, find the ordinate of P.

FIG. 12

If the co-ordinates of Q are (α, β), by using the Theorem of Pythagoras for the △QON obtain a relation connecting α and β. Hence write down the equation of the circle.

13. Prove that the line $y = x$ does not meet the curve $y = x^2 + 2x + 6$ in real points.

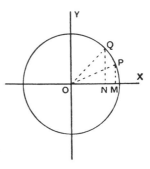

14. Find the points of intersection of the line $y = 4(x - 1)$ and the curve $y = x^2$. What does the result tell us about the line and the curve?

FIG. 13

Functions

The area of a circle A is connected with the radius r by the formula $A = \pi r^2$, where π is a constant.

The value of A *depends* on the value of r, and A is said to *be a function of r*.

A is called the *dependent* variable and r the *independent* variable.

In the same way, the volume of a circular cone V, which is equal to $\frac{1}{3}\pi r^2 h$, where r is the radius of the base and h the height, is a function of two independent variables r and h.

In general, when variables are related so that the value of one depends on the values of the others, the one variable is said to be a function of the others.

Implicit and Explicit Functions

Consider the equations

$$y = 3x^3 - x^2 + 2, \tag{1}$$

and

$$y^2 x + x^3 + y + 16 = 0. \tag{2}$$

In both cases y is a function of x, but in case (1) y is given *directly* in terms of x, and in case (2) y is given *indirectly* in terms of x.

Case (1) is an example of an *explicit function*,

and　Case (2) an example of an *implicit function*.

Ex. Write down equations in which

(a) y is given as an explicit function of x.

(b) p is given as an implicit function of q.

(c) V is given as an explicit function of s and t.

(d) v is given as an implicit function of s and t.

CHAPTER 2

THE STRAIGHT LINE

The Gradient of a Straight Line

Consider any two points P, Q on the straight line (Fig. 14).

Let PR, QR be parallel to OX and OY respectively. Then the ratio $\dfrac{QR}{PR}$ is constant for all pairs of points on the line. For if we take any other points P', Q', the $\triangle$'s PQR, P'Q'R' are similar, and hence

$$\frac{QR}{PR} = \frac{Q'R'}{P'R'}.$$

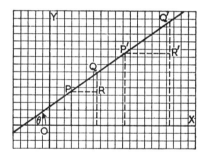

FIG. 14

This ratio $\dfrac{QR}{PR}$, constant for any two points on the straight line, is defined as the *gradient of the straight line*.

I.e. Gradient of straight line $= \dfrac{QR}{PR} = \dfrac{\text{increase in } y}{\text{increase in } x}$

$\qquad = $ rate of increase of y with respect to x.

If *the units on the axes are the same,* $\dfrac{QR}{PR} = \tan \theta$, where θ is the angle between the line and the positive direction of the x axis.

9

Sign of the Gradient

When y increases as x increases, both QR and PR are positive, and hence the gradient is positive. Or, when θ is acute, $\tan\theta$ is positive, and so the gradient of the line is positive.

If y decreases as x increases, QR is negative and PR is positive, and hence the gradient is negative. Or, when θ is obtuse, $\tan\theta$ is negative and hence the gradient is negative.

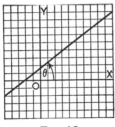

Fig. 15

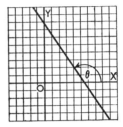

Fig. 16

In Fig. 15,
y increases as x increases,
θ is acute,
gradient of line is positive.

In Fig. 16,
y decreases as x increases,
θ is obtuse,
gradient of line is negative.

Gradient of the Straight Line joining two points whose co-ordinates are given

Ex. 1. *To find the gradient of the straight line joining the points* $A(1, 2), B(4, 3)$.

$$\text{Gradient of } AB = \frac{\text{increase in } y}{\text{increase in } x} = \frac{BC}{AC} = \frac{1}{3}.$$

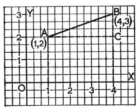

Fig. 17

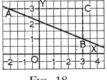

Fig. 18

Ex. 2. *To find the gradient of the straight line joining the points* $A(-2, 3), B(3, 1)$. *In this case* y *decreases as* x *increases, and the gradient is negative.*

$$\text{Gradient of } AB = \frac{\text{increase in } y}{\text{increase in } x} = \frac{-2}{5} = -\frac{2}{5}.$$

General Case

To find the gradient of the straight line joining the points $A(x_1, y_1)$, $B(x_2, y_2)$.

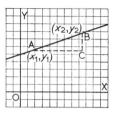

$$\textbf{Gradient of AB} = \frac{\textbf{BC}}{\textbf{AC}} = \frac{y_2 - y_1}{x_2 - x_1}$$

$$= \frac{\textbf{Difference in ordinates}}{\textbf{Difference in abscissae}}.$$

FIG. 19

Ex. 3. *Gradient of the line joining* $(-3, 2)$ *and* $(2, -1)$

$$= \frac{\text{Difference in ordinates}}{\text{Difference in abscissae}} = \frac{-1-2}{2-(-3)} = \frac{-3}{5} = -\frac{3}{5}.$$

To interpret the equations $y = mx$ **and** $y = mx + c$, **where** m **and** c **are constants.**

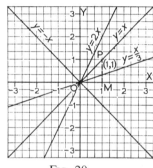

FIG. 20

Consider the equation $y = x$. This is the equation of a straight line passing through the origin and through the points $(1, 1)$, $(2, 2)$, etc.

$$\therefore \text{Gradient of line} = \frac{\textbf{PM}}{\textbf{OM}} = 1.$$

Similarly the equation $y = 2x$ represents a straight line passing through the origin and having a gradient equal to 2, and so on for different values of m.

Generally, $y = mx$ represents a straight line passing through the origin and having gradient m.

Now consider the equation $y = 2x + 2$. This represents a straight line parallel to the line $y = 2x$ and cutting the axis of y at distance 2 units from the origin.

Similarly, the equation $y = -3x - 1$ represents a straight line parallel to the line $y = -3x$ and cutting the axis of y at distance -1 unit from the origin.

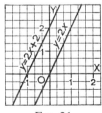

FIG. 21

Generally, $y = mx + c$ is the equation of a straight line, gradient m and cutting the y axis at distance c from the origin.

Ex. 4. *Show that the equation* $3y - 2x = 6$ *represents a straight line, and find its gradient and position.*

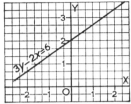

FIG. 22

$$3y - 2x = 6.$$

$$\therefore 3y = 2x + 6,$$

$$i.e. \quad y = \frac{2x}{3} + 2.$$

This is the equation of a straight line, gradient $\frac{2}{3}$ and cutting the y axis at distance 2 units from the origin. Using these results, the position of the line can be indicated roughly as shown in Fig. 22.

An equation of the first degree represents a straight line

Consider the general equation of the first degree,

$$lx + my = n \quad (l, m \text{ and } n \text{ constants}).$$

$$i.e. \quad y = -\frac{lx}{m} + \frac{n}{m}.$$

This is the equation of a straight line, gradient $\left(-\dfrac{l}{m}\right)$ and cutting the y axis at distance $\dfrac{n}{m}$ from the origin.

(A) EXAMPLES 2a

1. Find the gradients of the lines joining the following pairs of points:

(i) $(1, 2), (4, 5)$;   (ii) $(2, -6), (3, -2)$;
(iii) $(-1, 2), (0, -4)$;   (iv) $(-4, -2), (-1, 3)$;
(v) $(0, a), (a, 0)$.

2. Find the gradient of the line joining the points on the curve $y = x^3$ whose abscissae are 1 and 3.

3. What are the gradients of the lines joining the origin to the points of intersection of the curve $y = x^2$ and the line $y = x + 2$?

4. Find the gradients of the sides of the triangle whose vertices are the points $A(0, 1), B(3, -2), C(-1, 3)$.

5. Interpret the following equations

(i) $5x - 2y = 3$;   (ii) $x = 2y - 7$;

(iii) $\dfrac{x}{2} - \dfrac{y}{3} = 1$;   (iv) $7x - 4y = 8$;

(v) $ax - by = c$.

6. The straight line $\dfrac{x}{3}+\dfrac{y}{4}=1$ meets the axes of x and y in the points **A** and **B** respectively. Find the length and gradient of **AB**.

7. Find the gradients of the lines $y=3x-1$ and $x=\tfrac{1}{3}y+4$. What can be said about these two lines?

8. Points **P**, **Q**, whose abscissae are 2 and $2+h$, are taken on the curve $y=2x^2+1$. Find the gradient of the chord **PQ**. To what value does this gradient approach as h decreases towards the value zero?

9. Find the gradient of the chord of the curve $y=\dfrac{1}{x}$, joining the points whose abscissae are $1+h$ and $1-h$.

10. The points **P** and **Q** on the curve $y=x^2-x$ have abscissae a and $a+h$. Find the ordinates of **P** and **Q** and the gradient of the chord joining them. To what value does this gradient approach as h approaches zero?

To find the equation of a straight line passing through a given point and having a given gradient.

Ex. 5. *Find the equation of the straight line passing through the point* $(2,-1)$ *and having gradient* 3.

The general form of the equation of a straight line having gradient 3 is $y=3x+c$, where c is a constant.

As the line passes through the point $(2,-1)$ we have,

$$-1=6+c. \qquad i.e. \quad c=-7.$$

∴ Equation of the straight line is $y=3x-7$.

GENERAL CASE. *To find the equation of the straight line passing through the point* **A**(h,k) *and having gradient* m.

Take a variable point **P**(x,y) on the line.

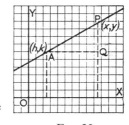

Then gradient $=m=\dfrac{\text{PQ}}{\text{AQ}}=\dfrac{y-k}{x-h}$.

$$\therefore \quad \boxed{y-k=m(x-h).}$$

This result is of importance and should be memorised.

Fig. 23

Ex. 6. *Equation of the line passing through the point* $(-3,2)$ *and having gradient* -2 *is*

$$y-2=-2(x+3),$$
$$i.e. \quad y+2x+4=0.$$

Equation of the Straight Line passing through two given points

Ex. 7. Find the equation of the straight line passing through the points $(-1, 2), (2, 4)$.

$$\text{Gradient of line} = \frac{4-2}{2-(-1)} = \frac{2}{3}.$$

∴ Equation of line is

$$y - 2 = \tfrac{2}{3}(x + 1),$$

$$i.e. \quad 3y - 2x = 8.$$

GENERAL CASE. *Equation of the straight line passing through the points* $A(x_1, y_1), B(x_2, y_2)$.

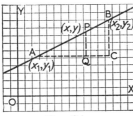

FIG. 24

Take any point $P(x, y)$ on the line.

By similar triangles, $\dfrac{PQ}{BC} = \dfrac{AQ}{AC}$,

$i.e.$ $\boxed{\dfrac{y - y_1}{y_2 - y_1} = \dfrac{x - x_1}{x_2 - x_1}}$

Ex. 8. Equation of the line passing through the points $(1, -2), (-3, 0)$ *is*

$$\frac{y+2}{0+2} = \frac{x-1}{-3-1},$$

$$i.e. \quad 2y + x + 3 = 0.$$

(A) EXAMPLES 2b

1. Find the equations of the following straight lines

   (i) gradient 2, passing through $(1, 3)$;
   (ii) gradient -1, passing through $(0, 0)$;
   (iii) gradient $\tfrac{1}{3}$, passing through $(-2, 3)$;
   (iv) gradient $-a$, passing through $(a, -a)$.

2. Find the equations of the straight lines joining the following pairs of points

   (i) $(1, 3), (2, 5)$;   (ii) $(-2, 0), (1, -3)$;   (iii) $(4, 6), (-3, -1)$;
   (iv) $(11, 7), (0, 1)$;   (v) $(a, 0), (0, a)$;   (vi) $(a, b), (b, a)$.

3. What are the equations of the straight lines passing through the point of intersection of the lines $y = x$ and $y = 2x - 3$ and

   (a) having gradient -2,   (b) passing through the point $(0, -1)$?

4. The vertices of a quadrilateral ABCD are $A(0, 2)$, $B(3, 1)$, $C(2, 5)$,

D$(-1, 4)$. Find the equations of the diagonals **AC** and **BD**, and hence find the co-ordinates of their point of intersection.

5. The sides of a quadrilateral consist of the following straight lines

AB, $y - 2x = 0$;	**BD**, $2y + 5x = 8$;
CD, $y = 2x + 4$;	**AD**, $2y + 5x = 0$.

Find the co-ordinates of the vertices of the quadrilateral and the equations of its diagonals.

If the diagonals intersect at **O**, find the co-ordinates of **O** and deduce that the diagonals bisect each other.

6. The vertices of the $\triangle$**PQR** are P$(-1, 2)$, Q$(2, 1)$, R$(0, 4)$. Find
 (*a*) the co-ordinates of the mid-points of **PQ**, **QR**, **RP**;
 (*b*) the equations of the medians of the triangle.
Deduce that the medians are concurrent and find the point of concurrence.

7. Find the equation of the chord joining the points on the curve $y = x^3 - 2$ whose abscissae are 0 and 2.

8. Find the equation of the chord joining the points on the curve $xy = 2$ whose ordinates are -1 and 1.

9. The straight line joining the point **P** on the curve $y^2 = 16x$ whose y co-ordinate is 4 to the point $(4, 0)$, meets the curve again in **Q**. Find the equation of **PQ** and the co-ordinates of **Q**.

10. **AB** is a chord of the curve $\dfrac{x^2}{4} + \dfrac{y^2}{9} = 2$ which passes through the origin **O**. If the x co-ordinate of **A** is 2, find the co-ordinates of **B**. Deduce that **O** is the mid-point of **AB**.

Parallel Lines

Lines are parallel when they have the same gradient.

Ex. 9. Find the equation of the straight line parallel to the line $3y - x = 2$ and passing through the point $(3, -2)$.

The equation $3y - x = 2$ can be written in the form

$$y = \frac{x}{3} + \frac{2}{3}.$$

$$\therefore \text{ Gradient of line} = \tfrac{1}{3}.$$

$\therefore$ Equation of required line is

$$y + 2 = \tfrac{1}{3}(x - 3), \quad [y - k = m(x - h)].$$
  i.e. $3y - x + 9 = 0$.

Perpendicular Lines

Consider two perpendicular lines with equations

$$y = mx + c, \quad y = m'x + c'.$$

Gradient of the line $y = mx + c$ is

$$m = \tan \theta.$$

Gradient of the line $y = m'x + c'$ is

$$m' = \tan \theta'$$
$$= \tan (90° + \theta) = -\cot \theta.$$

$$\therefore \ mm' = (\tan \theta) \cdot (-\cot \theta) = -1.$$

i.e. $\ m' = -\dfrac{1}{m}, \quad$ or $\ mm' = -1.$

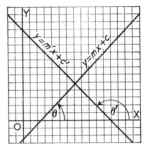

Fig. 25

Ex. 10. Find the equation of the straight line perpendicular to the line $y - 3x = 2$ and passing through the origin.

Gradient of the line $y = 3x + 2$ is 3.
$\therefore$ gradient of perpendicular line is $-\frac{1}{3}$.
$\therefore$ equation of required line is $y = -\frac{1}{3}x$,
or $3y + x = 0$.

To find the circumcentre of a given triangle

As an example, take the triangle whose vertices are $A(1, 2)$, $B(3, 4)$, $C(0, 3)$.

In order to find the circumcentre we require the point of intersection of the perpendicular bisectors of two sides of the triangle.

The co-ordinates of P, the mid-point of AB, are

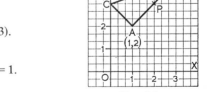

$$\left(\frac{1+3}{2}, \frac{2+4}{2}\right), \ i.e. \ (2, 3).$$

$$\text{Gradient of } AB = \frac{4-2}{3-1} = 1.$$

$\therefore$ Gradient of line perpendicular to $AB = -1$.

Fig. 26

$\therefore$ Equation of the perpendicular bisector of AB is

$$y - 3 = -1(x - 2),$$
$$i.e. \ \ y + x = 5.$$

The co-ordinates of the mid-point of AC are $(\frac{1}{2}, \frac{5}{2})$.

$$\text{Gradient of } AC = \frac{2-3}{1-0} = -1.$$

$\therefore$ Gradient of line perpendicular to $AC = 1$.

$\therefore$ Equation of perpendicular bisector of AC is

$$y - \tfrac{5}{2} = 1(x - \tfrac{1}{2}),$$

$$i.e. \quad y - x = 2.$$

The perpendicular bisectors meet in the point whose co-ordinates are the solutions of the equations

$$\left.\begin{array}{l} y - x = 2, \\ y + x = 5. \end{array}\right\}$$

Adding, we get $\qquad 2y = 7,$

$$\left.\begin{array}{l} y = 3\tfrac{1}{2}. \\ \therefore \ x = 1\tfrac{1}{2}. \end{array}\right\}$$

$\therefore$ The circumcentre is the point $(1\tfrac{1}{2}, 3\tfrac{1}{2})$.

(A) EXAMPLES 2c

1. Determine whether the following pairs of lines are parallel

(i) $2y - 3x + 6 = 0,$ $\qquad 4y = 6x + 3;$

(ii) $3y - 4x = 2,$ $\qquad 4x - 3y = 4;$

(iii) $3y - 2x - 1 = 0,$ $\qquad 4x = 6y + 7;$

(iv) $7y - 11x + 18 = 0,$ $\qquad 21y = 33x + 1;$

(v) $y + ax = 0,$ $\qquad a^2x - ay = \dfrac{1}{a}.$

2. Test whether the following lines are perpendicular

(i) $y - 7x = 0,$ $\qquad 7y = x - 11;$

(ii) $3y + x + 1 = 0,$ $\qquad 2y = 6x - 5;$

(iii) $7y = 3x + 4,$ $\qquad 14x + 6y = 0;$

(iv) $py - p^2x = 0,$ $\qquad p^2y + px = 0.$

3. What is the gradient of the line $3y - 4x = 7$?

Find the equation of the line which passes through the origin and is parallel to the given line.

4. Find the equation of the line parallel to the line $7x - y = 6$ which passes through the point $(-2, 3)$.

5. Find the gradient of all lines perpendicular to the line $y - 2x = 1$.

Find the equation of the perpendicular which passes through the origin.

6. What is the equation of the perpendicular bisector of the line joining the points $(0, 0)$ and $(2, 2)$?

7. Find the equation of the perpendicular to the line $2x - 3y = 4$ drawn from the point $(1, 3)$. Also obtain the co-ordinates of the foot of the perpendicular.

8. A$(0, 0)$, B$(1, 2)$ and C$(2, 3)$ are three vertices of a parallelogram ABCD. Write down
(a) the gradients of **AB** and **BC**;
(b) the equation of the line through **A** parallel to **BC**;
(c) the equation of the line through **C** parallel to **AB**.
Deduce the co-ordinates of **D**.

9. Prove that the lines joining the points $(-2, -1)$, $(-7, 4)$ to the point $(-1, 1)$ are at right angles.

10. The points A$(-2, -1)$, C$(-7, 4)$, and D$(-1, 1)$ are three vertices of a rectangle ABCD. Find
(a) the equation of the line through **C** perpendicular to **CD**;
(b) the equation of the line through **A** perpendicular to **AD**.
Hence deduce the co-ordinates of **B**.

11. The vertices of a triangle ABC are A$(1, 1)$, B$(-2, 2)$, C$(0, 5)$. Find the equation of the altitude of the triangle passing through **A**.

12. Prove that the figure bounded by the lines $y = 2x + 1$, $y - 3x = 6$, $2y + 6x = 7$, $y = 3x + 1$ is a trapezium, and find the co-ordinates of its vertices.

The Straight Line Law

In experimental work it is often required to find an algebraic law connecting two sets of numbers. If we can, using the numbers, obtain by a graph a series of points lying on a straight line, it is easy to obtain the law.

For we have seen that an equation of the form $y = a + bx$ where a and b are constants, is the equation of a straight line gradient b and cutting the y axis at distance a from the origin. Conversely, if, on plotting the values of one variable, y, against the values of a second variable, x, we obtain a straight line, then y and x must be connected by an equation of the form $y = a + bx$.

Ex. 11. The following are observed values of two quantities x and y. Show that there is an approximate linear law connecting x and y, and find it.

x	15·0	22·6	30·0	37·0	45·0	50·0	60·0
y	2·76	5·60	8·60	11·2	14·0	16·2	20·0

Take suitable axes and plot the points as shown below. We see that the plotted points lie approximately on a straight line. Therefore y and x are connected approximately by a law of the form $y = a + bx$.

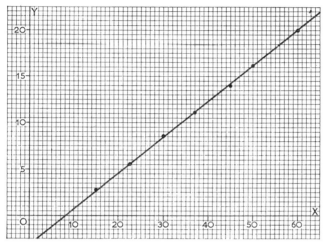

FIG. 27

To find the best values of a and b, take two points *on the graph* and read off the corresponding values of x and y.

We obtain $\qquad\qquad x = 20, y = 4\cdot6$⎫
and $\qquad\qquad\qquad\quad x = 60, y = 20.$⎭

Substituting in the equation $y = a + bx$ we get,

$$4.6 = a + 20b$$
and $\qquad\qquad 20 = a + 60b.$
$$\therefore \ a = -3\cdot1,$$⎫
$$b = 0\cdot385.$$⎭

$\therefore$ The approximate law connecting x and y is

$$y = -3\cdot1 + 0\cdot38x.$$

Laws reducible to a linear form

Laws, which are not linear, can sometimes be reduced to the linear form. For example, if the law is $y = a + bx^2$, the graph of y against x will be a curve not easily identified, but if we plot y against x^2, as there is a linear law connecting y and x^2, the graph will be a straight line.

3. A steam-engine, when its load is **P**, uses **W** kg of coal per hour. The following measurements were made

P	100	80	60	50	35	20
W	2240	1840	1440	1240	940	640

Show that there is a law of the form $W = a + bP$ and find the best values of a and b.

4. In an experiment with a screw jack, the following results were obtained

Load, W, in kg	100	120	140	160	180	200
Effort applied, P, in kgf	12·6	13·8	15·7	17·6	19·6	21·5

Plot these observations and find from your graph an approximate law of the form $W = a + bP$.

5. The following corresponding results for the resistance **R**, in newtons, and the weight **W**, in kg, of a body moving with constant speed were made

R	4·3	6·6	9·5	12·4	15·3	21·1	32·7
W	10	50	100	150	200	300	500

Verify that **R** and **W** are approximately connected by a linear law and find the law.

6. For a certain machine, the following table shows corresponding values of the effort **P** kgf and the load **W** kg

W	50	100	200	300	400	500	600
P	100	122	152	184	216	246	278

Plot these values and find the law connecting **P** and **W**.

7. A motor-bus company finds that the following table gives the average daily cost, **C** pounds, of running a motor-bus when **N** passengers are carried per day.

C	5·50	6·50	7·00	7·80	7·90	8·40	8·90
N	250	500	600	750	800	900	1000

Show that there is an approximate law of the form $C = a + bN$, and find a and b.

If, on the average, each passenger's fare is $1\frac{1}{2}$p., what is the least number of passengers which must be carried in order that there may be a profit of 15 per cent?

8. The velocity, V m s^{-1}, of an aeroplane t s after starting is given by the following table

V	26·3	31·4	42·6	51	65·5	107	135
t	6	8	12	15	20	35	45

By plotting V against t, obtain the law connecting V and t.

9. An examiner has given marks to papers; the highest number of marks is 85, the lowest 22. He desires to change all his marks according to a linear law, converting the highest number to 100 and the lowest to 40. Show how this can be done graphically and state the converted marks for papers originally marked 78, 64 and 36.

10. By plotting y against x^2 show that the following values of y and x are connected by laws of the form $y = a + bx^2$ and find a and b in each case.

(i)

y	4	2	−4	−14	−28	−46
x	0	1	2	3	4	5

(ii)

y	·65	1·85	3·85	6·65	10·25	14·65
x	2	4	6	8	10	12

(iii)

y	−1·60	−8·35	−19·60	−35·35	−55·60
x	1	2	3	4	5

11. The following are corresponding values of the resistance R g kg^{-1}, of a train, and its velocity V km h^{-1}. It is thought that there may be an approximate law of the form $R = a + bV^2$. Test if this is so, and find the best values of a and b.

V km h^{-1}	10	15	20	25	30	35	40
R g kg^{-1}	201	487	650	990	1400	1890	2440

12. The following table gives corresponding values of the pressure p cm of mercury, and the volume v cm^3 of a given mass of gas at constant temperature. By plotting p against $\dfrac{1}{v}$, obtain a simple relation between p and v.

p	89·9	113·4	177·9	195·6	210·2
v	22·1	17·6	11·2	10·2	9·5

13. For the following table, plot y against $\dfrac{1}{x}$ and deduce that y and x are approximately connected by a law of the form $y = a + \dfrac{b}{x}$. Find a and b.

x	10	12	15	18	20	22	25
y	3·120	2·985	2·850	2·755	2·710	2·675	2·628

14. The following corresponding values of P and V are taken from a table. There may be a law of the form $P = a + bV^3$ connecting them. Test if this is so, and find the best values of a and b.

P	290	560	1144	1810	2300
V	7·5	10·5	13·5	16·5	18

15. The following corresponding values of x and y were measured

x	1	4	9	16	25	36	49
y	0·25	0·37	0·49	0·61	0·69	0·82	0·93

By plotting y against $\sqrt{x}$, obtain an approximate law of the form $y = a + b\sqrt{x}$.

N.B. For examples on the use of logarithms to reduce given laws to a linear form see Examples 22b, (p. 312).

(B) MISCELLANEOUS EXAMPLES

1. Find the equation of the straight line passing through the point of intersection of the lines $3y - x = 2$, $2y - 5x = 1$ and perpendicular to the line $y - 2x = 7$.

2. The co-ordinates of **A, B, C**, the vertices of a triangle **ABC**, are $(1, 2), (3, 4), (4, -2)$. Find the equation of the line joining the mid-points of **AB** and **BC**. Deduce that this line is parallel to **AC** and of half the length.

3. Prove that the quadrilateral with vertices $(2, 1), (2, 3), (5, 6), (5, 4)$ is a parallelogram, and find the point of intersection of its diagonals.

4. Find the equations of the perpendicular bisectors of the lines joining the points $(2, 1), (6, 3)$ and $(6, 3), (8, 1)$. Deduce that the point of intersection of these bisectors is on the x-axis.

5. The co-ordinates of **P, Q, R**, the vertices of the triangle **PQR**, are $(1, 1), (5, 4), (4, 0)$. If the altitude through **P** meets **QR** in **X**, find

> (*a*) the equation of **PX**;
> (*b*) the co-ordinates of **X**;
> (*c*) the length **PX**.

Using these results, obtain the area of the triangle and check your result by a second method.

6. The co-ordinates of the vertices of a triangle are $(0, 4), (2, 0), (4, 2)$. Prove that the triangle is isosceles and find its area.

7. Find the equation of the lines passing through the point **A**$(3, -4)$, and (i) parallel, (ii) perpendicular to the line $4x + 3y = 8$. If these lines meet the line $2x + y = 1$ in points **B** and **C**, find the area of $\triangle$**ABC**.

8. Find the length of the perpendicular from the origin on to the line $2y + x = 4$.

9. The co-ordinates of **A, B, C**, three vertices of a rectangle **ABCD**, are $(0, 2), (-2, 0), (1, -3)$. Find the co-ordinates of **D**.

10. The co-ordinates of the points **A, B, C** are $(-3, -1), (11, 13), (-1, -3)$ respectively. Find the equations of the medians of the triangle **ABC** and the co-ordinates of their point of intersection.

11. For the triangle in Question 10 find the equations of the perpendicular bisectors of the sides **AB** and **AC**. Deduce the co-ordinates of the centre of the circumcircle of the triangle.

12. Find the radius and centre of the circumcircle of the triangle whose vertices are $(-1, 8), (-1, -2), (2, 4)$.

13. Prove that the quadrilateral whose vertices are $(1, 3), (1, -1), (3, 1), (-1, 1)$ is cyclic.

14. Find the equations of the altitudes of the triangle whose vertices are the points $(0, 0), (2, 3), (1, 4)$. Deduce the co-ordinates of their point of intersection (*i.e.* the orthocentre of the triangle).

15. Show that the points $(1, 3), (3, 4), (4, -3)$ are three vertices of a rectangle and find the co-ordinates of the fourth vertex.

16. Prove that the curves $y = x^2 + 1$, $y^2 = 3x + 1$ meet at points whose abscissae are 0 and 1. Deduce the equation of the common chord of the two curves.

17. Find the points of intersection of the line $y = x - 2$ and the curve $y^2 = 4x$. Write down the equation of the perpendicular bisector of the line joining these points and find where it meets the axis of x.

18. Find the points of intersection of the common chords of the curve $x^2 + y^2 = 4$ and (i) the curve $y^2 = 3x$, (ii) the curve $x^2 - 4y + 1 = 0$.

19. Find the points of intersection of the curve $x^2 + y^2 + 10(x + y) + 25 = 0$ with the co-ordinate axes. Explain the result.

20. Find the points of intersection of the curves $x^2 - y^2 = 20$, $xy = 24$.

21. For a system of pulleys, the following values were obtained for the load W g and the effort P gf:

W	50	100	150	200	250	300
P	40	50	59·7	69·5	79	89.

Plot W against P, and obtain an approximate law connecting W and P.

22. The following table shows the deflection $\theta°$ of the needle of a tangent galvanometer when a current of C amperes passes through the galvanometer. Show that there is a relation of the form $C = k \tan \theta$ and find the best value of k.

C	0·04	0·061	0·083	0·106	0·131	0·158	0·190
θ	10°	15°	20°	25°	30°	35°	40°

CHAPTER 3

GRADIENT OF A CURVE. DIFFERENTIATION

Tangent to a Curve

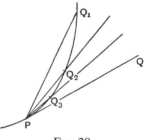

Consider a straight line cutting a curve in two points P and Q_1. PQ_1 is a *chord* of the curve. Consider points Q_2, Q_3, ... bisecting the arcs Q_1P, Q_2P... PQ_2, PQ_3, PQ_4, ... are nearer and nearer to the tangent at P to the curve. The tangent PQ at P is the limiting position of the chords PQ_1, PQ_2, PQ_3 ....

The points Q_2, Q_3, ... may be obtained from Q_1, Q_2 ... by any other simple law, *e.g.* arc PQ_2 may be one-tenth of arc PQ_1, arc PQ_3 one-tenth of arc PQ_2 ....

FIG. 29

Gradient of a Curve

The gradient of a curve at a given point is defined as the gradient of the tangent to the curve at that point.

In order to find the gradient at a point P, a neighbouring point Q is

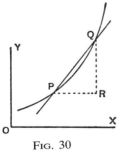

taken on the curve and the gradient of the chord PQ is obtained; *i.e.* the ratio $\dfrac{QR}{PR}$. Then Q is allowed to approach nearer and nearer to P and the limiting position of the chord PQ is the tangent at P. Hence the gradient of the curve at P will be the value to which the ratio $\dfrac{QR}{PR}$ approaches as Q approaches P.

FIG. 30

The value to which the ratio $\dfrac{QR}{PR}$ approaches as Q approaches P is called the *limit* of the ratio $\dfrac{QR}{PR}$ as Q tends to P, and is denoted symbolically as $\lim\limits_{Q \to P} \left(\dfrac{QR}{PR} \right)$.

To find the gradient of a curve at a given point

The numerical method of determining the gradient of a curve at a given point is illustrated by the following example.

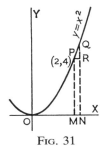

FIG. 31

Ex. 1. *To find the gradient of the curve* $y = x^2$ *at the point* P(2, 4). (*Fig.* 31.)

Take a point Q on the curve whose x co-ordinate is slightly greater than P.

Note.—The difference in the x co-ordinates of P and Q will be taken as ·1, ·01, ·001 .... Draw the ordinates PM, QN, and draw PR parallel to the x-axis. The portion of the curve PQ is shown on an enlarged diagram. (Fig. 32.)

Now take the distance PR (*i.e.* the difference in the x co-ordinates) as ·1.

∴ as the x co-ordinate of P is 2, the x co-ordinate of Q
(*i.e.* ON) = 2·1.

As Q lies on the curve $y = x^2$, the y co-ordinate of
$Q = (2·1)^2 = 4·41$.

Hence $QR = QN - RN = QN - PM = 4·41 - 4$.

∴ the gradient of the chord $PQ = \dfrac{QR}{PR}$,

$$= \frac{·41}{·1} = 4·1.$$

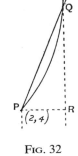

FIG. 32

Now let Q approach P by taking smaller and smaller values for PR, and in each case obtain the gradient of the chord PQ as above. For simplicity the results can be tabulated as follows

PR	·1	·01	·001	·0001	·00001	·000001
ON	2·1	2·01	2·001	2·0001	2·00001	2·000001
QN	4·41	4·0401	4·004001	4·00040001	4·0000400001	4·000004000001
QR	·41	·0401	·004001	·00040001	·0000400001	·000004000001
$\dfrac{QR}{PR}$	4·1	4·01	4·001	4·0001	4·00001	4·000001

It is clear that as PR decreases and approaches the value zero, the ratio $\dfrac{QR}{PR}$ approaches closer and closer to the value 4.

∴ the gradient of the curve at $P = \lim\limits_{Q \to P} \left(\dfrac{QR}{PR}\right)$,

$$= 4.$$

Gradient of a curve at any point

The student will have realised that the previous method only gives the gradient at one point on a given curve and must be repeated if the gradients at different points are required. This involves a considerable amount of work, and so it is desirable to try to obtain an expression which will give the gradient at any point on a given curve.

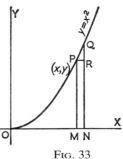

Ex. 2. To find the gradient at the point P (x, y) *on the curve* $y = x^2$.

The method is exactly as before. A point Q is taken on the curve whose abscissa is slightly greater than that of P.

FIG. 33

At this stage we introduce a new notation—a symbol to stand for a small increase or increment in a variable.

A small increment in the variable x is denoted by the symbol δx, called 'delta x'.

Similarly, a small increment in y is denoted by δy.

So we can let the small increase in x, PR, be δx and the corresponding small increase in y, QR, be δy.

The x co-ordinate of Q is $x + \delta x$ and the y co-ordinate is $y + \delta y$.

As P lies on the curve, $y = x^2$,
and as Q lies on the curve,

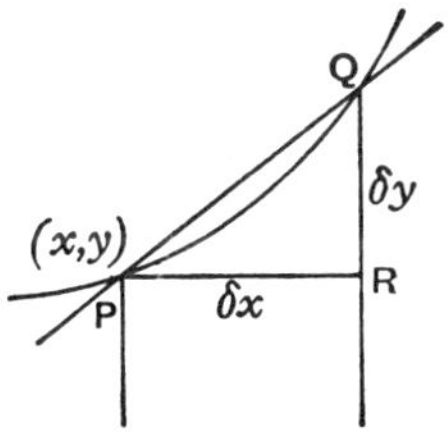

FIG. 34

$$y + \delta y = (x + \delta x)^2 = x^2 + 2x\delta x + (\delta x)^2.$$

Subtracting, $\delta y = 2x\delta x + (\delta x)^2$.

∴ The gradient of the chord $PQ = \dfrac{QR}{PR} = \dfrac{\delta y}{\delta x} = \dfrac{2x\delta x + (\delta x)^2}{\delta x}$,

$$= 2x + \delta x.$$

Now let $Q \to P$, *i.e.* let $\delta x \to 0$.

Then the gradient of the chord **PQ** becomes the gradient of the curve at **P**.

Hence the gradient of the curve at **P** is the value to which the fraction $\frac{\delta y}{\delta x}$ approaches as δx approaches zero.

$$i.e.\ \text{Gradient of curve at } \mathbf{P} = \lim_{\delta x \to 0} \frac{\delta y}{\delta x},$$

$$= \lim_{\delta x \to 0} (2x + \delta x),$$

$$= 2x,$$

where x is the abscissa of **P**.

e.g. when $x = 2$,   gradient at the point $(2, 4)\ = 2 \times 2 = 4$,

when $x = -1$, gradient at the point $(-1, 1) = -2$.

Using this result, we can write down the gradient at any point on the curve.

Ex. 3. Sketch roughly the curve $y = x^2 - 2x$ and find the gradient at any point (x, y) on it. Write down the equation of the tangent at the point $(2, 0)$.

Let **P** be the point (x, y) (Fig. 35).

Take a neighbouring point **Q** whose abscissa $x + \delta x$ is slightly greater than that of **P**.

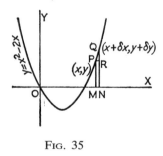

I.e. in moving from P to Q, x increases by δx.

Let the corresponding increase in y be δy.

The gradient of the chord **PQ**

$$= \frac{\text{difference in ordinates}}{\text{difference in abscissae}},$$

$$= \frac{\delta y}{\delta x}.$$

Fig. 35

As **P** lies on the curve, $y = x^2 - 2x$,

and as **Q** lies on the curve, $y + \delta y = (x + \delta x)^2 - 2(x + \delta x)$,

$$= x^2 + 2x\delta x + (\delta x)^2 - 2x - 2\delta x.$$

Subtracting, $\delta y = 2x\delta x + (\delta x)^2 - 2\delta x.$

Hence $\frac{\delta y}{\delta x} = 2x + \delta x - 2.$

The gradient of the tangent at $P = \lim\limits_{Q \to P}$ (gradient of chord PQ),

$$= \lim_{\delta x \to 0} \frac{\delta y}{\delta x},$$
$$= 2x - 2.$$

∴ gradient of the curve at the point $(2, 0) = 2 \cdot 2 - 2 = 2$.

∴ Equation of tangent is $\quad y - 0 = 2(x - 2)$,

i.e. $\quad y = 2x - 4$.

NOTATION. If y is expressed as a function of x and, with the usual notation, δy and δx represent corresponding small changes in y and x, then the limit of the fraction $\dfrac{\delta y}{\delta x}$ as δx decreases and approaches zero is called *the differential coefficient of y with respect to x* or *the derivative of y with respect to x*. This limit is denoted by the symbol $\dfrac{dy}{dx}$ [dy by dx].

As we have seen in the previous examples, the gradient of a curve is given by the $\lim\limits_{\delta x \to 0} \dfrac{\delta y}{\delta x}$, *i.e.* by the derivative $\dfrac{dy}{dx}$.

The process of obtaining the differential coefficient or derivative of a function is called *differentiation*.

Differentiation from First Principles

The general method of differentiation is illustrated by the following examples; to obtain a differential coefficient, we proceed in exactly the same way as we did to obtain the gradient of a given curve.

Ex. 4. If $y = 3x^2 - 2x + 1$, find $\dfrac{dy}{dx}$ from first principles.

Let x increase by a small amount δx and let the corresponding increase in y be δy.
Then as values of x and y are connected by the relation

$$y = 3x^2 - 2x + 1,$$

we have $\quad y + \delta y = 3(x + \delta x)^2 - 2(x + \delta x) + 1$,

$$= 3x^2 + 6x\delta x + 3(\delta x)^2 - 2x - 2\delta x + 1,$$

and $\quad y = 3x^2 - 2x + 1$.

Subtracting, $\quad \delta y = 6x\delta x + 3(\delta x)^2 - 2\delta x$.

Divide by δx, $\quad \dfrac{\delta y}{\delta x} = 6x + 3(\delta x) - 2$.

In order to obtain $\dfrac{dy}{dx}$, let $\delta x \to 0$.

Then $\dfrac{\delta y}{\delta x} \to \dfrac{dy}{dx}$ and $6x + 3(\delta x) - 2 \to 6x - 2$.

$$\therefore \dfrac{dy}{dx} = 6x - 2.$$

Ex. 5. Differentiate $(1 - x)^2$.

As a second variable is not mentioned, we introduce one.

Let $\qquad\qquad y = (1 - x)^2 = 1 - 2x + x^2$.

Let x increase by a small amount δx, and let the corresponding increase in y be δy.

$$\therefore y + \delta y = 1 - 2(x + \delta x) + (x + \delta x)^2,$$

$$= 1 - 2x - 2\delta x + x^2 + 2x\delta x + (\delta x)^2,$$

and $\qquad\qquad y = 1 - 2x + x^2$.

Subtracting $\qquad \delta y = -2\delta x + 2x\delta x + (\delta x)^2,$

$$\therefore \dfrac{\delta y}{\delta x} = -2 + 2x + \delta x.$$

Let $\delta x \to 0$.

$$\therefore \dfrac{dy}{dx} = \lim_{\delta x \to 0} \left(\dfrac{\delta y}{\delta x} \right) = -2 + 2x.$$

Ex. 6. If $s = 3t^2 - 4$, *find the value of* $\dfrac{ds}{dt}$ *when* $t = 2$.

We first obtain the differential coefficient and then substitute $t = 2$ in the result.

Let t increase by a small amount δt and let the corresponding increase in s be δs.

Then $\qquad s + \delta s = 3(t + \delta t)^2 - 4 = 3t^2 + 6t\delta t + 3(\delta t)^2 - 4,$

and $\qquad\qquad s = 3t^2 - 4$.

$$\therefore \delta s = 6t\delta t + 3(\delta t)^2.$$

$$\therefore \dfrac{\delta s}{\delta t} = 6t + 3\delta t.$$

Let $\delta t \to 0$.

Then $\qquad\qquad \dfrac{ds}{dt} = \lim_{\delta t \to 0} \dfrac{\delta s}{\delta t} = 6t.$

$$\therefore \text{ when } t = 2, \dfrac{ds}{dt} = 12.$$

(A) EXAMPLES 3

1. P is the point $(1, 2)$ on the curve $y = 2x^2$ and Q is a neighbouring point (Fig. 36). Complete the following table

PR	·1	·01	·001	·0001	·00001
QN					
QR					
$\dfrac{QR}{PR}$					

Hence deduce the gradient of the curve $y = 2x^2$ at the point $(1, 2)$.

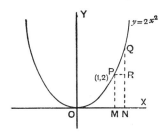

FIG. 36

2. Repeat question (1) for the cases

(i) P the point $(1, 3)$ on the curve $y = 3x^3$;

(ii) P the point $(2, 2)$ on the curve $y = x^2 - 2$;

(iii) P the point $(2, 6)$ on the curve $y = 2x^2 - x$.

3. P is the point $(3, 3)$ on the curve $y = x(x - 2)$ (Fig. 37). The abscissa of a neighbouring point Q is $3 + h$. Find the values of QN, QR, $\dfrac{QR}{PR}$. To what value does $\dfrac{QR}{PR}$ approach as h decreases and approaches zero? What is the gradient of the tangent at P?

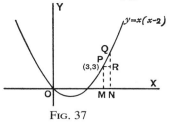

FIG. 37

4. Use the method of the previous example to obtain the gradients of the following curves at the points specified.

(i) The point $(1, 1)$ on the curve $y = x^2 + x - 1$;

(ii) The point $(1, 1)$ on the curve $y = \dfrac{1}{x}$.

5. P and Q are neighbouring points on the curve $y = x^2 + 2$. If the abscissae of P and Q are x and $x + \delta x$ respectively, find PM, QN, PR and QR in terms of x and δx. Hence obtain the ratio $\dfrac{QR}{PR}$ in terms of x and δx.

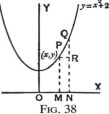

What is the gradient of the chord PQ?
Deduce the gradient of the curve at P.
What is the gradient of the curve at the point where $x = 3$?

FIG. 38

6. P and Q are neighbouring points on the curve $y = 6x - x^2$. P is the point (x, y), Q is the point $(x + \delta x, y + \delta y)$ (Fig. 39). Find the values of δy and $\dfrac{\delta y}{\delta x}$ in terms of x and δx.

Hence determine the gradient of the curve at P.

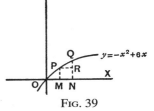

7. Sketch the curve $y = 1 - 2x^2$. Obtain the gradient of the curve at the point whose

FIG. 39

abscissa is x and deduce the values of the gradients at the points where $x = 1$ and $x = -2$.

8. Find the gradient of the tangent to the curve $y = 3x^2 - x$ at the point where $x = 2$. Hence write down the equation of the tangent at this point.

9. Find the gradients of the following loci at the point where $x = 0$:—

(i) $y = x^2$; (ii) $y = 1 - 3x$; (iii) $y = x(x + 2)$.

10. Find the equation of the tangent to the curve $y = 2x^2 - 1$ at the point where $x = -2$.

11. For the following functions of y, δy is the small increase in y due to a small increase δx in x. Find the values of δy, $\dfrac{\delta y}{\delta x}$ and $\dfrac{dy}{dx}$ in each case.

(i) $y = 3x^2$; (ii) $y = x^2 - x$; (iii) $y = 4x - 1$;
(iv) $y = x(4 - x)$; (v) $y = 2x^3$; (vi) $y = 3x^2 - 2x + 6$;
(vii) $y = (2x - 1)(x + 1)$.

12. If $s = 1 - t + 3t^2$, find $\dfrac{ds}{dt}$.

13. If $v = 3r^2 + 2r$, find $\dfrac{dv}{dr}$.

14. If $y = 4x^2$, find the value of $\dfrac{dy}{dx}$ when $x = 1$ and when $x = -1$.

15. Differentiate the following functions of x from first principles

(i) $2-3x$;   (ii) $4x-\dfrac{x^2}{2}$;   (iii) $2x^3-x$;   (iv) $(1+x)^2$.

16. A point P moves along a straight line such that after t seconds it is at a distance s metres from the origin O. In a further small time, δt seconds, the point moves a further distance δs metres to P'. If s and t are connected by the relation $s=2t^2+1$, find δs and $\dfrac{\delta s}{\delta t}$ in terms of t

Fig. 40

and δt. What is the average velocity of the point in the interval PP' ? Obtain the value of $\dfrac{ds}{dt}$. What does $\dfrac{ds}{dt}$ represent?

CHAPTER 4

GENERAL DIFFERENTIATION
THE DIFFERENTIAL COEFFICIENT OF x^n
TANGENTS AND NORMALS

The Differential Coefficient of a constant

Let $y = c$, where c is a constant. Then any change in x, say δx, will not affect y; *i.e.* $\delta y = 0$.

$$\therefore \quad \frac{\delta y}{\delta x} = 0.$$

$$i.e. \quad \frac{dy}{dx} = 0.$$

I.e. the differential coefficient of a constant is zero,

or
$$\frac{d}{dx}(c) = 0.$$

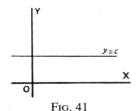

Fig. 41

This result has a simple geometrical interpretation.

The equation, $y = c$, represents a straight line parallel to the x-axis. The gradient of this line is zero, that is, $\frac{dy}{dx} = 0.$

The Differential Coefficient of x^n where n is a positive integer

We will lead to the general case of $\frac{d}{dx}(x^n)$ by means of the special cases,

(i) $\frac{d}{dx}(x)$, (ii) $\frac{d}{dx}(x^2)$, (iii) $\frac{d}{dx}(x^3)$.

(i) Let $y = x$.

Let x increase by a small amount δx and let the corresponding increase in y be δy.

36

Then $\qquad\qquad y + \delta y = x + \delta x,$

and $\qquad\qquad\qquad y = x.$

$$\therefore \; \delta y = \delta x,$$

and so $\qquad\qquad\qquad \dfrac{\delta y}{\delta x} = 1.$

Let $\delta x \to 0.$

$$\therefore \; \dfrac{dy}{dx} = 1.$$

$$i.e. \quad \dfrac{\mathbf{d}}{\mathbf{dx}}(\mathbf{x}) = \mathbf{1}.$$

(ii) Let $y = x^2.$

Let x increase by a small amount δx and let the corresponding increase in y be $\delta y.$

Then $\qquad\qquad y + \delta y = (x + \delta x)^2 = x^2 + 2x\delta x + (\delta x)^2,$

and $\qquad\qquad\qquad y = x^2.$

$$\therefore \; \delta y = 2x\delta x + (\delta x)^2,$$

and so $\qquad\qquad\qquad \dfrac{\delta y}{\delta x} = 2x + \delta x.$

Let $\delta x \to 0.$

$$\therefore \; \dfrac{dy}{dx} = 2x.$$

$$i.e. \quad \dfrac{\mathbf{d}}{\mathbf{dx}}(\mathbf{x}^2) = \mathbf{2x}.$$

(iii) Let $y = x^3.$

Let x increase by a small amount δx and let the corresponding increase in y be $\delta y.$

Then $\qquad\qquad y + \delta y = (x + \delta x)^3 = x^3 + 3x^2\delta x + 3x(\delta x)^2 + (\delta x)^3,$

and $\qquad\qquad\qquad y = x^3.$

$$\therefore \; \delta y = 3x^2\delta x + 3x(\delta x)^2 + (\delta x)^3,$$

and so $\qquad\qquad\qquad \dfrac{\delta y}{\delta x} = 3x^2 + 3x(\delta x) + (\delta x)^2.$

Let $\delta x \to 0.$

$$\therefore \; \dfrac{dy}{dx} = 3x^2.$$

$$i.e. \quad \dfrac{\mathbf{d}}{\mathbf{dx}}(\mathbf{x}^3) = \mathbf{3x}^2.$$

In the same way, we obtain the results,

$$\frac{d}{dx}(x^4) = 4x^3.$$

$$\frac{d}{dx}(x^5) = 5x^4, \text{ etc.}$$

The above results suggest the general result that when n is a positive integer,

$$\frac{d}{dx}(x^n) = nx^{n-1}.$$

Differential Coefficient of x^n when n is a negative integer

Again we use special cases to suggest a general result.
e.g. Differentiate (i) x^{-1}, (ii) x^{-2}.

(i) Let $y = x^{-1} = \dfrac{1}{x}$.

Let x increase by a small amount δx and let the corresponding increase in y be δy.

Then
$$y + \delta y = \frac{1}{x + \delta x},$$

and
$$\delta y = \frac{1}{x + \delta x} - \frac{1}{x} = \frac{-\delta x}{x(x + \delta x)}.$$

$$\therefore \frac{\delta y}{\delta x} = -\frac{1}{x^2 + x\delta x}.$$

Let $\delta x \to 0$.

$$\therefore \frac{dy}{dx} = -\frac{1}{x^2}.$$

i.e. $\dfrac{d}{dx}(x^{-1}) = (-1)x^{-2}.$

(ii) Let $y = x^{-2} = \dfrac{1}{x^2}$.

Then
$$y + \delta y = \frac{1}{(x + \delta x)^2},$$

and
$$\delta y = \frac{1}{(x + \delta x)^2} - \frac{1}{x^2},$$

$$= \frac{-2x\delta x - (\delta x)^2}{x^2(x + \delta x)^2}.$$

$$\therefore \frac{\delta y}{\delta x} = \frac{-2x - \delta x}{x^2(x + \delta x)^2}.$$

Let $\delta x \to 0$.

$$\therefore \frac{dy}{dx} = \frac{-2x}{x^4} = -\frac{2}{x^3}.$$

i.e. $\quad \dfrac{\mathbf{d}}{\mathbf{dx}} (\mathbf{x^{-2}}) = -\mathbf{2x^{-3}}.$

Similarly $\qquad \dfrac{\mathbf{d}}{\mathbf{dx}} (\mathbf{x^{-3}}) = -\mathbf{3x^{-4}}.$

These results suggest that the rule for differentiating x^n where n is a positive integer holds good for n a negative integer.

For all integral values of n,

i.e. $\quad \dfrac{\mathbf{d}}{\mathbf{dx}} (\mathbf{x^n}) = \mathbf{nx^{n-1}}.$

Ex. 1. *Differentiate with respect to* x (i) x^{10}; (ii) $\dfrac{1}{x^5}$.

We have $\qquad \dfrac{d}{dx} (x^{10}) = 10x^{10-1} = 10x^9,$

and $\qquad \dfrac{d}{dx} \left(\dfrac{1}{x^5} \right) = \dfrac{d}{dx} (x^{-5}) = -5x^{-5-1},$

$$= -5x^{-6} = -\frac{5}{x^6}.$$

The Differential Coefficient of cx^n where c is a constant

e.g. Differentiate $4x^2$ with respect to x.

Let $y = 4x^2$.

Let x increase by a small amount δx and let the corresponding increase in y be δy.

Then $\qquad y + \delta y = 4(x + \delta x)^2 = 4x^2 + 8x\delta x + 4(\delta x)^2,$

and $\qquad y = 4x^2.$

$$\therefore \ \delta y = 8x\delta x + 4(\delta x)^2.$$

$$\therefore \ \frac{\delta y}{\delta x} = 8x + 4\delta x.$$

Let $\delta x \to 0$.

$$\frac{dy}{dx} = 8x = 4(2x).$$

$$\therefore \ \frac{d}{dx} (4x^2) = 4 \frac{d}{dx} (x^2).$$

In the same way we can show that

$$\frac{d}{dx}(5x^3) = 5\frac{d}{dx}(x^3),$$

$$\frac{d}{dx}\left(\frac{2}{x}\right) = 2\frac{d}{dx}\left(\frac{1}{x}\right) \text{ etc.,}$$

and in general if c is a constant,

$$\frac{d}{dx}(cx^n) = c\frac{d}{dx}(x^n).$$

Differentiation of Sums and Differences

Referring to p. 31 we see that the differential coefficient of $3x^2 - 2x + 1$ is equal to $6x - 2$.

$$i.e. \quad \frac{d}{dx}(3x^2 - 2x + 1) = \frac{d}{dx}(3x^2) - \frac{d}{dx}(2x) + \frac{d}{dx}(1).$$

This result illustrates the fact that the differential coefficient of a sum, or difference, of several terms is equal to the sum, or difference, of the differential coefficients of the various terms.

Ex. 2. Differentiate $3x^4 - 2x + \dfrac{1}{x} + 6$ *with respect to x.*

Let

$$y = 3x^4 - 2x + \frac{1}{x} + 6,$$

then

$$\frac{dy}{dx} = 12x^3 - 2 - \frac{1}{x^2}.$$

(A) EXAMPLES 4a

1. In the following cases obtain $\dfrac{dy}{dx}$ from first principles

(i) $y = 3x^3$; (ii) $y = 2x^2 - 7x + 6$;

(iii) $y = \dfrac{2}{x}$; (iv) $y = \dfrac{3}{x^2}$; (v) $y(x+1) = 1$.

2. Differentiate the following functions with respect to x using the rules obtained in the chapter

(i) $3x^4$; (ii) $\dfrac{x^5}{2}$; (iii) $\dfrac{7}{x}$; (iv) $3x^{10}$;

(v) $\dfrac{1}{x^6}$; (vi) $\dfrac{1}{4x^3}$; (vii) $3x - x^4$; (viii) $2 + x + x^2$;

(ix) $(1+2x)^2$;   (x) $2+\dfrac{1}{x}$;   (xi) $4x^3-\dfrac{1}{x^2}$;   (xii) $7x^{11}+3x^4-4$;

(xiii) $\dfrac{1}{x^2}+\dfrac{1}{x}+1$;   (xiv) $\left(x-\dfrac{1}{x}\right)^2$;   (xv) $\dfrac{x^3-x^2}{x}$;   (xvi) $\dfrac{(1-x)^2}{x^2}$

3. Find $\dfrac{ds}{dt}$ in the following cases

     (i) $s=16t-t^2$;   (ii) $s=t^3-2t^2$;   (iii) $s=ut+\tfrac{1}{2}ft^2$,

where u and f are constants.

4. If $V=\tfrac{4}{3}\pi r^3$, find the value of $\dfrac{dV}{dr}$ when $r=2$. Take π as $\tfrac{22}{7}$.

5. Given that $pv=100$, find $\dfrac{dp}{dv}$ when $v=10$.

6. In the following cases, find the values of x for which $\dfrac{dy}{dx}$ is zero

     (i) $y=3x^2-x-2$;   (ii) $y=\tfrac{4}{3}x^3-2x^2-8x+2$;

     (iii) $y=2x+\dfrac{1}{2x}$;   (iv) $y=x^4-2x^3$.

7. Find the gradient of the curve $y=\left(x+\dfrac{1}{x}\right)^2$ at the point where $x=-1$.

8. Find the gradients of the curve $y=(1-2x)(1+x)$ at the points where it cuts the x-axis.

9. What are the gradients of the curve $y=3x^2+1$ at the ends of the chord whose equation is $y=3x+1$?

10. Find the co-ordinates of the points on the following curves where the tangent is parallel to the x-axis

     (i) $y=1-2x+x^2$;   (ii) $y=3x^3-x$;   (iii) $y=4x^3-3x^2-6x+1$.

11. Find the co-ordinates of the points on the curve $y=\tfrac{1}{3}x^3-\tfrac{1}{2}x^2-11x+2$ where the tangent makes an angle of $45°$ with the positive direction of the x-axis.

12. What is the gradient of the curve $y=x+\dfrac{1}{x}$ at the point $(2, 2\tfrac{1}{2})$? Prove that there is another point on the curve where the tangent is parallel to that at the point $(2, 2\tfrac{1}{2})$ and find its co-ordinates.

13. Sketch the curve $y=x^2-3x-4$. Find $\dfrac{dy}{dx}$ and determine the values of x for which (a) $\dfrac{dy}{dx}$ is positive, (b) $\dfrac{dy}{dx}$ is negative. Using your graph,

determine whether y is increasing or decreasing with x when (a) $\dfrac{dy}{dx}$ is

positive, (b) $\dfrac{dy}{dx}$ is negative.

14. Find the gradients of the curves $y = x^2 + 2$, $y = 4 - x^2$ at their points of intersection.

Tangents and Normals

The *normal* at any point on a curve is the line passing through the point which is perpendicular to the tangent at that point.

Ex. 3. Find the equations of the tangent and normal to the curve $y = 3x^2 - x^3$ at the point $(1, 2)$.

$$y = 3x^2 - x^3.$$

Then $$\frac{dy}{dx} = 6x - 3x^2.$$

Gradient of tangent at point $(1, 2)$

$$= \text{value of } \frac{dy}{dx} \text{ when } x = 1.$$

i.e. Gradient at $(1, 2) = 3$.

∴ Equation of tangent is

$$y - 2 = 3(x - 1) \qquad [y - k = m(x - h)],$$

or $$y = 3x - 1.$$

Gradient of normal is $-\frac{1}{3}$.

Equation of normal is

$$y - 2 = -\tfrac{1}{3}(x - 1),$$

or $$3y - 6 + (x - 1) = 0,$$

or $$3y + x = 7.$$

Ex. 4. For the curve $y = x^2 + 3$, show that $y = 2ax - a^2 + 3$ is the equation of the tangent at the point whose x co-ordinate is a.

Hence find the co-ordinates of the two points on the curve, the tangents at which pass through the point $(2, 6)$. (J.M.B.)

Curve $$y = x^2 + 3.$$

Gradient at any point $= \dfrac{dy}{dx} = 2x.$

∴ Gradient at point where $x = a$ is the value of $\dfrac{dy}{dx}$ at this point, *i.e.* $2a$.

∴ Gradient of tangent at the point whose x co-ordinate is $a = 2a$.

Ordinate of point whose x co-ordinate is $a = a^2 + 3$.

∴ Equation of tangent at point $(a, a^2 + 3)$ is

$$y - (a^2 + 3) = 2a(x - a),$$

$$i.e. \quad y = 2ax - a^2 + 3.$$

If this tangent passes through the point $(2, 6)$ we have

$$6 = 4a - a^2 + 3,$$

$$i.e. \quad a^2 - 4a + 3 = 0.$$

$$\therefore \quad a = 3 \text{ and } 1.$$

I.e. when $a = 3$ or 1, the tangent at the point $(a, a^2 + 3)$ passes through the point $(2, 6)$.

∴ required points are $(3, 12)$ and $(1, 4)$.

Subtangent and Subnormal

Let P be any point on a curve whose equation is given, and let PT and PN be the tangent and normal at P to the curve, and PM the ordinate.

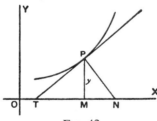

FIG. 42

In the figure, Fig. 42, T and N are on the axis of x; the length TM is called the *subtangent* and the length MN is called the *subnormal* at the point P.

$$\text{Gradient of tangent } \mathbf{PT} = \frac{dy}{dx}.$$

i.e.

$$\frac{\mathbf{PM}}{\mathbf{TM}} = \frac{dy}{dx}.$$

$$\therefore \mathbf{TM} = \frac{y}{\dfrac{dy}{dx}}.$$

Also
$$\frac{MN}{PM} = \tan M\hat{P}N = \tan P\hat{T}M = \frac{dy}{dx}.$$

$$\therefore MN = y\frac{dy}{dx}.$$

i.e. *Length of subtangent* $= \dfrac{y}{\dfrac{dy}{dx}}.$

and *Length of subnormal* $= y\dfrac{dy}{dx}.$

(A) EXAMPLES 4b

1. Find the equations of the tangents to the following curves at the points where $x = 2$

(i) $y = 2x^2$;

(ii) $y = 3x - x^2$;

(iii) $y = 3x^3 - 4x^2 + 2x - 2$;

(iv) $y = \dfrac{1}{x}$;

(v) $y = (2x - 1)^2$.

2. Find the equations of the normals to the curves in the previous example at the points where $x = -1$.

3. Find the equation of the tangent to the curve $y = 8x - 3x^2$ at the point where $x = 1$, and find where this tangent meets the line $x = y$.

4. Find the co-ordinates of the point of intersection of the normal to the curve $y = x^3 - 2x^2$ at the point $x = 0$, and the line $y = 4$.

5. The tangents to the curve $y = 2 - 3x + x^2$ at the points whose abscissae are 1 and 3 meet at P. Find the co-ordinates of P.

6. Find the co-ordinates of the point of intersection of the normal at the point $(1, 1)$ to the curve $y = x^2$ and the normal at the point $(1, -1)$ to the curve $y + x^2 = 0$.

7. Find the lengths of the subtangent and subnormal at the point $(1, 3)$ on the curve $y = 3x^3$.

8. The tangent and normal at the point P $(2, 0)$ on the curve $y = x(2 - x)$ meet the axis of y in the points T and N respectively. Find the lengths PT and PN.

9. Find the equations of the tangents to the curve $y = (x - 1)(1 - 2x)$ at its points of intersection with the x-axis, and find the co-ordinates of the point of intersection of these tangents.

10. If the normal at the point P $(1, 4)$ on the curve $y = 2x^2 + x + 1$ meets the x-axis in Q and the y-axis in R, find the equation of the normal and the lengths OR, OQ and RQ, where O is the origin.

11. On the curve $y = x + \dfrac{1}{x}$, find the points at which the tangents to the curve are parallel to the x-axis, and write down the equations of these tangents.

12. Find the co-ordinates of the points of intersection of the curve $y = 3x^2$ and the straight line $y = 2x + 1$. Also find the equations of the tangents to the curve at these points and find their point of intersection.

13. P is the point $(2, 1)$ on the curve $y = \dfrac{2}{x}$. If the tangent at **P** meet the axes in **A** and **B**, find the equation of the tangent and the co-ordinates of **A** and **B**.

14. Show that, if the gradient of a curve at a point is equal to unity, the subtangent and subnormal are equal in length.

15. Find the co-ordinates of the point on the curve $y = 3x^3$ at which the subtangent and subnormal are equal in length.

16. P is the point $(4, 4)$ on the curve $x^2 = 4y$. **S** is the point $(0, 1)$. The tangent and normal at **P** meet the axis of y in **T** and **N** respectively. Find the co-ordinates of **T** and **N**. Deduce that $ST = SP = SN$.

REVISION PAPERS I

PAPER A (1)

1. Find the distance between the points whose co-ordinates are $(2, -3)$ and $(-3, 2)$. What is the gradient of the line joining the two points?

2. Determine the equation of the straight line passing through the point $(2, -3)$ which is perpendicular to the straight line $2x - 3y = 7$.

3. Differentiate with respect to x

(i) $4x^4 - 2x^2 + 3$;   (ii) $(2x^2 - 1)(1 - x)$.

4. Find the co-ordinates of the point on the curve $y = 1 + 2x - 3x^2$ at which the tangent makes an angle of $45°$ with the positive direction of the x-axis.

5. Find the equations of the tangent and the normal to the curve $y = 1 - 2x^2$ at the point whose abscissa is -1.

PAPER A (2)

1. Find the equation of the straight line joining the origin to the mid-point of the line joining the points $(3, 2)$, $(-2, 4)$.

2. Find the equation of the straight line parallel to the line $4x - 5y + 2 = 0$ and passing through the point $(3, -2)$.

3. Differentiate with respect to t

(i) $(1 + 2t)^2$;   (ii) $\dfrac{t-1}{t^3}$.

4. Find the gradient of the curve $y = x^3 - 3x$ at the point whose abscissa is 3. Deduce the equation of the tangent to the curve at this point.

5. Find the lengths of the subtangent and subnormal to the curve $y = \dfrac{1}{x}$ at the point $(1, 1)$.

PAPER A (3)

1. Find the area of the triangle whose vertices are the points $(1, 2)$, $(3, 1)$, $(-1, 4)$.

2. Prove that the straight lines $2y + 3x - 5 = 0$ and $6y - 4x + 9 = 0$ are perpendicular to one another and find their point of intersection.

3. If $P = 3s^3 - 3s + 2$, find the value of $\dfrac{dP}{ds}$ when $s = -1$.

4. Find the points on the curve $y = 2x^3 - 3x^2 - 12x - 4$ at which the tangent is parallel to the axis of x.

5. Find the points of intersection of the curve $y = x^2 - 4x$ with the x-axis. Obtain the equations of the tangents to the curve at these points and find their point of intersection.

PAPER A (4)

1. Find the gradient of the line joining the points $(2, -3)$, $(-3, -4)$, and write down the equation of the perpendicular bisector of this line.

2. Find the equation of the straight line passing through the origin and through the point of intersection of the lines $2y - x = 1$, $y + x = 2$.

3. Differentiate with respect to x

(i) $(2 + x)(3 + x^2)$;   (ii) $\left(2x - \dfrac{1}{x}\right)^2$.

4. Find the co-ordinates of the point on the curve $y = 2x - x^2$ at which the tangent is perpendicular to the line $y - 2x = 3$.

5. The straight line $y = 3x + 6$ meets the curve $y = 2 + x^2$ in the points P and Q. Find the point of intersection of the normals at P and Q to the curve.

PAPER A (5)

1. Find the co-ordinates of the foot of the perpendicular from the origin to the line $3x - 2y = 4$.

2. A, B, C are the points $(-3, 2)$, $(0, 6)$, $(5, -2)$ respectively. Find the equation of the join of the mid-points of AB and BC.

3. If $pv = 10$, find the values of (a) $\dfrac{dp}{dv}$,   (b) $\dfrac{dv}{dp}$ when $v = 2$.

4. Find the gradients of the curve $y = (x + 1)(x - 1)(x - 3)$ at the points where it cuts the x-axis.

5. Find the equation of the normal to the curve $y = x + \dfrac{2}{x}$ at the point $(2, 3)$. If this normal meets the axes in A and B, find the length AB.

PAPER A (6)

1. The vertices of a triangle are the points A $(0, 0)$, B $(1, 3)$, C $(-2, 4)$. Find the equations of (i) the side AC and (ii) the altitude through B.

2. The co-ordinates of **A, B, C, D** are $(4, 2)$, $(2, 4)$, $(1, 3)$, $(4, 1)$ respectively. Find the gradient of each side of the quadrilateral **ABCD** and prove that one of its angles is a right angle.

3. Find the values of x for which the derivative of the function $4x^3 - 27x + 5$ is zero.

4. Obtain the ordinate of the point on the curve $y = 2x + 4x^2 + 3x^3$ whose abscissa is -1. Write down the equation of the tangent at this point, and obtain the co-ordinates of another point on the curve at which the tangent is parallel to that already found.

5. Find the lengths of the subtangent and subnormal to the curve $2y = x^3 - 2x$ at the point whose abscissa is -2.

PAPER B (1)

1. The co-ordinates of the points **A, B, C** are $(-3, -1)$, $(11, 13)$, $(-1, -3)$ respectively. Find the co-ordinates of the point of intersection of the medians of the triangle **ABC**. (J.M.B.)

2. A thin rod 6 cm long rests with one end **A** on the x-axis and the other end **B** on the y-axis. It passes through the point $(3, 1)$. If **O** is the origin and $\textbf{OA} = x$, find **OB** in terms of x and hence show that $x^2 + \dfrac{x^2}{(x-3)^2} = 36$. (J.M.B.)

3. Differentiate from first principles

(i) $(2x + 1)^2$;   (ii) $\dfrac{1}{x - 2}$.

4. The tangent to the curve $y = x^3$ at a point **P**, abscissa n, meets the axis of x in **T**, and the line through **P** perpendicular to **PT** meets the y-axis in **N**. Show that $\textbf{OT} = \dfrac{2n}{3}$, $\textbf{ON} = \dfrac{3n^4 + 1}{3n}$, and deduce that $\textbf{NT}^2 = \dfrac{(n^4 + 1)(9n^4 + 1)}{9n^2}$. (J.M.B.)

5. The following table shows the length to which a wire was stretched by weights hung on it

Weight W in kg	10	20	30	40	50	60
length l in m	4·14	4·30	4·46	4·61	4·76	4·92

By plotting **W** against l obtain the approximate law connecting **W** and l and deduce the unstretched length of the wire.

PAPER B (2)

1. Through the point **A** $(1, 5)$ is drawn a line parallel to the x-axis to meet at **B**, the line **PQ** whose equation is $3y = 2x - 5$. Find the length **AB** and the sine of the acute angle between **PQ** and **AB**. Deduce that the length of the perpendicular from **A** to **PQ** is $\dfrac{18}{\sqrt{13}}$.

If **PQ** meets the axes in **C** and **D**, find the area of triangle **OCD** where **O** is the origin. (J.M.B.)

2. Find the orthocentre of the triangle formed by the straight lines $y = x + 1$, $2y = 4x + 1$, $3y = 9x + 1$.

3. On the curve $y = 2x^2 - x$ are taken the points **P**, **R**, **Q** whose abscissae are $2 - h$, 2 and $2 + h$ respectively. Find the gradient of the chord **PQ** and show that it is equal to the gradient of the curve at the point **R**.

4. (a) Differentiate with respect to x

$$\text{(i) } \left(\sqrt{x} + \frac{1}{\sqrt{x}}\right)^2; \qquad \text{(ii) } x^p - \frac{1}{x^p}.$$

(b) Find the points at which the gradient of the curve $y = 2x^3 - 15x^2 + 36x - 20$ is equal to 12.

5. **P** is the point $(2a, a^2)$ on the curve $4y = x^2$ and **PM** is drawn parallel to the x-axis to meet the y-axis in **M**. The normal at **P** to the curve meets the axis of y in **N**. Find the gradient of the curve at **P** and the length **MN**. (J.M.B.)

PAPER B (3)

1. Prove that the straight lines joining the point $(1, -2)$ to the points $(-3, 0)$ and $(5, 6)$ are perpendicular. Calculate the co-ordinates of the fourth vertex of the rectangle of which these points are three vertices.

2. The vertices of a triangle are **A** $(-2, 3)$, **B** $(3, 7)$ and **C** $(4, 0)$. Find the co-ordinates of the point **P** on the same side of **AC** as **B** such that triangle **ACP** is right-angled at **C** and equal in area to triangle **ABC**.

3. Describe the changes in the gradient of the curve $y = x^3 - 6x^2 + 12x - 7$ as x increases from 1 to 3, giving a sketch of this portion of the curve. Find the value of x at each point of the curve where the gradient is $\frac{3}{4}$. (J.M.B.)

4. The tangent to the curve $y = \dfrac{1}{x}$ at a point **P**, abscissa n, meets the axes of x and y in **M** and **N** respectively, and the ordinate at **P** meets the x-axis in **K**. Find the equation of the tangent at **P** and show that $KM = n$. Deduce that **P** is the mid-point of **MN**. (J.M.B.)

5. In an experiment, two quantities **W** and **P** are found to have corresponding values given by the following table

W	11	20	27	36	45
P	7·2	9·4	11·6	14·2	16·2

Plot **P** against **W** and deduce an approximate law of the form **P** = a**W** + b.

PAPER B (4)

1. Find the area of the triangle whose sides are the lines $2y + x = 0$, $3y + 2x + 4 = 0$, $2y + 3x + 9 = 0$.

2. Find the co-ordinates of the circumcentre and the radius of the circumcircle of the triangle whose vertices are $(-3, 3)$, $(-1, -5)$, $(4, 2)$.

3. Write down the equation of the straight line passing through the point $(1, 2)$ and having gradient m. If this line meets the axes **OX**, **OY** at **H**, **K** respectively, and the parallelogram **OHKL** is completed, find the co-ordinates of **L**. Verify that **L** lies on the curve $xy = 2x - y$.

4. Find the co-ordinates of the points **A**, **B** on the curve $y = -\frac{1}{9}(2x^3 - 3x^2 - 21x + 11)$ where the tangents make an angle of $45°$ with the x-axis. Show that, if the tangents at **A**, **B** meet the x-axis in **C**, **D** respectively, then **BC**, **AD** are perpendicular to the x-axis. (J.M.B.)

5. Find, from first principles, the differential coefficient of $3x^3 - 4x + 7$ with respect to x.
Find the equation of the tangent to the curve $y = 3x^3 - 4x + 7$ at the point $(1, 6)$. Prove that this curve has two tangents parallel to $3x + y = 0$.
(J.M.B.)

PAPER B (5)

1. Show that the equation of the straight line passing through the point (a, b) and making an angle θ with the axis of x may be written $\dfrac{x - a}{\cos \theta} = \dfrac{y - b}{\sin \theta}$.
The co-ordinates of two points **A** and **B** are $(-3, 2)$ and $(1, 0)$ respectively; **C** is the mid-point of **AB**. Through **C** a straight line **DCE** is drawn at right angles to **AB** such that **DC** = **CE** = $2\sqrt{5}$. Find the co-ordinates of **D** and **E**. (J.M.B.)

2. The lines **AB**, **BC**, **CA** have equations $y + x = 0$, $3y = x + 6$, $y = 2x - 3$ respectively. Find the equations of the lines through **C** parallel and perpendicular to **AB**. Show that the distance of **C** from **AB** is $3\sqrt{2}$ and hence find the area of the triangle **ABC**.

3. The points **A** and **B** on the curve $y = \dfrac{3}{x+2}$ have abscissae 1 and $1 + h$. Show that the gradient of the line **AB** is $\dfrac{-1}{3+h}$. Hence deduce the gradient of the tangent to the curve at **A**, and write down the equation of the normal to the curve at the point **A**. (J.M.B.)

4. For the curve $y = 4x^2$ show that $m^2y = 8mx - 4$ is the equation of the tangent at the point whose x co-ordinate is $\dfrac{1}{m}$.

Hence find the co-ordinates of the two points on the curve, the tangents at which pass through the point $(\frac{3}{8}, -1)$, and write down the equations of the tangents.

5. Corresponding values of x and y are given by the following table

x	1·6	3·0	4·8	7·0	9·6
y	2·0	3·0	4·0	5·0	6·0

Writing p for $\dfrac{x}{y}$, plot p against y and deduce a simple relation connecting p and y. Hence express x in terms of y. (J.M.B.)

PAPER B (6)

1. Show that the points $(1, 3)$, $(3, 4)$, $(4, -3)$ are three of the vertices of a rectangle. Find the co-ordinates of the fourth vertex and the area of the rectangle.

2. Show that the equation of the perpendicular bisector of the straight line joining the points (a, b), (c, d) is $(a-c)x + (b-d)y = \frac{1}{2}(a^2 + b^2 - c^2 - d^2)$. Find the co-ordinates of the centre of the circle which passes through the points $(1, 2)$, $(-1, -1)$, $(2, -2)$.

3. A point with coordinates (x, y) is such that the sum of the squares of its distances from the points $(\pm a, 0)$ is $2k^2$ where $k > a$. Find the equation connecting x, y, a and k.

4. Determine the gradient of the curve $5y = (x+2)(x-1)^2$ at the points where (i) $x = -\frac{3}{2}$; (ii) $x = -2$.

If the tangent at the point $x = -2$ meets the axes at **A** and **B**, find the length **AB**. (J.M.B.)

5. Differentiate from first principles: $\dfrac{1}{2x+1}$.

Obtain an expression for the gradient of the curve $y = x + \dfrac{2}{2x+1}$ at any point. Calculate the gradient at the point where $x = 0$, and find the co-ordinates of another point where the gradient is equal to that at $x = 0$. Sketch the graph for values of x between -2 and $+2$.   (J.M.B.)

CHAPTER 5

DERIVATIVE AS A RATE MEASURER
APPROXIMATIONS

Velocity

If a body moves a distance of 88 m in 4 s, we say that its average velocity is $\frac{88}{4}$ m s^{-1}, *i.e.* total distance travelled divided by total time taken.

Unless the body maintains a constant velocity, we cannot say what was its velocity, say, after 2 s. In order to find this velocity, we should measure the distance travelled in a small time interval containing the instant at which the velocity is required, and work out the average velocity in this interval. By decreasing the interval, this average velocity will approach closer and closer to the actual velocity at the instant required.

Ex. 1. *A body moves on a straight line so that the distance travelled* s m, *in* t s, *is given by the equation* $s = 3t^2 - t$. *Find its velocity at any moment.*

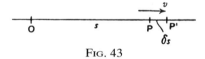

Fig. 43

After t s let the body be at **P** and let its velocity be v m s^{-1}. Let t increase by a small amount δt. Then s will also increase; let this increase be δs.

∴ In $(t + \delta t)$ s the body moves $(s + \delta s)$ m.

$$\therefore \ s + \delta s = 3(t + \delta t)^2 - (t + \delta t),$$
$$= 3t^2 + 6t\delta t + 3(\delta t)^2 - t - \delta t.$$

But $$s = 3t^2 - t.$$

$$\therefore \ \delta s = 6t\delta t + 3(\delta t)^2 - \delta t.$$

∴ Average velocity in the time interval t to $t + \delta t$

$$= \frac{\delta s}{\delta t} = 6t + 3(\delta t) - 1.$$

53

∴ Velocity after t s = value towards which the average velocity approaches, as the interval decreases, *i.e.* as $\delta t \to 0$,

$$= \lim_{\delta t \to 0} \left(\frac{\delta s}{\delta t}\right) = 6t - 1 \text{ m s}^{-1}$$

We can see that the velocity after t s

$$= \lim_{\delta t \to 0} \left(\frac{\delta s}{\delta t}\right).$$

But this limit $= \dfrac{ds}{dt}$ by our previous definition of a differential coefficient.

∴ **the velocity of a moving body at any instant** $= \dfrac{ds}{dt}$**, where s is the distance travelled in time t.**

Ex. 2. For a moving body, $s = 3t^3 - t^2$ where s m is the distance travelled in t s. Find the velocity of the body after 2 s.

Velocity after t s $= \dfrac{ds}{dt} = 9t^2 - 2t \text{ m s}^{-1}.$

∴ Velocity after 2 s $= 9 \cdot 2^2 - 2 \cdot 2 = 36 - 4 = 32 \text{ m s}^{-1}.$

Acceleration

The acceleration of a body is defined as the rate of change of velocity. If in 2 s, the velocity of a moving body increases by 20 m s^{-1}, we say that the average acceleration $= \frac{20}{2} = 10 \text{ m s}^{-2}$. To find the acceleration at any instant, we find the average acceleration in a small time interval containing the instant and find the value towards which this average acceleration approaches, as the interval decreases.

Ex. 3. The velocity v m s^{-1} of a moving point is connected to the time t s by the equation $v = 1 - 3t^2$. Find the acceleration at any time.

We have $v = 1 - 3t^2$ where v is the velocity after t s.

Now in a further small interval of time δt, let the velocity increase by a small amount δv.

∴ $v + \delta v = 1 - 3(t + \delta t)^2 = 1 - 3t^2 - 6t\delta t - 3(\delta t)^2.$

∴ $\delta v = -6t\delta t - 3(\delta t)^2.$

∴ The average acceleration in the interval $(t, t + \delta t)$

$$= \frac{\delta v}{\delta t}$$

$$= -6t - 3(\delta t) \text{ m s}^{-2}.$$

$\therefore$ Actual acceleration after t s $= \lim\limits_{\delta t \to 0} \dfrac{\delta v}{\delta t} = -6t$ m s^{-2}.

In general, as $\lim\limits_{\delta t \to 0} \left(\dfrac{\delta v}{\delta t}\right) = \dfrac{dv}{dt}$, **we see that the acceleration of a moving point is given by the differential coefficient of** v **with respect to** t.

Ex. 4. *The distance* (s) *moved by a point is connected to the time taken* (t), *by the equation*

$$s = \frac{2}{3}t^3 - \frac{9t^2}{2} + 10t.$$

Find the velocity and acceleration in terms of t and deduce that the point is at rest twice.

$$s = \frac{2}{3}t^3 - \frac{9t^2}{2} + 10t.$$

$$\text{Velocity } v = \frac{ds}{dt} = 2t^2 - 9t + 10.$$

$$\text{Acceleration } f = \frac{dv}{dt} = 4t - 9.$$

The point is at rest when $v = 0$;

$$i.e. \quad \text{when } 2t^2 - 9t + 10 = 0.$$

$$i.e. \quad (2t - 5)(t - 2) = 0.$$

$$i.e. \quad \text{when } t = 2\tfrac{1}{2} \text{ and } 2.$$

Rates of Change

If s denotes the distance of a moving point from a fixed origin and t the time, then the velocity of the point is the rate at which s is increasing with respect to t. We have just found, that the velocity is given by the differential coefficient of s with respect to t, i.e. $\dfrac{ds}{dt}$.

Hence $\dfrac{ds}{dt} =$ rate of increase of s with respect to t.

Similarly, the acceleration is the rate of increase of the velocity (v) with respect to the time (t). As the acceleration is given by $\dfrac{dv}{dt}$, we see that $\dfrac{dv}{dt} =$ rate of increase of v with respect to t.

In general, if we have any variable y given in terms of another variable x, then $\dfrac{dy}{dx}$ *not only gives the gradient of the* (x, y) *curve but also gives the rate of increase of y with respect to x.*

Ex. 5. The area A *cm*^{2} *of a blot of ink is growing so that after t s,* A $= 3t^2 + \frac{1}{5}t$. *Find the rate at which the area is increasing after 2 s.*

$$A = 3t^2 + \tfrac{1}{5}t.$$

The rate of increase of A with respect to t is given by $\dfrac{dA}{dt}$.

$$\frac{dA}{dt} = 6t + \tfrac{1}{5}.$$

∴ Rate of increase of A, when $t = 2$, is $12\frac{1}{5}$ cm^2 s^{-1}.

Approximations

When y is given as a function of x, $\dfrac{dy}{dx}$ by definition is equal to $\lim\limits_{\delta x \to 0} \dfrac{\delta y}{\delta x}$, where δy is the small increase in y due to a small increase δx in x.

Now if δx is very small, the fraction $\dfrac{\delta y}{\delta x}$ will be a close approximation to $\dfrac{dy}{dx}$.

$$\textit{i.e.} \quad \frac{\delta y}{\delta x} \simeq \frac{dy}{dx},$$

$$\therefore \quad \delta y \simeq \frac{dy}{dx} \cdot \delta x.$$

This result is useful in determining a small change in one variable produced by a given small change in a second variable.

Ex. 6. If the radius of a sphere increases from 2 *cm to* 2·01 *cm, find the approximate increase in volume.*

We have V $= \frac{4}{3}\pi r^3$, *where* V *is the volume and* r *the radius.*

$$\therefore \quad \frac{dV}{dr} = 4\pi r^2.$$

Let δr represent the small increase in r and δV the corresponding small change in V.

Then approximately $\quad \delta V = \dfrac{dV}{dr} \cdot \delta r,$

$$= 4\pi r^2 \cdot \delta r.$$

∴ when $r = 2$ cm and $\delta r = \cdot 01$ cm,

$$\delta V = 4\pi \cdot 4(\cdot 01) = \cdot 16\pi \text{ cm}^3.$$

Ex. 7. Given that $y = 3x^3$, *find the approximate % increase in y due to an increase of* ·1% *in x.*

We have $$y = 3x^3.$$

$$\therefore \; \delta y \simeq \frac{dy}{dx} \, . \, \delta x \simeq 9x^2 \, . \, \delta x.$$

Hence approximately $\dfrac{\delta y}{y} = \dfrac{9x^2}{3x^3} \, . \, \delta x = 3 \, . \, \dfrac{\delta x}{x} \, .$

But $\left(\dfrac{\delta x}{x} \times 100 \right) = \cdot 1$, as the % increase in x is $\cdot 1$.

$\therefore$ Approximate % increase in $y = \dfrac{\delta y}{y} \times 100 = 3 \times \cdot 1 = \cdot 3.$

(A) EXAMPLES 5a

1. A point moves along a straight line such that the distance travelled $s = 4t^2$, where t is the time taken. If δs is the distance travelled in the small time δt, find the values of δs and $\dfrac{\delta s}{\delta t}$ in terms of t and δt. What does $\dfrac{\delta s}{\delta t}$ represent? Find the velocity of the moving point after time t and deduce the velocity when $t = 2$.

2. For a moving body, the velocity $v = 1 + 2t^2$, where t is the time. Find δv and $\dfrac{\delta v}{\delta t}$ in terms of t and δt. What is the acceleration of the body when $t = 3$?

3. The volume of water in a vessel, $V \text{ cm}^3$, is connected with the time t by the relation $V = 1 + 2t - t^2$. Find the increase in volume δV due to a small increase δt in t, and obtain the value of $\dfrac{\delta V}{\delta t}$. What does $\dfrac{dV}{dt}$ represent and what is its value when $t = 3$?

4. The area A of a blot of ink is increasing so that $A = 3t + t^3$. Find the rate of increase of the area when $t = 1$.

5. If s, the distance travelled in metres by a point moving on a straight line, is connected to the time taken, t seconds, by the relation $s = 3t - t^3$, find the velocity and acceleration of the point when $t = \frac{1}{2}$ s. At what time is the point at rest?

6. A stone thrown up into the air rises s metres in t seconds where $s = 10t - 4 \cdot 9t^2$. Find the velocity of the stone when $t = 1$ s and $t = 2$ s. What is the meaning of the negative velocity?

7. The velocity, $v \text{ m s}^{-1}$, of a moving body is equal to $(1 - 2t)^2$, where t is the time taken in seconds. Find the acceleration of the body after t s. When is the body at rest? What is its acceleration at this instant?

8. The angle turned through by a rotating body, θ, in time t, is given by the relation $\theta = 4t^3 - t^2$. Find the rate of increase of θ when $t = 2$.

9. Express the following statements symbolically
   (i) the rate of increase of x with respect to $t = 2$;
   (ii) the rate of decrease of V with respect to $t = 3$;
   (iii) the rate of decrease of y with respect to x is proportional to x;
   (iv) the rate of increase of p with respect to q is inversely proportional to p^2.

10. If $y = 3x^3$, find the approximate increase in y when x increases from 2 to 2·01.

11. The volume of liquid in a vessel $V = 2t^4$, where t is the time. Find the approximate increase in the volume when t increases from 4 to 4·02.

12. If the radius of a circle increases from 5 cm to 5·1 cm, find the increase in the area.

13. When the radius of a sphere decreases from 4 cm to 3·95 cm, find the decrease in the volume and in the surface area.

$$\left[\begin{array}{c} \text{Area of surface} = 4\pi r^2 \\ \text{Volume} = \tfrac{4}{3}\pi r^3 \end{array} \right].$$

14. The volume of water in a spherical vessel of depth h cm, is given by the expression $V = \dfrac{\pi h^2}{3}(15 - h)$. Find the amount of water which must be poured into the vessel to increase its depth from 4 cm to 4·1 cm.

15. If the relation between the pressure p gf cm^{-2} and the volume v cm^3, of a given mass of gas, is $pv = 500$, find $\dfrac{dp}{dv}$ and hence obtain an approximate expression giving δp in terms of v and δv. What is the change in pressure if the volume expands from 50 to 51 cm^3?

(B) MISCELLANEOUS EXAMPLES

1. Differentiate from first principles

   (i) $1 - 3x + 4x^2$;   (ii) $(1 - 2x)^2$;

   (iii) $\dfrac{1}{1-x}$;   (iv) $\dfrac{1}{x^2 - x}$.

2. Differentiate the following functions with respect to t

   (i) $4t^5 - 3t^4 + 2t^2 - 7$,   (ii) $(2t^2 - 1)(1 - t^3)$;

   (iii) $\left(t^2 - \dfrac{1}{t^2} \right)^2$;   (iv) $\dfrac{1 - t^2}{3t^3}$.

3. Find the gradient of the curve $y = x(x-2)(x-3)$ at the points where it cuts the x-axis. Also find the abscissae of the points on the curve where the gradient is zero. Sketch the curve.

4. If the abscissae of three points P, Q, R on the curve $y = 3x^2$ are x, $x-h$, $x+h$ find the gradient of the chord QR, and show that it is equal to the gradient of the curve at P.

5. Find the point of intersection of the normals to the curve $x^2 = 16y$ at the ends of the chord $x = 2y - 6$.

6. The distance s m, which a body has travelled in t s, is given by $s = 4t^3 - 15t^2 - 18t + 12$. Find when the body is at rest, and the distance from the origin and the acceleration at this time.

7. If the radius of a circular cylinder of constant height increases by $\cdot2\%$, find the percentage increase in the volume and the curved surface area.

8. The path traced out by a shell is given by the equation $y = 4x - \dfrac{x^2}{3\cdot2}$, the x- and y-axes being the horizontal and vertical through the point of projection. Find the angle of projection, and the direction of motion when $x = 2$.

9. If $y = x - \dfrac{1}{x}$, prove that y always increases with x.

10. If $y = 4x^3 - 3x^2 - 6x + 1$, find the range of values of x for which y decreases as x increases.

11. Find the equations of the tangent and normal to the curve $y = x(2-x)$ at the point whose abscissa is a.

12. The tangent at the point where $x = t$ on the curve $y = x^3 - 2x$, meets the axes of x and y in A and B respectively. Find the co-ordinates of A and B, and the length AB.

13. Find the equation of the normal to the curve $4y = x^2$ at the point $(2t, t^2)$. Also find the co-ordinates of the second point of intersection of this normal with the curve.

14. A point moves along a straight line so that its distance from a fixed point O is given by $s = t - t^3$, where t is the time from the start. If v is the velocity and a the acceleration, at any instant, prove that $v = 1 + \frac{1}{2}at$.

15. The work done by a force P newtons, in moving through a distance s m in its own direction, is equal to Ps joules. If $P = 3s^2 - s$ and $s = 2t$, t being the time in s, find the rate of working when (a) $t = 2$ s; (b) $s = 10$ m.

16. The tangent at the point P on the curve $y = x(x-1)(x-2)$ whose x co-ordinate is equal to p meets the axis of y in T. Prove that the length $OT = 3p^2 - 2p^3$.

17. Find the equation of the tangent to the curve $y = \dfrac{c^2}{x}$, where c is a constant, at the point $\left(ct, \dfrac{c}{t}\right)$. If the tangent meets the axes of x and y in A and B, prove that the triangle AOB is of constant area.

18. Show that the equation of the tangent to the curve $y = 3x(1-x)$ at the point where $x = a$, is $y - x(3-6a) - 3a^2 = 0$. If this tangent passes through the point $(1, 3)$, find the possible values of a. Hence write down the equations of the tangents to the curve $y = 3x(1-x)$ which pass through the point $(1, 3)$.

19. Prove that the equation of the tangent to the curve $x^2 = y$ at the point whose x co-ordinate is equal to m, is $y - 2mx + m^2 = 0$. Find the values of m for which this line passes through the point $(2, 0)$. Deduce the equations of the tangents from the point $(2, 0)$ to the curve $x^2 = y$.

20. The chord PQ of the curve $x^2 = 4y$ passes through the point $(0, 1)$. if the co-ordinates of P are $(2t, t^2)$, find the co-ordinates of Q. Prove that the tangents at P and Q are at right angles and find their point of intersection.

CHAPTER 6

HIGHER DERIVATIVES. SIGN OF DERIVATIVE MAXIMA AND MINIMA

Suppose we are given y as some function of x, for example,

$$y = 3x^5 - x^3.$$

Then on differentiating, we get

$$\frac{dy}{dx} = 15x^4 - 3x^2. \tag{1}$$

Now, denote $\dfrac{dy}{dx}$ by y_1,

so $\qquad\qquad y_1 = 15x^4 - 3x^2.$

That is, the variable y_1 is a definite function of x, and in the ordinary way, on differentiating y_1 with respect to x, we have

$$\frac{dy_1}{dx} = 60x^3 - 6x. \tag{2}$$

$\dfrac{dy_1}{dx}$ is the result of differentiating $\dfrac{dy}{dx}$ with respect to x and is called *the second differential coefficient of y with respect to x.*

This second differential coefficient is denoted by the symbol $\dfrac{d^2y}{dx^2}$. If we differentiate $\dfrac{d^2y}{dx^2}$ with respect to x we get the third differential coefficient, denoted by $\dfrac{d^3y}{dx^3}$.

From (2), we have $\qquad \dfrac{d^3y}{dx^3} = 180x^2 - 6. \tag{3}$

The fourth differential coefficient is obtained by differentiating the third differential coefficient with respect to x;

$$i.e. \quad \frac{d^4y}{dx^4} = 360x, \text{ and so on.}$$

Ex. 1. *If* $v = 14t^3 - 8t$, *find the value of* $\dfrac{d^2v}{dt^2}$ *when* $t = 1$.

$$v = 14t^3 - 8t.$$

$$\therefore \frac{dv}{dt} = 42t^2 - 8.$$

$$\therefore \frac{d^2v}{dt^2} = 84t.$$

So, when $t = 1$, $\qquad \dfrac{d^2v}{dt^2} = 84.$

Ex. 2. *For the function* $y = x^3 - 3x$, *find the values of* x *for which* $\dfrac{dy}{dx}$ *is zero, and obtain the values of* $\dfrac{d^2y}{dx^2}$ *when* $\dfrac{dy}{dx}$ *is zero.*

$$\frac{dy}{dx} = 3x^2 - 3.$$

So $\dfrac{dy}{dx}$ is zero when $3x^2 - 3 = 0$.

$$\textit{i.e.} \quad \text{when } x = \pm 1.$$

$$\frac{d^2y}{dx^2} = 6x.$$

$\therefore$ when $x = 1$, $\qquad \dfrac{d^2y}{dx^2} = 6.$

and when $x = -1$, $\qquad \dfrac{d^2y}{dx^2} = -6.$

Sign of the Derivative

When y is given as a function of x, we have seen that $\dfrac{dy}{dx}$ gives the rate of increase of y with respect to x.

Hence, if $\dfrac{dy}{dx}$ *is positive,* y *is increasing as* x *increases, and if* $\dfrac{dy}{dx}$ *is negative,* y *is decreasing as* x *increases.*

In the same way, as $\dfrac{d^2y}{dx^2}$ is the differential coefficient of $\dfrac{dy}{dx}$ with respect to x, if $\dfrac{d^2y}{dx^2}$ is positive, $\dfrac{dy}{dx}$ is increasing as x increases, and if $\dfrac{d^2y}{dx^2}$ is negative, $\dfrac{dy}{dx}$ is decreasing as x increases.

(A) EXAMPLES 6a

1. In the following cases, find $\dfrac{dy}{dx}, \dfrac{d^2y}{dx^2}$ and $\dfrac{d^3y}{dx^3}$

(i) $y = 4x^6$; (ii) $y = 2x - \dfrac{x^3}{2}$; (iii) $y = \dfrac{2}{x}$;

(iv) $y = x^2 - \dfrac{1}{x^2}$; (v) $y = (1 - x^2)^2$

2. Find $\dfrac{d^2s}{dt^2}$ when

(i) $s = t - 3t^3$; (ii) $s = t + \dfrac{1}{t}$;

(iii) $s = ut + \frac{1}{2}ft^2$ where u and f are constants.

3. Find the values of $\dfrac{d^2v}{dh^2}$ for $h = 2$ and $h = -1$, when $v = 4h - 3h^3$.

4. If $y = x(3 - x)$, find the value of x for which $\dfrac{dy}{dx}$ is zero. What is the value of $\dfrac{d^2y}{dx^2}$ when $\dfrac{dy}{dx}$ is zero?

5. If $y = 2x^2(1 - x)$, find the value of x for which $\dfrac{d^2y}{dx^2}$ is zero. What is the value of $\dfrac{dy}{dx}$ when $\dfrac{d^2y}{dx^2}$ is zero?

6. For the curve $y = 4x^3 - 15x^2 - 18x + 12$, determine the signs of $\dfrac{d^2y}{dx^2}$ when $\dfrac{dy}{dx}$ is zero.

7. If $y = x^5 - 3x^4$, find the value of the first derivative which does not vanish when $x = 0$.

8. Sketch the curve $y = x^2 - 2x - 8$. Find the value of $\dfrac{dy}{dx}$. Mark on the curve the portions where (a) $\dfrac{dy}{dx}$ is positive; (b) $\dfrac{dy}{dx}$ is negative. Find a point on the curve where $\dfrac{dy}{dx}$ vanishes and changes sign from negative to positive.

9. If $y = x^3 - x^2$, for what values of x is y increasing as x increases? For what values of x is $\dfrac{dy}{dx}$ increasing as x increases?

10. Show that the function $3x^2 + 2x + 4$ increases as x increases for values of x greater than $-\frac{1}{3}$.

11. If $y = 3x^2 + 6x$, find $\dfrac{dy}{dx}$ and $\dfrac{d^2y}{dx^2}$.

Verify that $x^2\dfrac{d^2y}{dx^2} - 2x\dfrac{dy}{dx} + 2y = 0$.

12. If $y = x^2 + \dfrac{1}{x^2}$, find $\dfrac{dy}{dx}$ and $\dfrac{d^2y}{dx^2}$.

Verify that $x^2\dfrac{d^2y}{dx^2} + 4x\dfrac{dy}{dx} + 2y = 12x^2$.

Maximum and Minimum Points

Consider the graph shown in the diagram (Fig. 44).

For the portion **PA** of the curve, y is increasing as x increases and so $\dfrac{dy}{dx}$ is positive.

At the point **A**, $\dfrac{dy}{dx}$ is zero.

Along **AB**, $\dfrac{dy}{dx}$ is negative and is again

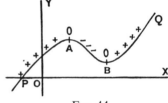

zero at **B**. For the portion **BQ**, $\dfrac{dy}{dx}$ is again positive.

Fig. 44

The points **A** and **B** on the curve are called *turning points*.

The ordinate at **A** is greater than any *neighbouring* ordinate and **A** is called a *maximum* point on the curve.

The ordinate at **B** is less than any *neighbouring* ordinate and **B** is called a *minimum* point on the curve.

Method of determining Maximum and Minimum Points

MAXIMUM POINT

MINIMUM POINT

Fig. 45

Maximum Point

(1) $\dfrac{dy}{dx} = 0$.

(2) $\dfrac{dy}{dx}$ changes from positive to negative in passing through the point.

i.e. $\dfrac{dy}{dx}$ is decreasing,

$\therefore \dfrac{d^2y}{dx^2}$ is negative.

Minimum Point

(1) $\dfrac{dy}{dx} = 0$.

(2) $\dfrac{dy}{dx}$ changes from negative to positive in passing through the point.

$$i.e. \quad \frac{dy}{dx} \text{ is increasing, and}$$

$$\therefore \quad \frac{d^2y}{dx^2} \text{ is positive.}$$

The method of determining the turning points on any curve is illustrated by the following examples:

Ex. 3. Determine the turning point on the curve $y = 3x^2 - x + 1$, and find whether it is a maximum or a minimum point. Illustrate by a rough graph.

$$y = 3x^2 - x + 1.$$

$$\frac{dy}{dx} = 6x - 1.$$

But for a turning point, $\qquad \dfrac{dy}{dx} = 0.$

$$\therefore \ 6x - 1 = 0.$$

$$i.e. \quad x = \tfrac{1}{6}.$$

So when $x = \tfrac{1}{6}$ there is a possible turning point.

To find whether this is a maximum or minimum point, we consider the sign of $\dfrac{d^2y}{dx^2}$ at this point.

We have $\qquad \dfrac{d^2y}{dx^2} = 6.$

$$\therefore \text{ when } x = \tfrac{1}{6}, \quad \frac{dy}{dx} = 0$$

and $\dfrac{d^2y}{dx^2}$ is positive.

$\therefore$ the point at which $x = \tfrac{1}{6}$ is a minimum point on the curve.

$\therefore$ Minimum value of

$$y = 3(\tfrac{1}{6})^2 - \tfrac{1}{6} + 1 = \tfrac{11}{12}.$$

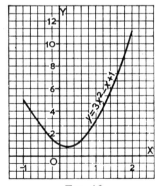

Fig. 46

Ex. 4. Find the turning points on the curve

$$y = 4x^3 - 3x^2 - 18x + 6$$

and distinguish between them.

$$y = 4x^3 - 3x^2 - 18x + 6.$$

$$\therefore \frac{dy}{dx} = 12x^2 - 6x - 18.$$

For a turning point, $\dfrac{dy}{dx} = 0.$

$$\therefore \ 12x^2 - 6x - 18 = 0.$$

$$\therefore \ 2x^2 - x - 3 = 0.$$

$$i.e. \ (2x - 3)(x + 1) = 0.$$

$$i.e. \ x = \tfrac{3}{2} \text{ and } -1.$$

Now differentiating again, $\dfrac{d^2y}{dx^2} = 24x - 6.$

$\therefore$ when $x = \tfrac{3}{2}$, $\qquad \dfrac{d^2y}{dx^2} = 30$, *i.e.* positive.

$\therefore$ the point at which $x = \tfrac{3}{2}$ is a minimum point.

When $x = -1$, $\qquad \dfrac{d^2y}{dx^2} = -30$, *i.e.* negative.

$\therefore$ the point at which $x = -1$ is a maximum point.
$\therefore$ the point $(\tfrac{3}{2}, -\tfrac{57}{4})$ is a minimum point $\Big\}$
and the point $(-1, 17)$ is a maximum point.

Exceptional Cases

In certain exceptional cases, the rule for discriminating between maximum and minimum points—*i.e.* the consideration of the sign of $\dfrac{d^2y}{dx^2}$—breaks down, as $\dfrac{d^2y}{dx^2}$ is zero. In cases of this sort we proceed as in the following examples.

Ex. 5. *Determine the maximum and minimum points on the curve* $y = x^4 - 4x^3$. *Sketch the curve.*

$$y = x^4 - 4x^3;$$

$$\frac{dy}{dx} = 4x^3 - 12x^2.$$

$$\therefore \frac{dy}{dx} = 0 \text{ when } 4x^3 - 12x^2 = 0;$$

$$i.e. \ \text{ when } x = 0 \text{ and } x = 3.$$

We have $\dfrac{d^2y}{dx^2} = 12x^2 - 24x.$

$\therefore$ When $x = 3$, $\dfrac{d^2y}{dx^2} = 108 - 72 = 36.$ [Positive.]

$\therefore$ $x = 3$ gives a minimum point on the curve.

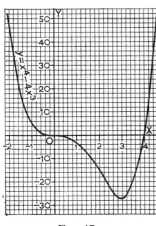

FIG. 47

When $x = 0$, $\dfrac{d^2y}{dx^2} = 0.$

In this case we consider the sign of $\dfrac{dy}{dx}$ immediately before and after the point $x = 0$.

If $\dfrac{dy}{dx}$ changes sign from negative to positive, the point is a minimum point.

If $\dfrac{dy}{dx}$ changes sign from positive to negative, the point is a maximum point.

If $\dfrac{dy}{dx}$ does not change sign, the point is neither maximum nor minimum.

In this case $\dfrac{dy}{dx} = 4x^2(x - 3).$

$\therefore$ when x is slightly <0, $\dfrac{dy}{dx}$ is negative,

and when x is slightly >0, $\dfrac{dy}{dx}$ is also negative.

$\therefore$ the point $x = 0$ is *not* a turning point. It is called a *point of inflexion.* (The graph (Fig. 47) shows the shape of the curve.)

Ex. 6. Determine the turning point on the curve $y = x^4$.

$$\frac{dy}{dx} = 4x^3.$$

$$\therefore \frac{dy}{dx} = 0 \text{ when } x = 0.$$

$$\frac{d^2y}{dx^2} = 12x^2.$$

So when $x = 0$, $\dfrac{d^2y}{dx^2} = 0$, and we cannot say whether the point is maximum, minimum or inflexion.

Considering a value of x slightly <0, $\dfrac{dy}{dx}$ is negative,

and considering a value of x slightly >0, $\dfrac{dy}{dx}$ is positive.

$\therefore$ The point where $x = 0$ is a minimum point on the curve.

To find the maximum and minimum values of a function of a given variable

The method employed to find the turning points on a given curve, can be used also to obtain the maximum and minimum values of a function of a given variable.

For example, suppose we require to find the greatest value of the function $x(2-x)$. If we introduce a second variable y, such that $y = x(2-x)$, the problem reduces itself to finding the maximum point on the curve $y = x(2-x)$.

Proceeding as before we have

$$y = x(2-x), \tag{1}$$

$$\therefore \frac{dy}{dx} = 2 - 2x.$$

$$\therefore \frac{dy}{dx} = 0, \text{ when } x = 1.$$

Also $$\frac{d^2y}{dx^2} = -2.$$

$\therefore$ when $x = 1$, $\quad \dfrac{dy}{dx} = 0$ and $\dfrac{d^2y}{dx^2}$ is negative.

$\therefore$ When $x = 1$, there is a maximum point on the curve $y = x(2-x)$.

The maximum value of y, that is, of $x(2-x)$, is found by putting $x = 1$ in equation (1).

$\therefore$ Maximum value of $x(2-x)$ is 1.

The following examples illustrate further the method of determining the maximum and minimum values of a function of one variable.

Ex. 7. Find the minimum value of $2x^2 - x + 3$.

Let $$y = 2x^2 - x + 3.$$

Then $\dfrac{dy}{dx} = 4x - 1.$

                              $\therefore \dfrac{dy}{dx} = 0$ when $x = \frac{1}{4}.$

Also $\dfrac{d^2y}{dx^2} = 4.$

$\therefore$ when $x = \frac{1}{4},$ $\dfrac{dy}{dx} = 0$ and $\dfrac{d^2y}{dx^2}$ is positive.

Hence when $x = \frac{1}{4},$ y is a minimum.

$\therefore$ Minimum value of $y(i.e.$ of $2x^2 - x + 3) = 2(\frac{1}{4})^2 - \frac{1}{4} + 3$
$$= \tfrac{1}{8} - \tfrac{1}{4} + 3$$
$$= 2\tfrac{7}{8}.$$

Ex. 8. The rate of working of an engine is given by the expression $10v + \dfrac{4000}{v}$, *where v is the speed of the engine. Find the speed at which the rate of working is least.*

We require to find the value of v for which the expression $10v + \dfrac{4000}{v}$ is a minimum.

Let $H = 10v + \dfrac{4000}{v}.$

                              $\therefore \dfrac{dH}{dv} = 10 - \dfrac{4000}{v^2}.$

$\therefore \dfrac{dH}{dv} = 0,$ when $v^2 = 400,$ *i.e.* when $v = \pm 20.$

Also $\dfrac{d^2H}{dv^2} = \dfrac{8000}{v^3}.$

$\therefore$ when $v = 20,$ $\dfrac{dH}{dv} = 0$ and $\dfrac{d^2H}{dv^2} = 1$ (*i.e.* positive).

$\therefore$ The rate of working, H, is a minimum when $v = 20.$

Ex. 9. If x + y = 10, find the maximum value of xy.

In this case we have a function xy of *two* variables, and before the previous method can be employed we must reduce this to a function of *one* variable.

Let $z = xy.$

We use the relation $x + y = 10$ to eliminate y.

$$\therefore \ z = x(10 - x),$$

and
$$\frac{dz}{dx} = 10 - 2x.$$

$$\therefore \ \frac{dz}{dx} = 0, \text{ when } x = 5,$$

and
$$\frac{d^2z}{dx^2} = -2.$$

$\therefore$ When $x = 5$, $\dfrac{dz}{dx} = 0$ and $\dfrac{d^2z}{dx^2}$ is negative.

$\therefore$ z is a maximum when $x = 5$.

i.e. xy is a maximum when $x = 5$ and $y = 5$.

$\therefore$ Maximum value of $xy = 25$.

(A) EXAMPLES 6b

1. For the graphs in Fig. 48, state (*a*) the portions of the curves for which $\dfrac{dy}{dx}$ is positive, (*b*) the portions of the curves for which $\dfrac{dy}{dx}$ is negative, (*c*) the points where $\dfrac{dy}{dx}$ is zero, (*d*) the maximum and minimum points on the curves, (*e*) points on the curves which are not turning points but where $\dfrac{dy}{dx}$ is zero.

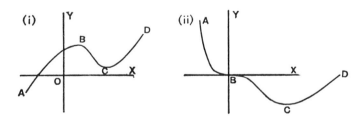

Fig. 48

2. Sketch the curve $y = x(3 - x)$. For what value of x is $\dfrac{dy}{dx}$ zero? What are the signs of $\dfrac{dy}{dx}$ immediately before and after this point? Is the point a maximum or a minimum point?

3. Sketch the graph $y = 1 + x^3$. What is the sign of $\dfrac{dy}{dx}$ when $x = 1$ and $x = -1$? For what value of x is $\dfrac{dy}{dx}$ zero? Does $\dfrac{dy}{dx}$ change sign as we pass through this point? Is the point a turning point?

4. Determine the turning points on the following curves, and discriminate between maximum and minimum points

(i) $y = 3x - x^2$;   (ii) $y = x^2 - 4x + 1$;   (iii) $y = (2 - x)^2$;

(iv) $y = (1 - x)(2x + 1)$;   (v) $y = x + \dfrac{1}{x}$;   (vi) $y = \dfrac{x^3}{3} - 4x$;

(vii) $y = 2x^3 - 9x^2 + 12x + 4$;   (viii) $y = x^2 + \dfrac{16}{x}$.

In each case illustrate your answer by making a rough sketch of the curve.

5. Find the points on the following curves where $\dfrac{dy}{dx}$ is zero. By considering the signs of $\dfrac{dy}{dx}$ before and after these points, determine whether the points are maximum, minimum or inflexion points. In each case illustrate your answer by a rough sketch of the curve.

(i) $y = x^3$;   (ii) $y = 2x^4$;   (iii) $y = 1 - x^3$;

(iv) $y = 3x^6$;   (v) $y = 2x^5 + 3$.

6. Find the maximum value of $3x(4 - x)$.

7. Find the least value of $(1 - 2x)(1 - x)$.

8. Find the value of p which makes $p + \dfrac{1}{p}$ a minimum.

9. What is the minimum value of $(1 - 2x)^2$?

10. Find the maximum and minimum values of the following functions of x

(i) $x^3 - x^2$;   (ii) $2x^3 - 3x^2 - 36x + 4$;

(iii) $2x - x^2 - 8x^3$;   (iv) $x^2 + \dfrac{1}{x^2}$.

11. The distance, s m, moved by a body in time t s, is given by the expression $s = 96t - 16t^2$. Find the maximum value of s.

12. The acceleration of a moving body is equal to $3t(t - 1)$, where t is the time. Find when the acceleration is a minimum and find its minimum value.

13. If $x + y = 5$, find the greatest value of $2xy$.

14. If $p + q = 6$, what is the least value of $p^2 + q^2$?

15. If $V = r^2h$ and $h + r = 6$, find the greatest and least values of V.

Harder Examples of Maxima and Minima

Ex. 10. A rectangular block with a square base has the total area of its surface equal to $150 \ cm^2$, *and the sides of the base are each* x *cm long. Prove that the volume of the block is* $\frac{1}{2}(75x - x^3) \ cm^3$, *and hence find the maximum volume of the block.*   (J.M.B.)

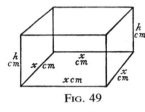

Fig. 49

In order to obtain the maximum volume of the block, say V, we must first obtain an expression giving V in terms of *one* other variable. As is indicated in the question, we will first show that $V = \frac{1}{2}(75x - x^3)$ where x cm is the length of the side of the base.

We have $V = x^2h \ cm^3$, where h cm is the height of the block.

To obtain h in terms of x, we use the fact that the surface area of the block is equal to $150 \ cm^2$.

$$\text{Surface area} = 2x^2 + 4xh = 150.$$

$$\therefore \ x^2 + 2xh = 75.$$

$$\therefore \ h = \frac{75 - x^2}{2x}$$

Hence $V = x^2h = x^2 \left\{ \dfrac{75 - x^2}{2x} \right\} = \frac{1}{2} \{ 75x - x^3 \} \ cm^3.$

Having obtained V in terms of one variable, we proceed to find its maximum value in the usual way.

We have
$$\frac{dV}{dx} = \frac{75}{2} - \frac{3x^2}{2}.$$

$$\therefore \ \frac{dV}{dx} = 0, \text{ when } \frac{75}{2} - \frac{3x^2}{2} = 0,$$

i.e. when $x^2 = 25$, or $x = \pm 5$.

In this case, the negative value for x has no meaning and we do not consider it.

$$\frac{d^2V}{dx^2} = -\frac{6x}{2} = -3x.$$

$\therefore$ when $x = 5$,
$$\frac{d^2V}{dx^2} = -15.$$

Hence, when $x = 5$, $\dfrac{dV}{dx} = 0$ and $\dfrac{d^2V}{dx^2}$ is negative.

∴ $x = 5$ cm makes V a maximum.

∴ Maximum value of $V = \frac{1}{2}\{375 - 125\} = 125$ cm³.

Ex. 11. *A wastepaper basket consists of an open circular cylinder. If the volume of the basket is to be 2000 cm³, find the radius of its base when the material used is a minimum.*

The material used in making the basket depends on the surface area of the basket.

Hence we require to find the radius of the base when the surface area is a minimum.

We first of all obtain an expression giving the surface area (say S cm²) in terms of the radius of the base (say *r* cm).

The total surface area $S = \pi r^2 + 2\pi rh$ cm², where *h* cm is the height of the cylinder.

FIG. 50

To obtain S in terms of *r* alone, *h* must be obtained in terms of *r*. This is done by using the fact that the volume of the basket = 2000 cm³.

We have,   volume $= 2000 = \pi r^2 h$.

$$\therefore h = \frac{2000}{\pi r^2}.$$

Hence $S = \pi r^2 + 2\pi rh,$

$$= \pi r^2 + 2\pi r . \frac{2000}{\pi r^2},$$

$$= \pi r^2 + \frac{4000}{r}.$$

i.e. $S = \pi r^2 + \dfrac{4000}{r}.$

$$\therefore \frac{dS}{dr} = 2\pi r - \frac{4000}{r^2}.$$

$$\therefore \frac{dS}{dr} = 0 \text{ when } 2\pi r - \frac{4000}{r^2} = 0.$$

I.e.   when $r^3 = \dfrac{2000}{\pi}$, or $r = \sqrt[3]{\dfrac{2000}{\pi}}.$

Also, $\dfrac{d^2S}{dr^2} = 2\pi + \dfrac{8000}{r^3}.$

∴ When $r = \sqrt[3]{\dfrac{2000}{\pi}}$, $\dfrac{d^2S}{dr^2}$ is positive.

∴ When $r = \sqrt[3]{\dfrac{2000}{\pi}}$, S is a minimum.

Or, the amount of material used will be a minimum when

$$r = \sqrt[3]{\frac{2000}{\pi}} \text{ cm} = 8 \cdot 60 \text{ cm}.$$

(B) EXAMPLES 6c

1. Find the points on the following curves where the gradient is zero, and determine the nature of these points

(i) $y = 4x^3 - 15x^2 - 18x + 79;$ (ii) $y = 3x^4 - x^3;$

(iii) $y = x^3 - 3x^2 + 3x;$ (iv) $y = 3x^4 - 8x^3 + 6x^2.$

2. Find the maximum and minimum values of $\dfrac{(x-1)(x-2)}{x}$, and illustrate your result by drawing the graph of the function between $x = \pm 3$.

3. Two quantities x and y are connected by the relation $x + y = 10$. Find the maximum and minimum values of $x^3 y^2$.

4. One side of a rectangular enclosure is formed by a hedge; the total length of fencing available for the other three sides is 200 m. Obtain an expression for the area of the enclosure, A m^2, in terms of its length x m, and hence deduce the maximum area of the enclosure.

5. If the volume of a circular cylindrical block is equal to 800 cm^3 prove that the total surface area is equal to $2\pi x^2 + \dfrac{1600}{x}$ cm^2, where x cm is the radius of the base. Hence obtain the value of x which makes the surface area a minimum.

6. It being given that the volume of a right circular cone is $\frac{1}{3}\pi r^2 h$, where r is the radius of the base and h the height of the cone, find the dimensions of the cone of greatest volume when the sum of the height and radius of the base is 6 cm.

7. The diagram represents the section of an open rectangular drain. The perimeter ABCD is equal to 3 m. Find the area of the section in terms of the length BC, and hence obtain the maximum area of the section.

FIG. 51

8. A closed rectangular box is made of sheet metal of negligible thickness, the length of the box being twice its width. If the box has a capacity of 243 cm^3, show that its surface area is equal to $4x^2 + \dfrac{729}{x}$. Hence obtain the dimensions of the box of least surface area.

9. A rectangular sheet of metal is 2 m by 1 m. Equal squares of side x cm are cut from each of the corners and the whole is folded up to form an open rectangular tray of depth x cm. Find the volume of the tray in terms of x, and find the value of x which makes the volume a maximum.

10. The running cost, C of a ship, in pounds per hour, is given by the formula $C = 4 + \dfrac{s^3}{1000}$ where s is the speed in km h^{-1}. Write down the total cost for a passage of 500 km and find the speed which makes this total cost a minimum.

11. The figure (Fig. 52) consists of a rectangle BCDE and an equilateral triangle ABE.
If the perimeter of the figure is 18 cm, find its dimensions when the area is a maximum.

12. A body moves a distance s m in t s where

$$s = t^4 - 2t^3 - 36t^2 + 2t + 1.$$

Find its velocity after t s and hence find its minimum velocity.

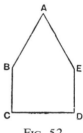

FIG. 52

13. An open rectangular basket has a square base. If the volume of the basket is 2000 cm^3, find its dimensions when the amount of material used is a minimum.

14. A point P, whose x co-ordinate is a, is taken on the line $y = 3x - 7$. If Q is the point (4, 1), prove that $PQ^2 = 10a^2 - 56a + 80$. Find the value of a for which this expression is a minimum. Hence deduce the co-ordinates of the foot of the perpendicular from Q to the line $y = 3x - 7$.   (J.M.B.)

15. Figure 53 represents a skeleton box on a square base ABCD. If the total length of wire used is 12 m, find the maximum volume of the box.

16. A point P is taken on the circumference of a semicircle with variable bounding diameter AB. If P moves such that $AP + PB = 10$ cm, find the maximum area of $\triangle ABP$.

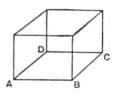

FIG. 53

17. An open cylindrical can has an area of surface equal to 100 cm^2. Find its maximum volume.

18. The base of a right circular cone is a section of a sphere, centre O, and of unit radius. The vertex of the cone is on the sphere and the centre of the sphere is within the cone. If the distance of O from the base of the cone is x, prove that the volume of the cone is $\dfrac{\pi}{3}(1 + x - x^2 - x^3)$. Hence find the value of x which makes the volume of the cone a maximum.
                                        (J.M.B.)

19. A right circular cone is cut from a solid sphere of radius a, the vertex and circumference of the base being on the surface of the sphere. If x is the distance of the base from the centre of the sphere, prove that the volume of the cone is $\frac{1}{3}\pi(a^2 - x^2)(a + x)$, and hence find the height of the cone when its volume is a maximum. (J.M.B.)

20. A solid circular cylinder is cut from a sphere radius R, the circumferences of the ends of the cylinder being on the surface of the sphere. Prove that the volume of the cylinder is $2\pi\sqrt{R^2 r^4 - r^6}$, where r is the radius of its circular ends. Obtain the value of r which makes $(R^2 r^4 - r^6)$ a maximum, and hence deduce the maximum volume of the cylinder.

CHAPTER 7

FUNCTION OF A FUNCTION

Suppose we are given y as some function of a variable u, say $4u^6$, and u as some function of x, say $(1-x^2)$, and we require to obtain $\dfrac{dy}{dx}$. In many cases it is of little assistance to obtain y directly in terms of x, and then attempt to differentiate. We need some method which does not involve the elimination of u. Considering the example given, we have $y = 4u^6$ and $u = 1 - x^2$. Proceeding with the usual method of differentiation from first principles, let x increase by a small amount δx. This will cause a small change in u, which in turn will cause a small change in y. Let the increases in u and y be δu and δy.

Now, by the ordinary laws of fractions,

$$\frac{\delta y}{\delta x} = \frac{\delta y}{\delta u} \times \frac{\delta u}{\delta x}.$$

As $\delta x \to 0$, δu also $\to 0$, $\dfrac{\delta y}{\delta x}$ becomes $\dfrac{dy}{dx}$, $\dfrac{\delta y}{\delta u}$ becomes $\dfrac{dy}{du}$, and $\dfrac{\delta u}{\delta x}$ becomes $\dfrac{du}{dx}$.

Hence, assuming that as $\dfrac{\delta y}{\delta u} \to \dfrac{dy}{du}$ and $\dfrac{\delta u}{\delta x} \to \dfrac{du}{dx}$,

then

$$\left(\frac{\delta y}{\delta u} \times \frac{\delta u}{\delta x}\right) \to \frac{dy}{du} \times \frac{du}{dx},$$

we have

$$\boxed{\frac{dy}{dx} = \frac{dy}{du} \times \frac{du}{dx}.}$$

In the given case, as $y = 4u^6$, $\dfrac{dy}{du} = 24u^5$,

and as $u = 1 - x^2$, $\dfrac{du}{dx} = -2x$.

$$\therefore \frac{dy}{dx} = -48x \cdot u^5 = -48x(1-x^2)^5.$$

Ex. 1.   Given $y = u^6$ and $u = 3x + 1$, to find $\dfrac{dy}{dx}$ in terms of x.

We have,
$$\frac{dy}{dx} = \frac{dy}{du} \times \frac{du}{dx},$$
$$= 6u^5 \times 3,$$
$$= 18u^5 = 18(3x+1)^5.$$

Ex. 2. *Differentiate* $(1-4x^2)^8$.

Let
$$y = (1-4x^2)^8,$$

and let
$$u = 1-4x^2 \atop \therefore\ y = u^8 \Bigg\}.$$

$$\therefore\ \frac{dy}{dx} = \frac{dy}{du} \times \frac{du}{dx},$$
$$= 8u^7 \times (-8x),$$
$$= -64x.\ u^7 = -64x(1-4x^2)^7.$$

Ex. 3. *Given* $v = 18r^2$, $\dfrac{dr}{dt} = 2$, *to find* $\dfrac{dv}{dt}$ *when* r = 1.

We have,
$$\frac{dv}{dt} = \frac{dv}{dr} \times \frac{dr}{dt},$$
$$= 36r \times 2 = 72r.$$

$$\therefore\ \text{when } r = 1, \qquad \frac{dv}{dt} = 72.$$

To prove that $\quad\dfrac{dy}{dx} = \dfrac{1}{\dfrac{dx}{dy}}$.

We have,
$$\frac{dy}{du} = \frac{dy}{dx} \times \frac{dx}{du},$$
Let u = y.
$$\therefore\ \frac{dy}{dy} = \frac{dy}{dx} \times \frac{dx}{dy},$$
$$\therefore\ \frac{dy}{dx} \times \frac{dx}{dy} = 1,$$
or
$$\frac{dy}{dx} = \frac{1}{\dfrac{dx}{dy}}.$$

Ex. 4. *If* $x = y^2$, *find* $\dfrac{dy}{dx}$.

Differentiate with respect to y, $\dfrac{dx}{dy} = 2y.$

$$\therefore \frac{dy}{dx} = \frac{1}{2y} = \pm \frac{1}{2\sqrt{x}}.$$

Ex. 5. Given $\left.\begin{array}{l} y = 3t^3 - t \\ x = 1 - 2t^2 \end{array}\right\}$, *to find* $\dfrac{dy}{dx}$ *when* $t = 2.$

We have $\qquad \dfrac{dy}{dx} = \dfrac{dy}{dt} \times \dfrac{dt}{dx} = \dfrac{dy}{dt} \Big/ \dfrac{dx}{dt}.$

$$\frac{dy}{dt} = 9t^2 - 1,$$

and $\qquad \dfrac{dx}{dt} = -4t.$

$$\therefore \frac{dy}{dx} = \frac{9t^2 - 1}{-4t}.$$

$\therefore$ when $t = 2,$ $\qquad \dfrac{dy}{dx} = -\dfrac{35}{8}.$

Differentiation of x^n where n is a fraction

The method of differentiating x^n, where n is a fraction, is illustrated by the following examples.

Ex. 6. Differentiate $x^{\frac{3}{2}}.$

Let $\qquad\qquad y = x^{\frac{3}{2}}$

and let $\qquad\qquad x^{\frac{1}{2}} = z.$

$$\therefore \left.\begin{array}{l} y = z^3 \\ x = z^2 \end{array}\right\}.$$

and

$$\therefore \frac{dy}{dx} = \frac{dy}{dz} \times \frac{dz}{dx} = \frac{3z^2}{2z} = \frac{3}{2} z = \frac{3}{2} x^{\frac{1}{2}}$$

i.e. $\qquad \dfrac{d}{dx}(x^{\frac{3}{2}}) = \dfrac{3}{2} x^{\frac{1}{2}}.$

Ex. 7. Differentiate $x^{-\frac{4}{3}}.$

Let $\qquad\qquad y = x^{-\frac{4}{3}}$

and let $\qquad z = x^{\frac{1}{3}}.$

$$\left. \begin{array}{c} \therefore \ y = z^{-4} \\ x = z^3 \end{array} \right\}.$$

$$\therefore \ \frac{dy}{dx} = \frac{dy}{dz} \times \frac{dz}{dx} = \frac{-4z^{-5}}{3z^2} = -\frac{4}{3}z^{-7} = -\frac{4}{3}x^{-\frac{7}{3}}.$$

i.e. $\quad \dfrac{d}{dx}(x^{\frac{4}{3}}) = -\dfrac{4}{3}x^{-\frac{7}{3}},$

The previous examples suggest that the rule used for differentiating x^n, when n is an integer, is also true when n is a fraction, and so we can state, for any value of n,

$$\frac{d}{dx}(x^n) = nx^{n-1}.$$

e.g. $\quad \dfrac{d}{dx}(\sqrt{x}) = \dfrac{d}{dx}(x^{\frac{1}{2}}) = \tfrac{1}{2}x^{-\frac{1}{2}} = \dfrac{1}{2\sqrt{x}}.$

Differentiation of x^n

A formal proof, that if $y = x^n$ then $\dfrac{dy}{dx} = nx^{n-1}$, for all real values of n, will now be given.

We have $\qquad y = x^n.$

Let x be increased by a small increment δx and as a consequence let y be increased by a small increment δy, so that

$$y + \delta y = (x + \delta x)^n.$$

$$\therefore \ \delta y = (x + \delta x)^n - x^n.$$

$$\therefore \ \frac{\delta y}{\delta x} = \frac{(x + \delta x)^n - x^n}{(x + \delta x) - x}.$$

Now $\qquad \dfrac{dy}{dx} = \lim_{\delta x \to 0} \dfrac{\delta y}{\delta x} = \lim_{\delta x \to 0} \dfrac{(x + \delta x)^n - x^n}{(x + \delta x) - x}.$

Let $x + \delta x = \mathsf{X}$, then this limit becomes

$$\lim_{\mathsf{X} \to x} \frac{\mathsf{X}^n - x^n}{\mathsf{X} - x},$$

When n is a positive integer, by direct division

$$\frac{\mathsf{X}^n - x^n}{\mathsf{X} - x} = \mathsf{X}^{n-1} + x\mathsf{X}^{n-2} + x^2\mathsf{X}^{n-3} \ldots + x^{n-1}$$

(containing n terms).

$$\therefore \ \lim_{\mathsf{X} \to x} \frac{\mathsf{X}^n - x^n}{\mathsf{X} - x} = x^{n-1} + x^{n-1} + x^{n-1} \ldots + x^{n-1} (n \text{ terms}),$$

$$= nx^{n-1}.$$

Let n be a positive fraction, say $\dfrac{p}{q}$, *where p and q are positive integers.*

Then
$$\lim_{X \to x} \frac{X^n - x^n}{X - x} = \lim_{X \to x} \frac{X^{p/q} - x^{p/q}}{X - x}.$$

Let $X^{1/q} = T$ and $x^{1/q} = t$.
Then the limit is

$$\lim_{T \to t} \frac{T^p - t^p}{T^q - t^q} = \lim_{T \to t} \left(\frac{T^p - t^p}{T - t}\right) \cdot \left(\frac{T - t}{T^q - t^q}\right),$$

$$= \frac{pt^{p-1}}{qt^{q-1}} \text{ (by first part)},$$

$$= \frac{p}{q} t^{p-q} = \frac{p}{q} (x^{1/q})^{p-q},$$

$$= \frac{p}{q} x^{(p/q)-1} = nx^{n-1}.$$

Now let n be a negative integer or fraction, say $-m$ *where m is positive,* then

$$\lim_{X \to x} \frac{X^n - x^n}{X - x} = \lim_{X \to x} \frac{X^{-m} - x^{-m}}{X - x},$$

$$= \lim_{X \to x} \frac{\dfrac{1}{X^m} - \dfrac{1}{x^m}}{X - x},$$

$$= -\lim_{X \to x} \frac{X^m - x^m}{X - x} \cdot \frac{1}{X^m x^m},$$

$$= -m . x^{m-1} . \frac{1}{x^{2m}} = (-m)x^{-m-1} = nx^{n-1}.$$

Thus for all real values of n, the limit is nx^{n-1}.

Hence $\dfrac{d}{dx} (x^n) = nx^{n-1}$ for all real values of n.

Ex. 8.　*Find* $\lim\limits_{x \to a} \dfrac{x^{\frac{2}{3}} - a^{\frac{2}{3}}}{x - a}$.

$$\text{Limit} = \lim_{x \to a} \frac{(x^{\frac{1}{3}})^2 - (a^{\frac{1}{3}})^2}{(x^{\frac{1}{3}})^3 - (a^{\frac{1}{3}})^3},$$

$$= \lim_{x \to a} \frac{(x^{\frac{1}{3}} - a^{\frac{1}{3}})(x^{\frac{1}{3}} + a^{\frac{1}{3}})}{(x^{\frac{1}{3}} - a^{\frac{1}{3}})(x^{\frac{2}{3}} + x^{\frac{1}{3}}a^{\frac{1}{3}} + a^{\frac{2}{3}})},$$

$$= \frac{2a^{\frac{1}{3}}}{3a^{\frac{2}{3}}} = \frac{2}{3} a^{-\frac{1}{3}}.$$

Ex. 9. If $y = \sqrt{(3x^2 - 4)}$, find $\dfrac{dy}{dx}$.

Let $$u = 3x^2 - 4.$$

$$\therefore \ y = \sqrt{u} = u^{\frac{1}{2}}.$$

$$\therefore \ \frac{dy}{dx} = \frac{dy}{du} \times \frac{du}{dx} = \tfrac{1}{2} u^{-\frac{1}{2}} \times 6x.$$

$$\therefore \ \frac{dy}{dx} = \frac{3x}{\sqrt{u}} = \frac{3x}{\sqrt{(3x^2 - 4)}}.$$

Ex. 10. If $y = \dfrac{1}{\sqrt{(1 - t^2)}}$ and $x = 3(1 - t^3)$, find $\dfrac{dy}{dx}$ *in terms of t.*

We have, $$\frac{dy}{dx} = \frac{dy}{dt} \times \frac{dt}{dx} = \frac{dy}{dt} \div \frac{dx}{dt}.$$

Now $$y = \frac{1}{\sqrt{(1 - t^2)}} = (1 - t^2)^{-\frac{1}{2}}.$$

Let $$u = (1 - t^2).$$

$$\therefore \ y = u^{-\frac{1}{2}}.$$

$$\therefore \ \frac{dy}{dt} = \frac{dy}{du} \times \frac{du}{dt} = -\tfrac{1}{2} u^{-\frac{3}{2}} \times (-2t),$$

$$= \frac{t}{(1 - t^2)^{\frac{3}{2}}},$$

and $$\frac{dx}{dt} = -9t^2,$$

Hence $$\frac{dy}{dx} = \frac{t}{(1 - t^2)^{\frac{3}{2}}} \div (-9t^2),$$

$$= -\frac{1}{9t(1 - t^2)^{\frac{3}{2}}}.$$

(A) EXAMPLES 7a

1. If $y = u^4$ and $u = 2x - 1$, find $\dfrac{dy}{du}$ and $\dfrac{du}{dx}$. Hence find the value of $\dfrac{dy}{dx}$.

2. Given that $V = 5u^6$ and $u = 1 - t^2$, write down the values of $\dfrac{dV}{du}, \dfrac{du}{dt}$ and hence find $\dfrac{dV}{dt}$.

3. If $y = 2t^2 - t$ and $x = 3t + 1$, find the values of $\dfrac{dy}{dt}$ and $\dfrac{dx}{dt}$. What is the value of $\dfrac{dt}{dx}$? Deduce the value of $\dfrac{dy}{dx}$.

4. If $y = u^{-3}$, where $x = u^4$, find $\dfrac{dy}{du}, \dfrac{dx}{du}$ and $\dfrac{du}{dx}$. Hence deduce the differential coefficient of $x^{-\frac{3}{4}}$.

5. Find $\dfrac{dy}{dx}$ in the following cases

$$\text{(i)} \left. \begin{array}{l} y = 5u^{11} \\[4pt] u = 4x + 3 \end{array} \right\}; \quad \text{(ii)} \left. \begin{array}{l} y = 3u^3 + u \\[4pt] u = 1 - 4x^2 \end{array} \right\}; \quad \text{(iii)} \left. \begin{array}{l} y = 4u^4 + \dfrac{1}{u} \\[4pt] u = 2x^2 - 1 \end{array} \right\}.$$

6. Differentiate with respect to x

$$(3x + 2)^3, \; (1 - x^2)^8, \; \frac{1}{(1 - 2x)^4}.$$

7. Find the values of $\dfrac{dy}{dt}, \dfrac{dx}{dt}$ and $\dfrac{dy}{dx}$ when $t = -1$, in the following cases

$$\text{(i)} \left. \begin{array}{l} y = 2t \\[4pt] x = t^2 \end{array} \right\}; \quad \text{(ii)} \left. \begin{array}{l} y = 4 - 3t^2 \\[4pt] x = t^2 - t + 1 \end{array} \right\}; \quad \text{(iii)} \left. \begin{array}{l} y = 1 - \dfrac{1}{t} \\[4pt] x = \dfrac{1}{t^2} \end{array} \right\}.$$

8. In the following cases find $\dfrac{dV}{dt}$, when $r = 4$

$$\text{(i)} \left. \begin{array}{l} V = 6r^2 \\[4pt] \dfrac{dr}{dt} = 6 \end{array} \right\}; \quad \text{(ii)} \left. \begin{array}{l} V = \tfrac{4}{3}\pi r^3 \\[4pt] \dfrac{dr}{dt} = -2 \end{array} \right\}; \quad \text{(iii)} \left. \begin{array}{l} V = 10\pi r^2 \\[4pt] \dfrac{dr}{dt} = \dfrac{1}{\pi} \end{array} \right\}.$$

9. Differentiate with respect to x

$$x^{\frac{3}{4}}, x^{\frac{7}{3}}, x^{-\frac{2}{3}}, x^{-1\cdot 5}, x^{3\cdot 6}, 4x^{\frac{1}{3}}, \frac{1}{2x^{\frac{1}{2}}}, \sqrt[3]{x^7}, \frac{1}{\sqrt[5]{x^2}}.$$

10. Find $\dfrac{dy}{dx}$ in the following cases

$$\text{(i)} \; y = \sqrt{x} + \frac{1}{\sqrt{x}}; \quad \text{(ii)} \; y = \sqrt{1 + x};$$

$$\text{(iii)} \; y = \frac{1}{\sqrt{x^2 + 1}}; \quad \text{(iv)} \; y = 3(3x + 1)^{\frac{2}{3}}.$$

11. Find the following limits

$$\lim_{x\to a}\frac{x^{15}-a^{15}}{x-a}\ ;\quad \lim_{x\to a}\frac{x^{\frac12}-a^{\frac12}}{x-a}\ ;$$

$$\lim_{x\to a}\frac{x^{\frac34}-a^{\frac34}}{x-a}\ ;\quad \lim_{x\to a}\frac{x^{-2}-a^{-2}}{x-a}\ .$$

Rates of Change of Connected Variables

If we say that the rate of change of a variable x is $2\,\text{cm s}^{-1}$, it is understood that we mean the rate of change of x with respect to the time t. We have seen previously that a differential coefficient gives a rate of change and the rate of increase of x with respect to t is given by the derivative $\frac{dx}{dt}$.

Consider the following examples on rates of change.

Ex. 11. The radius, r cm, of a blot of ink is increasing at the rate of $1\cdot5$ mm s^{-1}. Find the rate at which the area A is increasing after 4 s.

We require the rate of increase of area, i.e. $\frac{dA}{dt}$.

We know the rate of increase of r, i.e. $\frac{dr}{dt}$, and also, as we have a relation between A and r, we can find $\frac{dA}{dr}$. Hence $\frac{dA}{dt}$ is found using the relation $\frac{dA}{dt}=\frac{dA}{dr}\times\frac{dr}{dt}$.

Now
$$A=\pi r^2.$$
$$\therefore\ \frac{dA}{dr}=2\pi r.$$

Also, $\frac{dr}{dt}=$ rate of increase of $r=1\cdot5$ mm s^{-1},
$$=0\cdot15\ \text{cm s}^{-1}$$
$$\therefore\ \frac{dA}{dt}=\frac{dA}{dr}\times\frac{dr}{dt}=2\pi r\times\cdot15,$$
$$=\cdot3\pi r\ \text{cm}^2\,\text{s}^{-1}.$$

We require $\frac{dA}{dt}$ after 4 s. As r increases $\cdot15$ cm each second, the radius after 4 s is equal to $\cdot6$ cm.

$$\therefore\ \text{when }r=\cdot6\text{ cm},\ \frac{dA}{dt}=\cdot18\pi\ \text{cm}^2\,\text{s}^{-1}.$$

i.e. after 4 s, the rate of increase of area $=\cdot18\pi\ \text{cm}^2\,\text{s}^{-1}$.

Ex. 12. *Water is poured into a cone of semi-vertical angle* 30°, *at the rate of* 2 m³ min⁻¹. *At what rate is the surface rising when the depth is* 1 m?

At any instant, let the depth of the liquid be h m and its volume V m³.

Then $V = \frac{1}{3}\pi r^2 h$ and $r = h \tan 30°$.

$\therefore V = \frac{1}{3}\pi h^3 \tan^2 30° = \frac{1}{9}\pi h^3$.

i.e. $V = \frac{1}{9}\pi h^3$. . . . (1)

As the rate of increase of V is equal to 2 m³ min⁻¹,

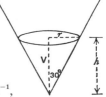

Fig. 54

$$\therefore \frac{dV}{dt} = 2 \text{ m}^3 \text{ min}^{-1}.$$

We require to know the rate of increase of h, *i.e.* $\frac{dh}{dt}$.

Now $$\frac{dV}{dt} = \frac{dV}{dh} \cdot \frac{dh}{dt},$$

$$\therefore \frac{dh}{dt} = \frac{2}{\frac{dV}{dh}},$$

and from (1), $$\frac{dV}{dh} = \frac{1}{3}\pi h^2.$$

$$\therefore \frac{dh}{dt} = \frac{6}{\pi h^2} \text{ m min}^{-1}$$

$\therefore$ when $h = 1$, $$\frac{dh}{dt} = \frac{6}{\pi} \text{ m min}^{-1}.$$

i.e. when the depth is 1 m, the surface is rising at a rate of $\frac{6}{\pi}$ m min⁻¹ = 1·91 m min⁻¹.

Ex. 13. *A kite is* 112 m *above the ground, and has* 130 m *of string out. If the kite is travelling horizontally at* 6 m s⁻¹ *directly away from the boy who is flying it, at what rate is the string being paid out?*

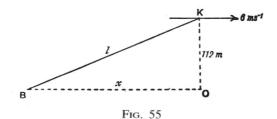

Fig. 55

Let the horizontal distance of kite from boy at any time be x m, and let the length of string out be l m.

Then horizontal velocity = rate of increase of $x = \dfrac{dx}{dt}$.

$$\therefore \frac{dx}{dt} = 6 \text{ m s}^{-1}.$$

By Pythagoras' Theorem, $\qquad l^2 = x^2 + (112)^2$

Differentiating with respect to t, $\dfrac{d}{dt}(l^2) = \dfrac{d}{dt}(x^2).$

But $\qquad\qquad \dfrac{d}{dt}(l^2) = \dfrac{d}{dl}(l^2) \times \dfrac{dl}{dt} = 2l\dfrac{dl}{dt},$

and $\qquad\qquad \dfrac{d}{dt}(x^2) = 2x\dfrac{dx}{dt}.$

$$\therefore \frac{dl}{dt} = \frac{x}{l} \cdot \frac{dx}{dt} = 6\frac{x}{l} \text{ m s}^{-1}.$$

But when $l = 130$ m, $\qquad x = 66$ m (Pythagoras).

$$\therefore \frac{dl}{dt} = \frac{66}{130} \times 6 = \frac{198}{65} = 3{\cdot}04 \text{ m s}^{-1}.$$

∴ Rate of increase of length of string, when length is 130 m
$$= 3{\cdot}04 \text{ m s}^{-1}$$

(B) EXAMPLES 7b

1. The area of the surface of a sphere $A = 4\pi r^2$, where r is the radius. What is the value of $\dfrac{dA}{dr}$ when $r = 3$ cm? The radius of the sphere is increasing at a rate of 2 cm s^{-1}. Express this statement symbolically in terms of a derivative, and find the rate of increase of A where $r = 3$ cm.

2. The volume, V, of a sphere of radius r m, is $\frac{4}{3}\pi r^3$ m^3. Find the value of $\dfrac{dV}{dr}$ when $r = 6$ m. If the radius of the sphere is increasing at a rate of 4 m s^{-1}, find the rate at which the volume is increasing when the radius is equal to 6 m.

3. The area of a circular blot of ink is increasing at a rate of $2 \text{ cm}^2 \text{ s}^{-1}$. Find the rate at which the radius is increasing when the area is equal to 4 cm^2.

4. Water is being poured into an open cylindrical tank of radius 2 m at a rate of $6 \text{ m}^3 \text{ min}^{-1}$. Find the relation between the volume of water V m^3 and its depth h m, and find the rate at which the depth is increasing.

5. Air is leaking from a spherical balloon at a rate of $2 \text{ cm}^3 \text{ s}^{-1}$. Find the rate of decrease of the radius and the surface area, when the radius is equal to 10 cm.

6. The point (x, y) lies on the curve $y = 3x^2$. If the abscissa x is increasing at 2 units per second, find the rate of increase of the ordinate y when $x = 4$.

7. The co-ordinates of a moving point at a time t, are given by $x - 3t^2$, $y = 2 + 3t$. Find the component velocities of the point in the directions of the x- and y-axes $\left(i.e. \dfrac{dx}{dt} \text{ and } \dfrac{dy}{dt} \right)$, and hence find the direction of motion of the point when $t = 2$ s.

8. A balloon is expanded so that its surface area is increasing at the rate of $2 \text{ m}^2 \text{ min}^{-1}$. At what rate is its volume increasing when its radius is 4 m?

9. A circular cone, with semi-vertical angle 45°, is fixed with its axis vertical and its vertex downwards. Water is poured into the cone at a rate of $2 \text{ cm}^3 \text{ s}^{-1}$. Find the rate at which the depth of the water is increasing when the depth is 4 cm.

10. A conical filter paper, vertical angle 60°, is held with its axis vertical and its vertex downwards. If the cone is filled with water which runs out at the rate of $3 \text{ cm}^3 \text{ s}^{-1}$, find the rate at which the level of water in the cone is falling when the depth is 3·5 cm.

11. Sand is falling on to the ground at the rate of $10 \text{ cm}^3 \text{ s}^{-1}$, and is forming a heap in the shape of a circular cone of vertical angle 90°. At what rate is the vertical height of the heap increasing after 2 s?

12. The section of a trough, 2 m long, is an isosceles triangle of height 30 cm and base (upwards) 24 cm. Water runs into the trough at the rate of $600 \text{ cm}^3 \text{ min}^{-1}$. Find the rate at which the level is rising after 2 min.

13. A glass vessel is constructed so that when the depth of liquid in it is x cm, the volume of the liquid V is $3x^2 - \frac{1}{3}x^3 \text{ cm}^3$. Liquid is poured into the vessel at a steady rate such that when its depth is 3 cm its level is rising at a rate of 2 cm s^{-1}. Find the rate at which the liquid is being poured in.

14. A man of height 1·8 m walks directly away from a lamp of height 3 m, on a level road, at 2 m s^{-1}. Find the rate at which the length of his shadow is increasing when he is 3 m from the foot of the lamp.

15. For a given mass of gas at a constant temperature, the pressure is inversely proportional to the volume. The volume of the gas is decreased at a rate of $2 \text{ cm}^3 \text{ s}^{-1}$. Find the rate at which the pressure is changing when the volume is reduced to 100 cm^3, if the initial volume of the gas was 250 cm^3, and its initial pressure 1000 gf cm^{-2}.

Chapter 8

INTEGRATION. INDEFINITE INTEGRALS

In differentiation, we are given, say, $y =$ some function of x, and it is required to find $\frac{dy}{dx}$. In the reverse process, called *integration*, we are given $\frac{dy}{dx}$ and require to find y.

Take the simple case, $\frac{dy}{dx} = 2x$. Find y.

Obviously *one* possible value of y is x^2.

In addition, we have solutions $y = x^2 + 1$, $y = x^2 - 6$ ... or more generally, $y = x^2 + c$ where c is any constant.

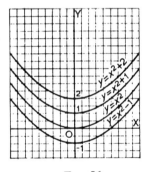

FIG. 56

We see at once, that on being given $\frac{dy}{dx}$ there is more than one possible value for y. It is interesting to consider the geometrical meaning of this result.

As the gradient of a curve is given by $\frac{dy}{dx}$, we require to find a curve which has a gradient equal to $2x$ at any point, where x is the abscissa of the point. We see that we can get any number of parallel curves all satisfying the condition that the gradient is equal to $2x$.

If, however, in addition to being given $\frac{dy}{dx}$, we are given *one* point on the curve, the solution becomes unique.

e.g. Given $\frac{dy}{dx} = 2x$, and $y = 1$ when $x = 2$,

we have $\qquad y = x^2 + c,$ where c is any constant.

And as $y = 1$ when $\qquad x = 2,$

$$1 = 4 + c, \text{ i.e. } c = -3.$$

$\therefore$ Required curve is $\qquad y = x^2 - 3.$

Ex. 1. *Find the equation of the curve, gradient* $3 - 2x^2$, *which passes through the origin.*

We have, $$\frac{dy}{dx} = 3 - 2x^2.$$

$$\therefore \ y = 3x - \tfrac{2}{3}x^3 + c, \qquad \text{where } c \text{ is constant.}$$

But as the curve passes through the origin, $y = 0$ when $x = 0$.

$$\therefore \ c = 0.$$

$\therefore$ Equation of curve is $y = 3x - \tfrac{2}{3}x^3$.

Ex. 2. *The velocity v of a body moving in a straight line is given by the relation $v = 2t - t^2$. Find s, the distance of the body from a fixed point O in the line, when $t = 2$.*

We have, $$v = \frac{ds}{dt} = 2t - t^2.$$

$$\therefore \ s = t^2 - \tfrac{1}{3}t^3 + c, \qquad \text{where } c \text{ is a constant.}$$

In this case c has a definite meaning, for it is the value of s when $t = 0$, *i.e.* the distance from the origin at which the body starts.
Suppose the body starts at **O**, then $s = 0$ when $t = 0$ and hence $c = 0$.

$$\therefore \ s = t^2 - \tfrac{1}{3}t^3.$$

So when $t = 2$, $\qquad\qquad s = 1\tfrac{1}{3}.$

Ex. 3. *The velocity, v, of a point moving on a straight line is given in terms of the time, t, by the equation $v = 2t^2 - 9t + 10$, the point being at the origin when t is zero. Find (a) the acceleration, (b) the distances from the origin when the point is at rest.*

Let s be the displacement from the origin at time t.

Then $$v = \frac{ds}{dt} = 2t^2 - 9t + 10.$$

Acceleration $$= \frac{dv}{dt} = 4t - 9.$$

As $$\frac{ds}{dt} = 2t^2 - 9t + 10,$$

$$\therefore \ s = \tfrac{2}{3}t^3 - \tfrac{9}{2}t^2 + 10t + c, \qquad \text{where } c \text{ is a constant.}$$

But, as the body is at the origin when t is zero, $s = 0$ when $t = 0$.

$$\therefore \ c = 0.$$

$$\therefore \ s = \tfrac{2}{3}t^3 - \tfrac{9}{2}t^2 + 10t.$$

The body is at rest when $v = 0$,

$\qquad$ *i.e.* when $2t^2 - 9t + 10 = 0$.

$\qquad\qquad$ *i.e.* when $t = 2$ and $2\tfrac{1}{2}$.

When $t = 2,$ $s = 7\frac{1}{3},$

and when $t = 2\frac{1}{2},$ $s = 7\frac{7}{24}.$

∴ The point is at rest at points distances $7\frac{1}{3}$ and $7\frac{7}{24}$ units from the origin.

Notation

If $\dfrac{dy}{dx} = 2x,$

we say that y is equal to the integral of $2x$ with respect to x and write the statement as follows

$$y = \int 2x\,dx,$$

and we have, $y = \displaystyle\int 2x\,dx = x^2 + c.$

Ex. 4.   Integrate, with respect to x, the following expressions $3x + 1$, $3x^2 - 2x$.

(i) $\displaystyle\int (3x + 1)\,dx = \tfrac{3}{2}x^2 + x + c,$

(ii) $\displaystyle\int (3x^2 - 2x)\,dx = x^3 - x^2 + c,$ where c is any constant.

Integration of x^n

We have $\dfrac{d}{dx}(x^{n+1}) = (n+1)x^n,$

and so $\dfrac{d}{dx}\left(\dfrac{x^{n+1}}{n+1}\right) = x^n,$ except for $n = -1.$

Hence $\displaystyle\int x^n dx = \dfrac{x^{n+1}}{n+1} + c$ where c is a constant,

the case $n = -1$ being excluded.

e.g. $\displaystyle\int x^8 dx = \tfrac{1}{9}x^9 + c.$

$\displaystyle\int 4x^5 dx = 4 \cdot \tfrac{1}{6}x^6 + c = \tfrac{2}{3}x^6 + c.$

Ex. 5.   Evaluate

(i) $\displaystyle\int \left(x^3 + \dfrac{1}{x^3}\right)dx;$ (ii) $\displaystyle\int \left(\sqrt{x} - \dfrac{1}{\sqrt{x}}\right)dx;$ (iii) $\displaystyle\int \left(x - \dfrac{1}{x}\right)^2 dx.$

(ii) $\int\left(x^3+\dfrac{1}{x^3}\right)dx = \int(x^3+x^{-3})dx = \dfrac{x^4}{4}+\dfrac{x^{-2}}{-2}+c,$

$$=\dfrac{x^4}{4}-\dfrac{1}{2x^2}+c.$$

(iii) $\int\left(\sqrt{x}-\dfrac{1}{\sqrt{x}}\right)dx = \int(x^{\frac{1}{2}}-x^{-\frac{1}{2}})dx = \dfrac{x^{\frac{3}{2}}}{\frac{3}{2}}-\dfrac{x^{\frac{1}{2}}}{\frac{1}{2}}+c,$

$$=\tfrac{2}{3}\sqrt{x^3}-2\sqrt{x}+c.$$

(iv) $\int\left(x-\dfrac{1}{x}\right)^2 dx = \int\left(x^2-2+\dfrac{1}{x^2}\right)dx,$

$$= \int(x^2-2+x^{-2})dx,$$

$$=\dfrac{x^3}{3}-2x+\dfrac{x^{-1}}{-1}+c,$$

$$=\dfrac{x^3}{3}-2x-\dfrac{1}{x}+c$$

(A) EXAMPLES 8

1. If $\dfrac{dy}{dx}=3x$, write down three possible values for y. What is the most general expression giving y in terms of x? Interpret your result graphically.

2. Prove that $y=x+\dfrac{1}{x}$ is a solution of the equation $\dfrac{dy}{dx}=1-\dfrac{1}{x^2}$. What is the most general solution?

3. If $\dfrac{dy}{dx}=2x^2$ find y, if $y=1$ when $x=1$.

4. If $\dfrac{dy}{dx}=4x^3-4x$, and $y=0$ when $x=2$, find y in terms of x.

5. Given that $\dfrac{ds}{dt}=3-t^2$, find s in terms of t, if $s=5$ when $t=0$. What is the value of s when $t=2$?

6. If $\dfrac{dp}{dv}=2v-\dfrac{v^3}{2}$, and $p=0$ when $v=0$, find the value of p when $v=1$.

7. At the point (x, y) on a curve, the gradient is equal to $(2x-1)$. If the curve passes through the point $(3, 4)$ what is its equation?

8. Find the equation of the curve with gradient $2x^2+3x-1$, which passes through the origin.

9. Integrate the following expressions with respect to x

$$x^4, \quad x^{14}, \quad \frac{1}{x^3}, \quad x^{\frac{3}{2}}, \quad 4x^3, \quad \frac{2}{x^2}, \quad \sqrt[3]{x^2}, \quad x^{-2}, \quad 4x^3-x^2, \quad 2-\frac{1}{x^3},$$

$$\sqrt[3]{x} \quad \frac{1}{\sqrt[3]{x}},(3x-1)(1-x).$$

10. Evaluate the following integrals

$$\int x^7 dx; \quad \int (x^2+x)dx; \quad \int \left(t+\frac{1}{t}\right)^2 dt;$$

$$\int (1-3x)(1+x)dx; \quad \int \frac{x^3-1}{x^2} dx.$$

11. The velocity of a moving point, v, is given in terms of the time, t, by the following equations

(a) $v=3t-1$;   (b) $v=t^2+t^3$;   (c) $v=1-2t-3t^2$.

In each case find the distance travelled in time t, assuming that this distance is measured from the position of the point when $t=0$.

12. A body moves along a line **OA**, so that its velocity t seconds after passing **O** is $2+3t^2$ m s^{-1}. The time from **O** to **A** is 2 s. Find the length **OA**.

13. A particle starts from rest and moves in a straight line with a speed given by $v=4t-2t^2$. Find how far it goes in the 2nd second.

14. Given that $\frac{d^2y}{dx^2}=4x$, and that $\frac{dy}{dx}=0$, $y=2$ when $x=0$. Find $\frac{dy}{dx}$ and y in terms of x.

15. The acceleration of a moving point is given in terms of the time t, by the following equations

(a) $f=2t$;   (b) $f=1-t^2$;   (c) $f=2t-t^3$.

In each case find the distance travelled s in time t, assuming that the point is initially at rest and that $s=0$ when $t=0$.

CHAPTER 9

AREA UNDER A CURVE
DEFINITE INTEGRATION
VOLUME OF REVOLUTION

Ex. 1. Find the area included between the curve $y = x^2$, the axis of x, and the ordinates $x = 1$, $x = 2$.

Referring to the diagram, Fig. 57, we see that it is required to find the area **ALMB**. Imagine a point **P**, co-ordinates (x, y), to move along the curve from the point **A** to the point **B**.

Then **PQ**, the ordinate of **P**, will sweep out the area **ALMB**.

Consider **P** to be at any point between **A** and **B**, and let the area **ALQP** be denoted by **A**.

Then as we vary x, the abscissa of **P** (*i.e.* change the position of **PQ**), the area **A** will also vary. Hence **A** must be some function of x. In order to obtain the required area, we must find this function of x.

Imagine x to increase by a small amount δx [*i.e.* let **PQ** move into a neighbouring position **P'Q'**].

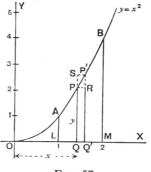

FIG. 57

Let the corresponding increase in y[*i.e.* **P'R**] be δy, and the corresponding increase in **A** [*i.e.* area **PQQ'P'**] be δA.

From the diagram, we see that

area **PQQ'P'** < area rect. **SQQ'P'**,

and   area **PQQ'P'** > area rect. **PQQ'R**.

i.e. $y\delta x < \delta A < (y + \delta y)\delta x.$

$$\therefore \ y < \frac{\delta A}{\delta x} < y + \delta y.$$

Now, let $\delta x \to 0$.

So $\dfrac{\delta A}{\delta x} \to \dfrac{dA}{dx}$ and $\delta y \to 0$.

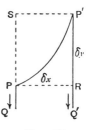

FIG. 58

93

$$\therefore \text{ we have, } \frac{dA}{dx} = y = x^2.$$

$\therefore$ Integrating, $A = \frac{1}{3}x^3 + c$, where c is a constant.

But when PQ is in the position AL, $A = 0$.

$$\therefore A = 0 \text{ when } x = 1.$$

$$\therefore 0 = \frac{1}{3} + c. \qquad i.e. \quad c = -\frac{1}{3}.$$

$$\therefore A = \frac{1}{3}x^3 - \frac{1}{3}.$$

As we require the area between BM and AL, we want the value of A when $x = 2$.

$$\text{When } x = 2, A = \frac{8}{3} - \frac{1}{3} = \frac{7}{3} \text{ unit}^2.$$

$$\therefore \text{ Required area} = \frac{7}{3} \text{ unit}^2.$$

General Case

To find the area bounded by the curve $y = f(x)$, the axis of x, and the ordinates $x = a$, $x = b$.

The diagrams illustrate the two possible cases
(i) $f(x)$ an increasing function of x (Fig. 59);
(ii) $f(x)$ a decreasing function of x (Fig. 60).

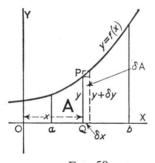

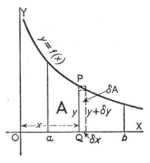

FIG. 59   FIG. 60

We adopt the previous method and notation.

A = area included between the moving ordinate PQ
and the fixed ordinate $x = a$.

δA = small increase in A due to a small increase δx in x.

δy = small increase in y due to a small increase δx in x.

N.B. In case (i), δy is positive and⎫
      in case (ii), δy is negative. ⎭

As before, the area δA lies between the areas of two rectangles, and we get,

in case (i) $y\delta x < \delta A < (y + \delta y)\delta x$,

in case (ii) $(y + \delta y)\delta x < \delta A < y\delta x$.

In both cases, δA lies between $y\delta x$ and $(y + \delta y)\delta x$.

$$\therefore \quad \frac{\delta A}{\delta x} \text{ lies between } y \text{ and } y + \delta y.$$

$\therefore$ Letting $\delta x \to 0$, $\quad \dfrac{dA}{dx} = y = f(x).$

$$\therefore \quad A = \int y dx \mid c = \int f(x) dx + c.$$

But $A = 0$ when $x = a$.

$$\therefore \quad c = -\left\{ \text{value of} \int y dx \text{ when } x = a \right\},$$

and the required area is obtained by taking $x = b$.

$$\therefore \quad \text{Area} = \left\{ \text{value of} \int y dx \text{ when } x = b \right\},$$

$$-\left\{ \text{value of} \int y dx \text{ when } x = a \right\}.$$

This result is written as:

area between $x = a$ and $x = b$ is equal to $\displaystyle\int_a^b y dx$.

i.e. The symbol $\displaystyle\int_a^b y dx$ is used to represent the value of $\displaystyle\int y dx$ when $x = b$, MINUS the value of $\displaystyle\int y dx$ when $x = a$. Such an expression is called a *definite integral*.

So we have the important result, that the area included between the curve $y = f(x)$, *the x-axis, and the ordinates* $x = a$ *and* $x = b$ *is equal to*

$$\int_a^b y dx = \int_a^b f(x) dx.$$

It follows by the same method that the area contained between a curve, the axis of y, *and the two abscissae* $y = y_1$ *and* $y = y_2$ *is*

$$\int_{y_1}^{y_2} x dy.$$

Definite Integrals

We have $\displaystyle\int_a^b f(x) dx = $ value of $\displaystyle\int f(x) dx$ when $x = b$,

$$- \text{value of} \int f(x) dx \text{ when } x = a.$$

We express this difference symbolically in the form $\left[\int f(x)dx\right]_a^b$

i.e. $\int_a^b f(x)dx = \left[\int f(x)dx\right]_a^b$.

Ex. 2. $\int_1^2 (2x-1)dx = \left[x^2-x+c\right]_1^2$

$= (value\ of\ x^2-x+c\ when\ x=2)$

$-(value\ of\ x^2-x+c\ when\ x=1),$

$= (4-2+c)-(1-1+c) = 2.$

We notice that the constant of integration cancels out and in future we will omit it in a definite integral.

Ex. 3. $\int_{-2}^2 (x^2-x)dx = \left[\dfrac{x^3}{3}-\dfrac{x^2}{2}\right]_{-2}^2 = (\tfrac{8}{3}-\tfrac{4}{2})-(-\tfrac{8}{3}-\tfrac{4}{2}) = \tfrac{16}{3}.$

Ex. 4. Evaluate $\int_1^3 \left(x-\dfrac{1}{x^2}\right)dx.$

$\int_1^3 \left(x-\dfrac{1}{x^2}\right)dx = \left[\dfrac{x^2}{2}+\dfrac{1}{x}\right]_1^3 = (\tfrac{9}{2}+\tfrac{1}{3})-(\tfrac{1}{2}+1) = 3\tfrac{1}{3}.$

Ex. 5. Find the area contained between the curve $y=2x^2$, and
(a) the axis of x and the ordinates $x=1$ and $x=3$;
(b) the axis of y and the abscissae $y=1$ and $y=4$.

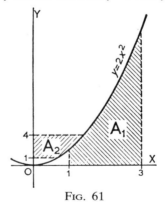

Fig. 61

(a) $A_1 = \int_a^b y\,dx,$

$= \int_1^3 2x^2 dx = \left[2\dfrac{x^3}{3}\right]_1^3,$

$= (2.9)-(2.\tfrac{1}{3}),$

$= 17\tfrac{1}{3}\ unit^2.$

(b) $$A_2 = \int_{y_1}^{y_2} x \, dy,$$

$$= \int_1^4 \sqrt{\frac{y}{2}} \cdot dy = \frac{1}{\sqrt{2}} \int_1^4 y^{\frac{1}{2}} dy,$$

$$= \frac{1}{\sqrt{2}} \left[\frac{y^{\frac{3}{2}}}{\frac{3}{2}} \right]_1^4 = \frac{1}{\sqrt{2}} (\tfrac{2}{3} \cdot 8) - \frac{1}{\sqrt{2}} (\tfrac{2}{3} \cdot 1),$$

$$= \frac{14}{3\sqrt{2}} = 3 \cdot 3 \text{ unit}^2.$$

Sign of Area

Referring to the previous work on area, we had the result that the element of area δA lies between $y\delta x$ and $(y + \delta y)\delta x$.

Now δx is positive, and so the sign of δA depends on the sign of y.

For the area A_1, Fig. 62, as y is positive, A_1 will be positive, but for the area A_2, as y is negative, this area will be calculated as negative,

i.e. $\int_a^b y \, dx$ will be negative.

If the area contained between the curve, the axis of x, and the ordinates $x = 0$ and $x = b$ is required, we must calculate $A_1 \left[i.e. \int_0^a y \, dx \right]$ and A_2 separately, and then neglecting the sign of A_2, the total area is found by addition.

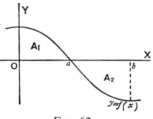

FIG. 62

The integral $\int_0^b f(x) \, dx$ will give the algebraic sum of A_1 and A_2, i.e. the difference between the two areas.

Ex. 6.   Find the area enclosed by the curve $y = x(4 - x)$, the axis of x, and the ordinates $x = 0$ and $x = 6$.

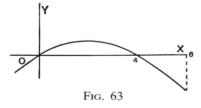

FIG. 63

Using our general result, it appears that the required area

$$= \int_0^6 (4x - x^2) dx = \left[2x^2 - \frac{x^3}{3} \right]_0^6 = 0.$$

In this case part of the area is above the axis and part below, and as the integral is zero, we observe that the portions above and below the axis are equal in magnitude, but because of their opposite signs the sum appears as zero.

To get over this difficulty it is always advisable to draw a rough diagram showing the area required.

The area included between $x = 0$ and $x = 4$

$$= \int_0^4 (4x - x^2)dx,$$

$$= \left[2x^2 - \frac{x^3}{3} \right]_0^4 = 32 - \frac{64}{3},$$

$$= 10\tfrac{2}{3} \text{ unit}^2.$$

The area included between $x = 4$ and $x = 6$

$$= \int_4^6 (4x - x^2)dx,$$

$$= \left[2x^2 - \frac{x^3}{3} \right]_4^6 = 0 - 10\tfrac{2}{3},$$

$$= -10\tfrac{2}{3} \text{ unit}^2.$$

∴ Total area between $x = 0$ and $x = 6$

$$= 10\tfrac{2}{3} + 10\tfrac{2}{3}$$

$$= 21\tfrac{1}{3} \text{ unit}^2.$$

Ex. 7.   Find the area of the segment cut off from the curve $y = x(2 - x)$ by the line $y = \dfrac{x}{2}$.

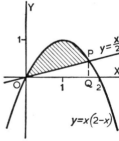

Fig. 64

The required area is shaded in the diagram (Fig. 64).

Required area = area under curve between O and Q

− area of △OPQ.

We require the co-ordinates of P.

For the points of intersection, $y = x(2-x)$,

and $$y = \frac{x}{2}.$$

$$\therefore \frac{x}{2} = x(2-x).$$

$$\therefore x = 4x - 2x^2.$$

i.e. $2x^2 - 3x = 0$; i.e. $x = 0$ and $x = \frac{3}{2}$.

$\therefore$ the point P has x co-ordinate $\frac{3}{2}$ and y co-ordinate $\frac{3}{4}$.

$$\therefore \text{ Area under curve} = \int_0^{\frac{3}{2}} x(2-x)dx = \int_0^{\frac{3}{2}} (2x - x^2)dx,$$

$$= \left[x^2 - \frac{x^3}{3} \right]_0^{\frac{3}{2}} = \frac{9}{4} - \frac{9}{8} = \frac{9}{8} \text{ unit}^2.$$

Area of $\triangle POQ = \frac{1}{2} \cdot \frac{3}{2} \cdot \frac{3}{4} = \frac{9}{16}$ unit^2.

$\therefore$ Area of segment $= \frac{9}{8} - \frac{9}{16} = \frac{9}{16}$ unit^2.

(A) EXAMPLES 9a

1. Evaluate the following definite integrals

(i) $\int_0^1 x^2 dx$; (ii) $\int_1^2 3x^4 dx$; (iii) $\int_1^4 \sqrt{x}dx$; (iv) $\int_{-2}^{-1} \frac{dx}{x^2}$;

(v) $\int_{-1}^1 (2x^2 - 1)dx$; (vi) $\int_0^2 (3x^3 - x)dx$; (vii) $\int_2^3 \left(x^2 - \frac{1}{x^2} \right)dx$;

(viii) $\int_{-1}^2 (3x-1)(2x+1)dx$; (ix) $\int_1^2 x(x-1)(x-2)dx$; (x) $\int_1^9 \frac{x+1}{\sqrt{x}}dx$.

2. Evaluate the integral $\int_1^4 P ds$, where

(a) $P = 2s^2 - s$; (b) $P = \left(s + \frac{1}{s} \right)^2$.

3. Sketch the curve $y = 4x^2$. Find the area included by the curve and
(a) the axis of x and the ordinates $x = 0$, $x = 3$;
(b) the axis of y and the abscissae $y = 1$, $y = 4$.

4. Find the areas bounded by the following curves, the axis of x, and
the ordinates $x = 1$, $x = 3$.

(i) $y = 2x^3$; (ii) $y = \frac{1}{x^2}$; (iii) $y = 2x^2 + x + 1$;

(iv) $y = (1-x)^2$; (v) $y = 2\sqrt{x}$.

5. The diagram (Fig. 65) represents the curve $y = 2x(2-x)$. What are the co-ordinates of P? Evaluate the area of the graph above the x-axis.

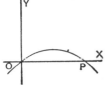

FIG. 65

6. Sketch the curve $y = (1-x)(2+x)$ and find the area of the portion of the curve above the x-axis.

7. For the curve $y = (3x-1)(x-2)$, find the range of values of x for which y is negative. Also find the area of the portion of the curve below the x-axis.

8. Find the areas of the segments cut off by the x-axis from the curves

(i) $y = x(x-1)(x-2)$;   (ii) $y = x(x+1)(x+2)$.

9. Sketch the curve $y = 3x - x^2$ for values of x between 0 and 5. Find the area contained between the curve and the ordinates $x = 0$ and $x = 5$. Evaluate $\int_0^5 (3x - x^2)dx$ and interpret the result.

10. Evaluate the integral $\int_0^3 (2x - x^2)dx$ and explain the result.

11. A curve passes through the point $(1, 1)$, and its gradient at any point (x, y) is $1 + 2x^2$. Find the area bounded by the curve, the axis of x, and the ordinates $x = 1$ and $x = 4$.

12. Find the area of the segment cut off on the curve $y = 5x - x^2$ by the line $y = 6$.

Approximate evaluation of definite integrals

As any area under a curve can be expressed as a definite integral, so any definite integral can be considered as the area under a certain curve. In many cases a definite integral cannot be easily evaluated, and in these cases an approximate result can be obtained by plotting a curve and obtaining the area under it.

Ex. 8. Obtain an approximate value for the integral $\int_1^3 \dfrac{1}{x+1} dx.$
This integral cannot be evaluated by any method we have so far considered. To obtain its approximate value, we plot the curve $y = \dfrac{1}{x+1}$ and actually obtain the area contained by the curve, the axis of x and the ordinates $x = 1$, $x = 3$. The area is obtained by counting squares.

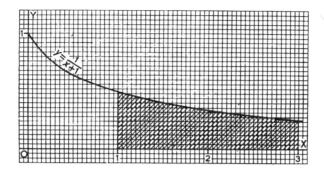

FIG. 66

Approximate area = 346 small squares.

But   1 sq. unit ≡ 500 small squares.

$$\therefore \int_1^3 \frac{1}{x+1}\, dx = \tfrac{346}{500} = \cdot 692, \text{ approximately.}$$

Volume of Revolution

Ex. 9.   *The portion of the curve* $y = x^2$ *between* $x = 1$ *and* $x = 3$ *is rotated about* OX. *Find the volume of the solid of revolution formed.* (*Fig.* 67).

The solid required is bounded by fixed planes perpendicular to OX, passing through A and B. Consider a plane parallel to these and free to move from A to B.

Let the volume contained between this plane PN and the fixed plane through A, be V.

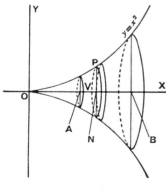

FIG. 67

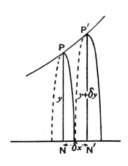

FIG. 68

Let $\mathbf{ON} = x$ and $\mathbf{NP} = y$. Then as $\mathbf{P}$ lies on the curve, $y = x^2$

As x varies, *i.e.* $\mathbf{N}$ varies, $\mathbf{V}$ will also change, and so $\mathbf{V}$ is some function of x. We need to obtain this function.

Let x increase by a small amount δx, and let $\delta \mathbf{V}$ be the small increase in $\mathbf{V}$ and δy the small increase in y.

$\delta \mathbf{V}$ is the volume contained between two circular planes, radii y and $y + \delta y$, and distance δx apart. (Fig. 68.)

$$\therefore \quad \delta \mathbf{V} \text{ lies between } \pi(y + \delta y)^2 \delta x \text{ and } \pi y^2 \delta x.$$

$$\therefore \quad \frac{\delta \mathbf{V}}{\delta x} \text{ lies between } \pi(y + \delta y)^2 \text{ and } \pi y^2.$$

Now let $\delta x \to 0$, and $\dfrac{d\mathbf{V}}{dx} = \pi y^2$.

$$\therefore \quad \mathbf{V} = \int \pi y^2 dx + c.$$

But $\mathbf{V} = 0$ when $x = 1$.

$$\therefore \quad c = -\left\{ \text{value of} \int \pi y^2 dx \text{ when } x = 1 \right\}.$$

$\therefore$ Required volume

$$= \left\{ \text{value of} \int \pi y^2 dx \text{ when } x = 3 \right\}$$

$$- \left\{ \text{value of} \int \pi y^2 dx \text{ when } x = 1 \right\},$$

$$= \int_1^3 \pi y^2 dx = \pi \int_1^3 x^4 dx,$$

$$= \pi \left[\frac{x^5}{5} \right]_1^3 = \pi \left[\tfrac{243}{5} - \tfrac{1}{5} \right],$$

$$= 48 \cdot 4 \pi \text{ unit}^3$$

In an exactly similar way, we find that the volume obtained by rotating the portion of the curve $y = f(x)$, between $x = a$ and $x = b$, about the x-axis is

$$\int_a^b \pi y^2 dx \quad \text{or} \quad \int_a^b \pi (f(x))^2 dx.$$

If the portion of curve between $y = y_1$ and $y = y_2$ were rotated about the y-axis, the corresponding result for the volume would be

$$\int_{y_1}^{y_2} \pi x^2 dy.$$

Ex. 10.   *The portion of the curve* $y^2 = 4x$ *between* $(0, 0)$ *and* $(1, 2)$ *is rotated* (i) *about the x-axis, and* (ii) *about the y-axis. Find the volumes generated in each case.*

(i) Volume generated $= \pi \int_0^1 y^2 dx = \pi \int_0^1 4x dx = \pi \left[2x^2 \right]_0^1$,

$$= 2\pi = 6 \cdot 28 \text{ unit}^3.$$

(ii) Volume generated $= \pi \int_0^2 x^2 dy = \pi \int_0^2 \frac{y^4}{16} dy = \frac{\pi}{16} \left[\frac{y^5}{5} \right]_0^2$,

$$= \frac{\pi}{16} \cdot \frac{32}{5} = \frac{2\pi}{5} = 1 \cdot 26 \text{ unit}^3.$$

Ex. 11.   *The loop of the curve* $y^2 = x(x-2)^2$ *is rotated about the axis of x; what is the volume of the solid so formed?*

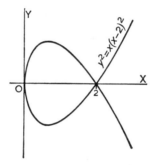

FIG. 69

The loop of the curve $y^2 = x(x-2)^2$ is the portion between $x = 0$ and $x = 2$.

∴ Volume of revolution $= \pi \int_a^b y^2 dx = \pi \int_0^2 x(x-2)^2 dx$,

$$= \pi \int_0^2 (x^3 - 4x^2 + 4x) dx = \pi \left[\frac{x^4}{4} - \frac{4x^3}{3} + 2x^2 \right]_0^2,$$

$$= \pi [4 - 10\tfrac{2}{3} + 8] = \frac{4\pi}{3},$$

$$= 4 \cdot 20 \text{ unit}^3.$$

(A) EXAMPLES 9b

In questions 1–10, find the volumes formed when the given areas are rotated about the axis of x.

1. The area contained between the curve $y = \sqrt{x}$ and the ordinates $x = 1$, $x = 4$.

2. The area contained between the line $y = 2x$ and the ordinates $x = 0$, $x = 2$.

3. The area contained between the curve $y = \dfrac{3}{x}$ and the ordinates $x = 1$, $x = 3$.

4. The area contained between the line $y = 2x + 1$ and the ordinates $x = -1$, $x = 1$.

5. The area contained between the curve $y^2 = 1 + x^2$ and the ordinates $x = -2$, $x = -1$.

6. The area contained between the curve $y = 1 + \sqrt{x}$ and the ordinates $x = 0$, $x = 1$.

7. The area contained between the curve $y = (1 + x)^2$ and the ordinates $x = 2$, $x = 3$.

8. The area contained between the curve $y\sqrt{x^3} = 2$ and the ordinates $x = 1$, $x = 5$.

9. The area contained between the curve $x^2 + y^2 = 16$ and the ordinates $x = 0$, $x = 4$.

10. The area contained between the curve $x^2 - y^2 = 1$ and the ordinates $x = 2$, $x = 4$.

11. Sketch the curve $xy = 4$. The portion of the curve from $y = 1$ to $y = 4$ is rotated about the y-axis. Find the volume of the solid generated.

12. The diagram (Fig. 70) represents the loop of the curve $y^2 = x(x - 6)^2$. Find the co-ordinates of P, and find the volume of the solid formed when the loop is rotated about the x-axis.

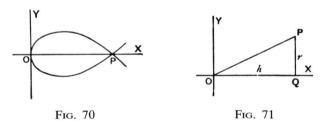

FIG. 70   FIG. 71

13. A vessel is formed by revolving the curve $y = \frac{1}{10}x^2$ about the y-axis. If the radius of the rim is 4 cm, find the depth and the volume of the vessel. [Unit on each axis 1 cm.]

14. Find the volume generated by revolving about the x-axis, the area included between that axis and the curve $y = (x - 1(x + 2)$.

15. OPQ is a right-angled triangle with OQ $= h$, PQ $= r$. (Fig. 71.) What is the equation of the line OP? If the area OPQ is rotated about OX, what solid is swept out? Use the previous results to obtain the volume of a right circular cone, base radius r and height h.

16. The diagram (Fig. 72) shows a quadrant of a circle centre O, radius r. If the co-ordinates of P, any point on the curve, are (x, y), what relation exists between x, y and r? By considering the rotation of the quadrant about OX, deduce the volume of a hemisphere of radius r.

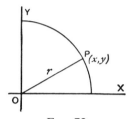

FIG. 72

(B) EXAMPLES 9c

1. Find the area bounded by the curve $y = 3x^2 + x$ and the ordinates $x = a$, $x = 2a$.

2. Find the area in the positive quadrant bounded by the curve $x^2 y = c^2$ and the ordinates $x = a$, $x = b$.

3. The portion of the curve $y^2 = 4ax$ between the ordinates $x = 0$, $x = a$ is rotated (i) about the x-axis, (ii) about the y-axis. Find the ratio of the volumes of the solids formed.

4. Find the area bounded by the curve $y = x^2 + 3x + 4$ and the lines $x = 2$, $x = 3$, $y = 1$.

5. What are the co-ordinates of the points of intersection of the straight line $y = 4x$ and the curve $x^2 = 2y$? Determine the area of the segment cut off from the curve by the straight line.

6. Find the area of the segment cut off on the curve $y = x(6 - x)$ by the line $y = 3x$.

7. Find the area bounded by the x-axis and the portion of the curve $y = 2(x - 1)(x - 3)$ which lies below it. If this area is rotated about the x-axis, find the volume swept out.

8. Sketch the curve $x = y(3 - y)$. Find (a) the area enclosed between the curve and the y-axis, and (b) the volume swept out when this area is rotated about the axis of y.

9. Sketch the curve $y = x(x - 1)(x - 3)$, and find the ratio of the areas of the two figures bounded by the curve and the axis of x.

10. The area enclosed between the curve $y = x(1-2x)$ and the line $y = \dfrac{x}{2}$, is rotated about the x-axis. Find the volume of the solid so formed.

11. Sketch the curve $y = \sqrt{4x}$. Find
    (a) the area bounded by the curve, the axis of x and the ordinate $x = 9$;
    (b) the value of a for which the ordinate $x = a$ divides this area into equal portions.

12. Prove that the curves $2y^2 = x$, $x^2 = 4y$ meet at the origin and at the point $(2, 1)$, and find the area intercepted between them.

13. The diagram (Fig. 73) represents the loop of the curve $y^2 = x(x-4)^2$. Find the area of this loop, and the volume of the solid formed when the loop is rotated about **OX**.

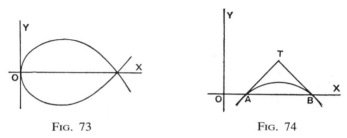

FIG. 73   FIG. 74

14. Figure 74 shows a portion of the curve $y = (x-1)(3-x)$ with the tangents to the curve at its points of intersection with the x-axis. Find the area included between the curve and the tangents **TA**, **TB**.

15. What area is represented by the integral $\displaystyle\int_1^4 \dfrac{dx}{x+2}$? By drawing the curve $y = \dfrac{1}{x+2}$ for values of x between 1 and 4, obtain an approximation to the value of this integral.

16. Interpret the following integrals geometrically, and obtain their values approximately

    (i) $\displaystyle\int_1^4 \dfrac{dx}{\sqrt{x+1}}$;   (ii) $\displaystyle\int_0^2 \dfrac{x\,dx}{x+1}$,   (iii) $\displaystyle\int_1^{10} \log_{10} x\,dx$;

    (iv) $\displaystyle\int_0^{90°} \sin x\,dx.$ [x in degrees.]

REVISION PAPERS II

PAPER A (1)

1. The speed of a body at time t s, is given by the formula $v = 3t - \frac{1}{4}t^2$ m s^{-1}. Find the acceleration of the body when $t = 4$ s. What is the maximum velocity?

2. (a) Differentiate $(2x + 1)^4$ with respect to x.

(b) If $y = x^3 - \frac{1}{x^3}$, find the value of $\frac{d^2y}{dx^2}$ when $x = 1$.

3. For the curve $y = 3x - x^2$, find (a) the greatest value of y; (b) the area above the x-axis between this axis and the curve.

4. Integrate   (i) $x^3 - 2x + 1$;   (ii) $(2x - 1)^2$.

Evaluate the integrals $\int_1^2 x(x - 1)dx$, $\int_1^9 \frac{dx}{\sqrt{x}}$.

5. A curve $y = f(x)$ has a gradient $3x^2 - 1$. If the curve passes through the origin, find its equation.

If the part of the curve between $x = 0$ and $x = 1$ is rotated about the x-axis, find the volume of the solid of revolution. [Take $\pi = 3 \cdot 14$.]

PAPER A (2)

1. Differentiate   (i) $(2x - 1)^2$;   (ii) $\sqrt{x + 2}$.

Integrate   (iii) $(3x + 2)^2$;   (iv) $\sqrt{x}(x + 2)$.

2. A curve passes through the origin, and its gradient function is $2x - \frac{x^2}{2}$. Find its ordinate when $x = 2$.

3. The distance s m travelled by a body in t s is given by $s = 30t - 16t^2$. Find (a) the velocity when $t = \frac{1}{2}$ s, (b) the distance travelled in the first $\cdot 501$ s.

4. Find the maximum and minimum values of the function $x(12 - x^2)$.

5. Find the area included between the axis of x and the portion of the curve $y = x^2 - 2x - 3$ which lies below that axis. If this area is rotated about the x-axis, find the volume of the solid formed.

PAPER A (3)

1. The radius of curvature, ρ, of the curve $y = f(x)$ at the point (x, y) is given by the formula

$$\rho = \frac{\left\{1 + \left(\dfrac{dy}{dx}\right)^2\right\}^{\frac{3}{2}}}{\dfrac{d^2 y}{dx^2}}.$$

Calculate the radius of curvature of the curve $y = x^2 + 2x$ at the origin.

2. The speed of a body after moving in a straight line for t s is $(12t - 3t^2)$ m s^{-1}. Find how far it goes in the third second.

3. Find the value of x for which the functions $2x^2 - x + 3$ and $x^2 - 3x + 1$ are increasing at the same rate.

4. The radius of a circular blot of ink is increasing at 5 cm min^{-1}. Find the rate of increase of the area when the radius is equal to 5 cm. Find also the approximate change in the area when the radius increases from 5 to 5·001 cm.

5. (i) Evaluate   (a) $\displaystyle\int \left(x^2 - \frac{1}{x^2}\right) dx$;   (b) $\displaystyle\int_4^9 \left(\sqrt{x} - \frac{1}{\sqrt{x}}\right) dx$.

     (ii) The line $y = 2x$ revolves about the axis of x and forms a circular cone. Find the volume of this cone between the planes $x = 0$, $x = 4$.

PAPER A (4)

1. If $x = 3t^2 - 1$, $y = t^3 - t$, find $\dfrac{dx}{dt}$, $\dfrac{dy}{dt}$ and $\dfrac{dy}{dx}$. For what values oɪ t is $\dfrac{dy}{dx}$ zero?

2. Find the minimum value of H, if $H = t^2(t - 27)$.

3. (i) Differentiate $\dfrac{1}{(2x - 1)^2}$.

     (ii) If $\dfrac{dy}{dx} = 2x - \dfrac{1}{x^2}$, and if $y = 1$ when $x = 1$, find y in terms of x.

4. The volume of a sphere, radius r cm, is $\frac{4}{3}\pi r^3$ cm^3. Find the approximate decrease in volume of a sphere when the radius decreases from 3 to 2·98 cm.

5. Find the area under the curve $y = 4\sqrt{x}$ between the ordinates $x = 0$ and $x = 9$. If this area is rotated about the x-axis, find the volume of the solid of revolution.

PAPER A (5)

1. If $V = \frac{4}{3}\pi r^3$ and $s = 4\pi r^2$, find the value of $\dfrac{ds}{dV}$ when $r = 2$.

2. A moving point covers s m in the first t s of its motion, where $s = 3 + 5t + t^2$. Find its velocity and acceleration after 3 s.

3. If $y = \sqrt{3x + 1}$, find (i) $\dfrac{dy}{dx}$, (ii) $\dfrac{d^2y}{dx^2}$.

4. (a) Evaluate (i) $\displaystyle\int (x^2 - 3x + 5)dx$; (ii) $\displaystyle\int \frac{2x - 1}{x^3}\, dx$.

(b) If $pv^{1.5} = 1$, evaluate $\displaystyle\int_1^4 p\, dv$.

5. A curve passes through the origin, and its gradient at any point (x, y) is $2 - \frac{1}{3}x^2$. Find the area bounded by the curve, the x-axis and the ordinates $x = 0$, $x = 1$.

PAPER A (6)

1. Find the gradient of the curve $y = (2x - 1)^4$ at the point $(1, 1)$. Write down the equations of the tangent and the normal to the curve at this point.

2. If $x + y = 9$, find the maximum value of $x^2 y$.

3. Differentiate $\dfrac{1}{4x - 3}$, and use your result to evaluate

$$\int_1^3 \frac{dx}{(4x - 3)^2}.$$

4. The velocity, v, of a moving point after time t is given by the relation $v = (t - 1)(2 - t)$. Find the distance between the two points at which the body is at rest.

FIG. 75

5. The diagram (Fig. 75) shows a rough sketch of the curve $y^2 = x(x - 1)^2$. Find the volume of the solid formed by rotating the loop about the axis of x.

PAPER B (1)

1. A particle is moving in a straight line, and at the end of t s is distant s m from a fixed point on the line. If $3s = 3 + 12t^2 - t^3$, show that at the end of 8 s the particle begins to retrace its path. Find also the greatest velocity attained during the 8 s. (J.M.B.)

2. The equation of a curve is $y = 2x^3 + 3x^2 + 1$. Sketch the curve and find the maximum and minimum values of y. If the tangent at the point where $x = 2$ meets the x-axis at T, calculate the distance of T from the origin.   (J.M.B.)

3. The contents, V cm^3, and the depth, x cm, of water in a vessel are connected by the relation $V = 5x^2 - \frac{1}{6}x^3$. If water is being poured into the vessel at the rate of 5 cm^3 s^{-1}, find the rate at which the depth is increasing when $x = 2$ cm.

4. A piece of lead 6 m long by 20 cm wide is to be bent to make an open rectangular gutter of length 6 m. Find the maximum area of cross-section of the gutter.

5. (a) Find the equation of the curve whose gradient is $4x(x^2 - 1)$ and which passes through the point $(2, 1)$.

(b) Find the area cut off between the x-axis and the line $y = 3$ by the curve $y = x(4 - x)$.   (J.M.B.)

PAPER B (2)

1. (a) Differentiate (i) $3x^4 - \dfrac{2}{x^3}$; (ii) $\dfrac{1}{(x+1)^3}$.

(b) The radius of a solid circular cylinder of fixed height 10 cm increases at the rate of 1 mm s^{-1}. At a certain instant the radius is 2 cm. Find the rate of increase, in cm^2 s^{-1}, of the *total* surface (i) at that instant; (ii) 10 s later.   (J.M.B.)

2. A point moves in a straight line so that its distance from a fixed point O of that line is $27t - t^3$ cm after t s. Prove that it will move outwards from O for 3 s, and that when it returns to O its speed will be 54 cm s^{-1}.

3. (a) Show that the greatest value of $x^3(4a - x)$ is $27a^4$.

(b) If 10 solid cubes of side x cm and 40 of side y cm are made, where $x + y = 12$, find the values of x and y that will make the total volume a minimum.   (J.M.B.)

4. (a) Given that $\dfrac{dy}{dx} = x^2 + \dfrac{1}{x^2}$, and that $y = 3$ when $x = 1$, find the value of y when $x = 3$.

(b) Find the area bounded by the axis of x and that part of the curve $y = (1 + x)(2 - x)$ for which y is positive.   (J.M.B.)

5. The part of the curve $y = \sqrt{(20x - x^2)}$ between $x = 0$ and $x = 10$ is a quarter of a circle. The arc of the curve from the origin to a point P whose x co-ordinate is a is revolved about the x-axis. If the volume of the bowl thus obtained is one-half of the bowl obtained by revolving the whole quadrant, show that $30a^2 - a^3 = 1000$.   (J.M.B.)

PAPER B (3)

1. Find the equation of the tangent to the curve $y = x(x - 2)^2 + 5$ at the point $(3, 8)$. Find also the maximum and minimum values of y. Draw a rough sketch of the curve.   (J.M.B.)

2. (a) Find, from first principles, the differential coefficient of x^3 with respect to x.

(b) A spherical soap bubble is of radius r and volume v. If r is subject to a slight variation, show that the percentage increase in r is approximately one-third of the percentage increase in v. If r increases from 1 cm to $1 \cdot 03$ cm, find, correct to two significant figures, the increase in v. (J.M.B.)

3. A closed rectangular box is made of sheet metal of negligible thickness, the length of the box being twice its width. Find the box of least surface that has a capacity of 243 cm^3. (J.M.B.)

4. A particle starting from **O** moves in a straight line, and its velocity, v m s^{-1}, after t s is given by the relation $v = 10 + 25t - 4t^2$. Show that its velocity will increase for the first $3\frac{1}{8}$ s, and find the distance travelled in this time.

5. (a) Find the area enclosed by the curve $y = x^2(x-1)^2$ and the x-axis.

(b) The portion of the curve $y = \dfrac{x^4 + 3}{x^2}$ between $x = 1$ and $x = 2$ is rotated about the x-axis. Find the volume enclosed by the resulting surface of revolution.

PAPER B (4)

1. If $x = (1 - 3t^2)^3$, $y = \dfrac{1}{\sqrt{2t+1}}$, find the value of $\dfrac{dy}{dx}$ when $t = 1$.

2. (a) If a cone has semi-vertical angle α and height h, prove that its volume is $\frac{1}{3}\pi h^3 \tan^2\alpha$.

(b) Water is being poured into a cone of semi-vertical angle $60°$ at $2 \text{ m}^3 \text{ min}^{-1}$. At what rate is the surface rising when the depth is 1 m?

3. A point **P** moves in a straight line so that its distance from a fixed point **A** in the line is given by $s = t - t^3$, where s is the distance in metres from **A** and t is the time in seconds.

(i) How far will **P** move in the 3rd second?

(ii) At what instant will it be at rest?

(iii) If v is the velocity and f the acceleration at any instant, prove that $v = \cdot 1 + \frac{1}{2}ft$.

4. If $\dfrac{dy}{dx} = x(x-1)^2$ and if $y = 1$ when $x = 1$, find y in terms of x. Show that the ordinate at the point $x = 1$ is neither a maximum nor a minimum, though $\dfrac{dy}{dx}$ is equal to 0 when $x = 1$. (J.M.B.)

5. Plot the curve $y = \dfrac{1}{5 - x^2}$, between $x = 0$ and $x = 2$, by calculating y when x is 0, $\frac{1}{2}$, 1, $1\frac{1}{2}$, 2. (*continued on next page*)

Hence evaluate approximately the integral $\int_0^2 \dfrac{dx}{5-x^2}$. (J.M.B.)

PAPER B (5)

1. (a) If $y\sqrt{1-x} = 1$, find $\dfrac{d^2y}{dx^2}$.

(b) Evaluate $\int_1^{10} P\,ds$ where $P = \dfrac{(1-s^2)^2}{s^2}$.

2. Prove that in the curve $y = x+5-\dfrac{4}{x+1}$, y increases as x increases.
Find the co-ordinates of the two points on the curve where the gradient is 2. If the tangents at these points meet the x-axis in P, Q, find the length of PQ. (J.M.B.)

3. (a) Find the maximum and minimum values of the expression $x^2(7-2x^5)$.

(b) A rectangular field is bounded on one side by a straight river, and on the other three sides by a fence whose total length is 160 m. Show that the area of the field cannot exceed 3200 m². (J.M.B.)

4. Sketch the curve $y = 2\sqrt{x}$ between $x = 0$ and $x = 9$. Find
(a) the area bounded by the x-axis, this portion of the curve, and the ordinate $x = 9$;
(b) where a line should be drawn parallel to the y-axis so as to bisect this area.

5. (a) If $\dfrac{dy}{dx} = \left(x+\dfrac{1}{x}\right)^2$ and $y = 7$ when $x = 1$, find y when $x = 2$.

(b) The part of the curve $y = x^2(1-x)$ between $x = -1$ and $x = 1$, and the ordinates at its extremities, are rotated about the axis OX. Find the volume swept out. (J.M.B.)

PAPER B (6)

1. The tangent to the curve $y(1+x^2) = 2$ at the point $(2, \tfrac{2}{5})$ meets the curve again in Q. Find the co-ordinates of Q.

2. A right circular cone is cut from a solid sphere of radius a, the vertex and the circumference of the base being on the surface of the sphere. If x is the distance of the base from the centre of the sphere, prove that the volume of the cone is $\tfrac{1}{3}\pi(a^2-x^2)(a+x)$. Hence find the height of the cone when its volume is a maximum. (J.M.B.)

3. The velocity, v, of a point moving along a straight line is given in terms of t by the equation $v = 2t^2 - 7t + 6$, the point being at the origin when t is 0. Find the equations, giving in terms of t, (a) the distance from the origin, and (b) the acceleration. Show that the point is at rest twice and find its distances from the origin when at rest.

4. (a) If $\dfrac{dy}{dx} = \dfrac{4}{(x+1)^2} - \dfrac{9}{(x+2)^2}$ and $y = 0$ when $x = 0$ find y when $x = 7$.

(b) Find the area in the first quadrant bounded by the x-axis and the curve $y = 2x - 8x^3$. (J.M.B.)

5. On the curve $y^2 = ax$, P is the point where $y = 2a$; PN is the perpendicular on the axis of x and PN the perpendicular on the axis of y. Find the volumes generated (i) by the revolution of the area enclosed by PN, NO and the arc OP about the axis of x, (ii) by the revolution of the area enclosed by PM, MO and the arc OP about the axis of y. (J.M.B.)

CHAPTER 10

THEORY OF THE QUADRATIC EQUATION.
QUADRATIC FUNCTIONS.

Sum and Product of the roots of a quadratic equation

Let the roots of the equation $ax^2 + bx + c = 0$ be α and β.
We can write the equation with roots α and β in the form

$$(x - \alpha)(x - \beta) = 0,$$

$$i.e. \quad x^2 - x(\alpha + \beta) + \alpha\beta = 0.$$

Hence the equations $\quad x^2 - x(\alpha + \beta) + \alpha\beta = 0,$

and $\qquad\qquad\qquad ax^2 + bx + c = 0$

have the same roots.

$\therefore$ Comparing coefficients, $\quad \dfrac{1}{a} = -\dfrac{(\alpha + \beta)}{b} = \dfrac{\alpha\beta}{c}.$

$$\left.\begin{array}{c} \therefore \ \alpha + \beta = -\dfrac{b}{a} \\[2mm] \alpha\beta = \dfrac{c}{a} \end{array}\right\}.$$

or

$i.e.$ **Sum of the roots** $= -\dfrac{\textbf{Coefficient of x}}{\textbf{Coefficient of x}^2}.$

Product of roots $= \dfrac{\textbf{Constant term}}{\textbf{Coefficient of x}^2}.$

Ex. 1. *If the roots of the equation* $4x^2 - 6x + 1 = 0$ *are* α *and* β, *find the value of* $\alpha^3 + \beta^3$, *and find the equation whose roots are* $\left(\alpha + \dfrac{1}{\beta}\right)$ *and* $\left(\beta + \dfrac{1}{\alpha}\right)$.

If the roots of the equation $4x^2 - 6x + 1 = 0$ are α and β,

$$\text{then } \alpha + \beta = -\left(\frac{-6}{4}\right) = \frac{3}{2},$$

$$\alpha\beta = \tfrac{1}{4}.$$

$$\therefore \; \alpha^3 + \beta^3 = (\alpha + \beta)(\alpha^2 - \alpha\beta + \beta^2),$$

$$= (\alpha + \beta)\{(\alpha + \beta)^2 - 3\alpha\beta\},$$

$$= \tfrac{3}{2} \cdot \{\tfrac{9}{4} - \tfrac{3}{4}\} = \tfrac{9}{4}.$$

To find the equation whose roots are $\alpha + \dfrac{1}{\beta}$ and $\beta + \dfrac{1}{\alpha}$.

Equation is $x^2 - x(\text{sum of roots}) + \text{product of roots} = 0$.

$$\text{Sum of roots} = \left(\alpha + \frac{1}{\beta}\right) + \left(\beta + \frac{1}{\alpha}\right) = \alpha + \beta + \frac{\alpha + \beta}{\alpha\beta},$$

$$= \frac{3}{2} + \frac{\tfrac{3}{2}}{\tfrac{1}{4}} = 7\tfrac{1}{2}.$$

$$\text{Product of roots} = \left(\alpha + \frac{1}{\beta}\right)\left(\beta + \frac{1}{\alpha}\right) = \alpha\beta + 2 + \frac{1}{\alpha\beta},$$

$$= \tfrac{1}{4} + 2 + 4 = 6\tfrac{1}{4}.$$

$$\therefore \; \text{Equation is} \quad x^2 - x \cdot \tfrac{15}{2} + \tfrac{25}{4} = 0,$$

$$\textit{i.e.} \quad 4x^2 - 30x + 25 = 0.$$

Ex. 2. *If α, β are the roots of the equation $ax^2 + bx + c = 0$, find the value of $(1 - \alpha^3)(1 - \beta^3)$.*

We have
$$\alpha + \beta = -\frac{b}{a}, \qquad \alpha\beta = \frac{c}{a}.$$

$$\therefore \; (1 - \alpha^3)(1 - \beta^3) = 1 - (\alpha^3 + \beta^3) + \alpha^3\beta^3,$$

$$= 1 - (\alpha + \beta)(\alpha^2 + \beta^2 - \alpha\beta) + \alpha^3\beta^3,$$

$$= 1 - (\alpha + \beta)\{(\alpha + \beta)^2 - 3\alpha\beta\} + \alpha^3\beta^3,$$

$$= 1 - \left(-\frac{b}{a}\right)\left\{\frac{b^2}{a^2} - \frac{3c}{a}\right\} + \frac{c^3}{a^3},$$

$$= 1 + \frac{b^3 - 3abc + c^3}{a^3},$$

$$= \frac{a^3 + b^3 + c^3 - 3abc}{a^3}.$$

Roots of a Quadratic Equation

Consider the equation

$$ax^2 + bx + c = 0,$$

$$\textit{i.e.} \quad x^2 + \frac{bx}{a} = -\frac{c}{a}.$$

Completing the square,

$$x^2 + \frac{bx}{a} + \left(\frac{b}{2a}\right)^2 = -\frac{c}{a} + \left(\frac{b}{2a}\right)^2,$$

$$\therefore \left(x + \frac{b}{2a}\right)^2 = \frac{b^2 - 4ac}{4a^2}.$$

$$\therefore x + \frac{b}{2a} = \pm\frac{\sqrt{b^2 - 4ac}}{2a},$$

i.e. $$x = \frac{-b \pm \sqrt{b^2 - 4ac}}{2a}.$$

i.e.   the roots of the equation $ax^2 + bx + c = 0$ are

$$\frac{-b \pm \sqrt{b^2 - 4ac}}{2a}.$$

Case   (i). **If $b^2 - 4ac > 0$, the roots are real and different.**
Case   (ii). **If $b^2 - 4ac = 0$, the roots are real and equal.**
Case   (iii). **If $b^2 - 4ac < 0$, the roots are imaginary.**

Ex. 3.   Discuss the nature of the roots of the equation

$$3x^2 - 7x + 2 = 0.$$

Here $a = 3, \quad b = -7, \quad c = 2.$

$$\therefore b^2 - 4ac = 49 - 24 = 25.$$

i.e. $b^2 - 4ac > 0$ and hence the roots of the equation are real and different.

Ex. 4.   For what values of k are the roots of the equation $x^2 + 2x + 1 = k(x + 2)$ equal?

Rewriting the equation with terms collected,

$$x^2 + x(2 - k) + 1 - 2k = 0,$$

For equal roots, $b^2 - 4ac = 0,$

$$\therefore (2 - k)^2 - 4(1)(1 - 2k) = 0,$$

$$k^2 + 4k = 0,$$

$$k(k + 4) = 0,$$

i.e. $k = 0$ and $-4.$

Ex. 5.   If x is real and $4x^2 - 2kx + 8 - k = 0$, show that k cannot lie between certain limits and find these limits.

As x is real, the roots of the given equation are real.

$$\therefore b^2 - 4ac \geqslant 0.$$

i.e. $(-2k)^2 - 4(4)(8-k) \geqslant 0,$

$$4k^2 + 16k - 128 \geqslant 0,$$

$$k^2 + 4k - 32 \geqslant 0.$$

Factorising, $(k+8)(k-4) \geqslant 0.$ . . . . (i)

Now the function $(k+8)(k-4)$ is zero when $k = -8$ and 4. For a value of k between these values, say $k = 0$, the function is negative.

Hence the inequality (i) is not satisfied when k takes values between -8 and 4.

Consequently if x is real, k cannot lie between -8 and 4.

The Quadratic Function. Sign of the Quadratic Function

The following examples show how the sign of a quadratic function can be determined.

Ex. 6. Determine the signs of the function $(x-3)(2x-1)$ *for real values of* x.

The function is zero when $x = 3$ and $x = \frac{1}{2}$.

For values of x between 3 and $\frac{1}{2}$, one factor is negative and the other positive, and hence the function is negative.

For values of x not lying between 3 and $\frac{1}{2}$, the function is positive, as each factor will be of the same sign.

Ex. 7. Determine the sign of the function $2x^2 - 4x + 3$ *for real values of* x.

We have $2x^2 - 4x + 3 = 2\{x^2 - 2x + \frac{3}{2}\},$

$$= 2\{(x-1)^2 + \frac{3}{2} - 1\}, \text{ (completing the square)}$$

$$= 2\{(x-1)^2 + \frac{1}{2}\}.$$

As x is real, the squared term is always positive, and hence the function is positive for all real values of x.

Ex. 8. Find the greatest value of c, *if there is no real value of* x *for which the function* $c - 8x - 3x^2$ *is positive.*

Expression $= -3x^2 - 8x + c = -3\left\{x^2 + \dfrac{8x}{3} - \dfrac{c}{3}\right\},$

$$= -3\left\{\left(x + \dfrac{4}{3}\right)^2 - \dfrac{c}{3} - \dfrac{16}{9}\right\} = -3\left\{\left(x + \dfrac{4}{3}\right)^2 - \dfrac{3c+16}{9}\right\}.$$

As the expression is to be negative for all real values of x, $\left\{(x + \frac{4}{3})^2 - \dfrac{3c+16}{9}\right\}$ must be positive for *all* real values of x.

Hence $\dfrac{3c+16}{9}$ must be $\leqslant 0$.

$i.e.$ $3c+16 \leqslant 0$,

or $c \leqslant -\dfrac{16}{3}$.

$\therefore$ Greatest possible value of c is $-\dfrac{16}{3}$.

Maximum or Minimum Values of a Quadratic Function

The maximum or minimum value of a quadratic function can be obtained without the use of the differential calculus in the following manner.

Ex. 9. Determine the minimum value of the function $2-3x+4x^2$.

We have
$$2-3x+4x^2 = 4\{x^2 - \tfrac{3}{4}x + \tfrac{1}{2}\},$$
$$= 4\{(x-\tfrac{3}{8})^2 + \tfrac{1}{2} - \tfrac{9}{64}\},$$
$$= 4\{(x-\tfrac{3}{8})^2 + \tfrac{23}{64}\}.$$

$\therefore$ The expression is a minimum when $x - \tfrac{3}{8}$ is zero,

$i.e.$ when $x = \tfrac{3}{8}$.

Minimum value $= 4 \cdot \tfrac{23}{64} = \tfrac{23}{16}$.

(A) EXAMPLES 10a

1. Write down the sums and the products of the roots of the following equations

(i) $2x^2 - 4x + 3 = 0$; (ii) $3x^2 - x = 7$;

(iii) $\dfrac{x^2}{2} = x - 4$; (iv) $(2x-1)^2 = 3$;

(v) $x + \dfrac{1}{x} = 2$; (vi) $x^2 + 2px - q = 0$;

(vii) $\dfrac{bx}{x^2 + ax + b} = 1$.

2. In each of the following cases find the quadratic equation the sum and product of whose roots have the stated values.

(i) sum 6, product 8; (ii) sum -2, product 4;
(iii) sum $\tfrac{3}{5}$, product $\tfrac{2}{5}$; (iv) sum $2a$, product a^2;
(v) sum $-3k$, product $5k^2$; (vi) sum $2a - b$, product $a + b$.

3. Find the sums of the squares of the roots of the following equations

(i) $x^2 - x + 6 = 0$; (ii) $2x^2 - 4x + 1 = 0$;
(iii) $3x^2 - 12x + 5 = 0$; (iv) $x^2 + px + q = 0$.

4. Find the sums of the cubes of the roots of the equations in the previous question.

5. If the roots of the equation $4x^2 - 6x + 1 = 0$ are α and β, find the values of

(i) $(\alpha - \beta)^2$; (ii) $\alpha^2 - \beta^2$; (iii) $\alpha^3 - \beta^3$; (iv) $\left(\alpha^2 + \dfrac{1}{\beta^2}\right)\left(\beta^2 + \dfrac{1}{\alpha^2}\right)$.

6. Determine whether the roots of the following equations are real and different, real and equal, or imaginary

(i) $3x^2 - 4x + 6 = 0$; (ii) $1 - 4x + 3x^2 = 0$; (iii) $x^2 - 2x + 8 = 0$;
(iv) $4x^2 - 5x = 1$; (v) $x^2 - ax + a^2 = 0$.

7. If the roots of the equation $3x^2 - 5x + q = 0$ are equal, find q.

8. If one root of the equation $3x^2 - px + 2 = 0$ is double the other, find the values of p.

9. Find q if the roots of the equation $3x^2 + qx - 12 = 0$ differ by 4.

10. If the roots of the equation $2x^2 - x + 6 = 0$ and α and β, find the equations whose roots are

(i) $2\alpha, 2\beta$; (ii) α^2, β^2; (iii) $\dfrac{1}{\alpha}, \dfrac{1}{\beta}$; (iv) α^3, β^3;

(v) $\alpha + \dfrac{1}{2\beta}, \beta + \dfrac{1}{2\alpha}$; (vi) $\alpha + 2\beta, \beta + 2\alpha$.

11. If k is negative, determine the nature of the roots of the equation $x^2 + 2x + 1 - k = 0$.

12. For what value of m are the roots of the equation $x^2 + 2x(2 - m) + (1 + m)^2 = 0$ equal?

13. For what values of x are the following functions negative?

(i) $(x - 2)(x - 3)$; (ii) $(3x - 4)(5x - 2)$; (iii) $x^2 - 3x - 28$;
(iv) $3x^2 - 10x + 3$; (v) $15 + x - 2x^2$; (vi) $6x^2 - 13x + 6$.

14. Find the signs of the following functions for real values of x

(i) $2x^2 + 5x + 4$; (ii) $2x^2 - 6x + 11$; (iii) $12x - 15 - 3x^2$;
(iv) $2x - 3 - 2x^2$; (v) $2x - x^2 - 8$.

15. Find, algebraically, the maximum or minimum values of the following expressions and check your results by differentiation.

(i) $3x^2 - 4x + 1$; (ii) $2 - x - 4x^2$;
(iii) $7 + 10x + 5x^2$; (iv) $4 + 3x - x^2$.

(B) EXAMPLES 10b

1. If α, β are the roots of the equation $x^2 - 5x + 2 = 0$, find the value of $\alpha^2 + \beta^2$ and hence find the value of $\alpha^4 + \beta^4$.

2. For what values of a are the roots of the equation $x^2 - 2x + a - 2 = 0$ real and different?

3. If the roots of the equation $px^2 + qx + r = 0$ are α, β, find the value of $\left(\alpha^3 + \dfrac{1}{\beta^3}\right)\left(\beta^3 + \dfrac{1}{\alpha^3}\right)$.

4. If α, β are the roots of the equation $x^2 - 3x - 7 = 0$, find the equation whose roots are $\alpha^2 + \dfrac{1}{\beta^2}$ and $\beta^2 + \dfrac{1}{\alpha^2}$.

5. If α, β are the roots of the equation $x^2 + 3x + 1 = 0$, prove that the equation whose roots are $\alpha + c\alpha^{-1}$, $\beta + c\beta^{-1}$ is $x^2 + 3(1 + c)x + 1 + 7c + c^2 = 0$.

6. If the roots of the equation $ax^2 + bx + c = 0$ are in the ratio $m : n$, prove that $(m^2 + n^2)ac = (b^2 - 2ac)_{\mid}mn$.

7. If α, β are the roots of the equation $x^2 - px + q = 0$, obtain the equation whose roots are $1/\alpha^2$ and $1/\beta^2$.

8. Show that the roots of the equation $(x + 1)(2x - 1) = kx$ are real and different for all real values of k.

9. The roots of the equation $x^2 + px + 1 = 0$ are α, β and the roots of the equation $x^2 - 9x + q = 0$ are $\alpha + 2\beta$, $\beta + 2\alpha$. Find the values of p and q.

10. One root of the equation $x^2 - px - 8 = 0$ is the square of the other root. Find the real value of p.

11. Given that α is a root of the equation $x^2 - 5x + 3 = 0$, show that $\alpha^2 - 5\alpha = -3$. If β is the other root write down the value of $\beta^2 - 5\beta$ and use these two results to evaluate $\alpha^2 + \beta^2$.

12. If $x - y = 2$, find the least value of $x^2 + xy - x - 9$.

13. If $a > b > 0$, prove that the equation $x(x - a) = k(x - b)$ has real roots whatever real value k may have.

14. Find the values of m for which the equation $2(x^2 - 3x + 4) = m(x^2 - x - 2)$ has equal roots.

15. Prove that the equation $4x^2 + 4x(2 - m) + 3m - 8 = 0$ has no real roots if $3 < m < 4$.

16. The roots of the equation $x^2 + px + 1 = 0$ are α, β and the roots of the equation $x^2 + qx + 1 = 0$ are y, δ. Prove that $(\alpha\delta - \beta y)(\alpha y - \beta\delta) = p^2 - q^2$.

17. Show that the roots of the equation $(2x - 5)(x + 1) = m(x - 1)$ are real for all real values of m.

18. If the roots of the equation $x^2 + 2x(1-k) + 3k - 5 = 0$ are real, show that k cannot take any value between 2 and 3.

19. If p and q are real, prove that the expression $x^2 + (p+q)x + p^2 - pq + q^2$ is always positive.

20. If $y(x^2 + x + 1) = x^2 - x + 1$ and x is real, show that y must take values from $\frac{1}{5}$ to 3

CHAPTER 11

ALGEBRAIC METHODS

The method of substitution

The method of substitution is of importance in mathematics and its applications in algebra to elimination and to the solution of simultaneous equations are illustrated in the following examples.

Ex. 1. *Eliminate t from the equations* $x = t^2 + t$, $y = 2t - 1$.

From the linear equation $\qquad y = 2t - 1$,

we have $\qquad\qquad\qquad t = \frac{1}{2}(y + 1)$.

Substituting this expression for t in the first equation

gives $\qquad\qquad\qquad x = [\frac{1}{2}(y + 1)]^2 + \frac{1}{2}(y + 1)$,

$$4x^2 = y^2 + 4y + 3,$$

or $\qquad\qquad\qquad 4x^2 - y^2 - 4y - 3 = 0.$

Ex. 2. *If* $x = at^3$, $y = bt^2$, *obtain the equation connecting* x, y *and the constants* a, b.

We have $\qquad\qquad\qquad x^2 = a^2 t^6; \; y^3 = b^3 t^6.$

$$\therefore \; t^6 = \frac{x^2}{a^2} = \frac{y^3}{b^3},$$

so $\qquad\qquad\qquad \dfrac{x^2}{a^2} = \dfrac{y^3}{b^3},$

or $\qquad\qquad\qquad b^3 x^2 = a^2 y^3.$

Ex. 3. *Eliminate x from the equations* $x^2 + a_1 x + b_1 = 0$, $x^2 + a_2 x + b_2 = 0$.

Subtracting the equations in order to obtain a linear equation in x,

$$x^2 + a_1 x + b_1 - (x^2 + a_2 x + b_2) = 0,$$

$$x(a_1 - a_2) - (b_2 - b_1) = 0,$$

$$x = \frac{b_2 - b_1}{a_1 - a_2}.$$

Now substitute for x in either equation, say the first.

$$\left(\frac{b_2-b_1}{a_1-a_2}\right)^2 + a_1\left(\frac{b_2-b_1}{a_1-a_2}\right) + b_1 = 0,$$

i.e. $(b_2-b_1)^2 + a_1(a_1-a_2)(b_2-b_1) + b_1(a_1-a_2)^2 = 0.$

Ex. 4. *Solve the equations* $2x+3y=1$, $3x^2+7xy+4y^2=0$.

From the linear equation, $x = \dfrac{1-3y}{2}$.

Substituting this expression for x in the quadratic equation,

$$3\left(\frac{1-3y}{2}\right)^2 + 7y\left(\frac{1-3y}{2}\right) + 4y^2 = 0.$$

Multiplying by 4 and collecting terms gives

$$y^2 - 4y + 3 = 0,$$

i.e. $(y-1)(y-3) = 0,$

$$\left.\begin{array}{l} y = 1, 3 \\ x = -1, -4 \end{array}\right\}.$$

and

Ex. 5. *Solve the equations* $2xy+x-y=13$, $xy-x+y=5$.

The neatest solution is obtained by first eliminating the xy term from the equations to give a linear equation in x and y.

Multiplying the second equation by 2 and subtracting from the first gives $3x-3y=3$ or $x-y=1$.

So $y = x-1,$

and substituting in either of the given equations, say the first, gives

$$2x(x-1)+x-(x-1) = 13,$$
$$2x^2-2x-12 = 0,$$

or $x^2-x-6 = 0,$
$$(x-3)(x+2) = 0,$$

$$\left.\begin{array}{l} x = 3, -2 \\ y = 2, -3 \end{array}\right\}.$$

and

(A) EXAMPLES 11a

1. Eliminate t from the equations $y = t^2$, $x = 2t$.

2. Given $x = 3t$, $y = 3/t$, obtain the equation connecting x and y.

3. Solve the equations $2x^2+y^2=6$, $y=2x$.

4. If $x = t^3$, $y = t^2$, find the relationship between x and y.

5. Solve the equations $xy = 12$, $2x-y = 5$.

6. Eliminate m if $x - 1 = 4m^2$ and $y = 8m$.

7. A point has co-ordinates $(4t, 4t^{-1})$. By writing $x = 4t$, $y = 4t^{-1}$ and eliminating t, find the (x, y) equation of the curve on which the point lies.

8. Solve (i) $x^2 + y^2 = 25, 2y + x = 5$;

(ii) $\dfrac{1}{x} + \dfrac{1}{y} = 3, 4x - y = 1$.

9. Solve the equations $3x^2 - 2xy - y^2 = 3$, $y = x - 3$.

10. Find the (x, y) equations of the curves on which the following points lie:—

(i) $(-8t, 4t^2)$; (ii) $(3 + t^2, 1 + 2t)$; (iii) $\left(1 + 4t, 2 + \dfrac{4}{t}\right)$;

(iv) $(3t + 2, 2t + 5)$.

11. Solve (i) $x^2 + 3xy - 2y^2 = 1, 3x - y = 3$;
(ii) $2y^2 + 3xy - x^2 = 1, 3y - x + 1 = 0$.

12. Eliminate y from the equations $ax + by + 1 = 0$, $bx + ay + 1 = 0$, where $a \neq b$.

13. Solve the equations $\dfrac{2}{x} + y = -4$, $3x - \dfrac{4}{y} = -3$.

14. If $x = t + \dfrac{1}{t}$, $y = t - \dfrac{1}{t}$, show that $\frac{1}{2}(x + y) = t$ and hence eliminate t from the equations.

15. If $x = \frac{1}{2}a\left(t + \dfrac{1}{t}\right)$, $y = \frac{1}{2}a\left(t - \dfrac{1}{t}\right)$ show that $x^2 - y^2 = a^2$.

16. Solve $12xy + 13y^2 = 25$, $4x - 3y = 1$.

17. Given $x = t^2 + t$, $y = t^2 - 2t$, eliminate t^2 and obtain an expression for t in terms of x and y. Hence eliminate t between the given equations.

18. Solve the equations

$$xy = 6, \qquad xy + x + y = 1.$$

Identities

An *identity* is a relationship which is true for *all* values of the variable or variables involved. It differs from an equation in that the latter is only true for a number of specific values of the variable involved—the roots of the equation.

E.g. $x^2 + 2x - 1 = (x + 1)^2 - 2$ is an identity whereas $x^2 + 2x - 1 = 0$ is an equation. In the case of an identity the symbol $\equiv$ is often used instead of the equality symbol $=$. Identities are handled either by equating coefficients or by giving particular values to the variable or variables involved. Both methods are illustrated below.

Ex. 6. Express $4x^2 + x - 1$ *in the form* $\mathsf{A} + \mathsf{B}(x+1) + \mathsf{C}x(x+1)$, *where* $\mathsf{A}, \mathsf{B}, \mathsf{C}$ *are constants.*

Let $\qquad\qquad 4x^2 + x - 1 \equiv \mathsf{A} + \mathsf{B}(x+1) + \mathsf{C}x(x+1)$.

Equating coefficients of x^2, $\qquad 4 = \mathsf{C}$

Equating coefficients of x, $\qquad 1 = \mathsf{B} + \mathsf{C}$,

∴ $\qquad\qquad\qquad\qquad\qquad \mathsf{B} = -3$.

Equating constant terms, $\qquad -1 = \mathsf{A} + \mathsf{B}$,

∴ $\qquad\qquad\qquad\qquad\qquad \mathsf{A} = 2$.

Hence $\qquad\qquad 4x^2 + x - 1 \equiv 2 - 3(x+1) + 4x(x+1)$.

Alternative method

Let $x = -1$, then $\qquad 4 - 1 - 1 = \mathsf{A}; \quad \mathsf{A} = 2$.

Let $x = 0$, then $\qquad\qquad -1 = \mathsf{A} + \mathsf{B}; \quad \mathsf{B} = -3$.

Let $x = 1$, then $\qquad 4 + 1 - 1 = \mathsf{A} + 2\mathsf{B} + 2\mathsf{C}$;

$\qquad\qquad\qquad\qquad \mathsf{C} = 4$ as before.

N.B. The values of x, -1 and 0, where used as they made terms on the R.H.S. vanish.

Ex. 7. Given that $2x^2 - 8x + 9 = p(x-q)^2 + r$ *for all values of* x, *find the values of the constants* p, q, r.

We have $\qquad\qquad 2x^2 - 8x + 9 \equiv px^2 - 2pqx + pq^2 + r$.

Equating coefficients, $\qquad 2 = p$.

$\qquad\qquad\qquad\qquad -8 = -2pq; \quad q = 2$.

$\qquad\qquad\qquad\qquad 9 = pq^2 + r; \quad r = 1$.

So $\qquad\qquad\qquad\qquad p = 2, q = 2, r = 1$.

Ex. 8. Express $\dfrac{4x - 13}{(x+2)(2x-3)}$ *in the form* $\dfrac{\mathsf{A}}{x+2} + \dfrac{\mathsf{B}}{2x-3}$ *where* A, B *are constants.*

We have $\qquad \dfrac{4x - 13}{(x+2)(2x-3)} \equiv \dfrac{\mathsf{A}}{x+2} + \dfrac{\mathsf{B}}{2x-3}$,

$\qquad\qquad\qquad\qquad \equiv \dfrac{\mathsf{A}(2x-3) + \mathsf{B}(x+2)}{(x+2)(2x-3)}$.

∴ $\qquad\qquad 4x - 13 \equiv \mathsf{A}(2x-3) + \mathsf{B}(x+2)$.

Let $x = \frac{3}{2}$, $\qquad\qquad -7 = \mathsf{B}(\frac{7}{2}); \quad \mathsf{B} = -2$.

Let $x = -2$, $\qquad\qquad -21 = \mathsf{A}(-7); \quad \mathsf{A} = 3$.

So
$$\frac{4x-13}{(x+2)(2x-3)} \equiv \frac{3}{x+2} - \frac{2}{2x-3}.$$

(A) EXAMPLES 11b

1. Find the values of the constants A, B in each of the following identities:—

(i) $2(x-1)^2 \equiv A(x^2+1) + Bx$;

(ii) $3x-1 \equiv A(x-1) + B(x+1)$;

(iii) $1 \equiv A(1-2x) + B(1-3x)$.

2. Express $1-10x+x^2$ in the form $A(1-x)^2 + B(1+x)^2$, where A, B are constants.

3. Find the constants p, q if x^2-4x+5 is expressed in the form $(x-p)^2+q$.

4. In each of the following identities find the values of the constants l, m, n:—

(i) $(2x-1)^2 \equiv lx^2 + mx + n$;

(ii) $2x^2-3x+2 \equiv l + m(x+1) + nx(x+1)$;

(iii) $x \equiv l(x-1)(x+1) + mx(x+1) + nx(x-1)$;

(iv) $3 \equiv (lx+m)(x-1) + n(x^2+1)$.

5. Express the function $2x^2+9x-13$ in the form

$$a(x-2)(x+3) + b(x+3)(x-1) + c(x-1)(x-2),$$

where a, b, c are constants.

6. In each of the following identities find the values of the constants A, B:—

(i) $\dfrac{x}{(x+1)(x-3)} \equiv \dfrac{A}{x+1} + \dfrac{B}{x-3}$;

(ii) $\dfrac{7}{(2x-1)(x+3)} \equiv \dfrac{A}{2x-1} + \dfrac{B}{x+3}$;

(iii) $\dfrac{2x-1}{x+1} \equiv A + \dfrac{B}{x+1}$;

(iv) $\dfrac{5x-1}{1-x^2} \equiv \dfrac{A}{1-x} + \dfrac{B}{1+x}$.

7. Express the function $4x^2-4x+9$ in the form $p(x-q)^2+r$ where p, q, r are constants.

8. Given $x^4+4x^2-14 \equiv a(x^2-1)^2 + b(x^2-4)^2$, find the values of the constants a, b.

9. Find the values of the constants a, b, c if

$$\frac{20x}{(x+1)(x-2)(x+3)} \equiv \frac{a}{x+1} + \frac{b}{x-2} + \frac{c}{x+3}.$$

10. If the factors of the function $x^3 + x^2 - 3x + 1$ are $x - a$ and $x^2 + bx - 1$, find the values of a, b.

11. Express $x^2 - 3x + 2$ in the form $Ax^2 + B(x+2) - Cx(x+2)$, where A, B, C are constants.

12. Express $x^4 + x^2 + 1$ in the form

$$(x^2 + ax - b)(x^2 + bx + a),$$

where a, b are constants.

13. Find the values of the constants p, q, r, if

$$\frac{3x^2}{2x-1} \equiv px + q + \frac{r}{2x-1}.$$

14. For what value of k is the expression $5x^2 + 4x + 2 + k(x^2 + 1)$ a perfect square of the form $(ax + b)^2$?

15. If $2x^2 - 12x + 21$ can be expressed in the form $p(x-q)^2 + r$, find the values of the constants p, q, r.

16. In each of the following identities find the values of the constants A, B, C:—

(i) $\dfrac{x+1}{x^2(x-1)} \equiv \dfrac{A}{x} + \dfrac{B}{x^2} + \dfrac{C}{x-1}$;

(ii) $\dfrac{x+2}{x(x^2+1)} \equiv \dfrac{A}{x} + \dfrac{Bx+C}{x^2+1}$.

Graphical Solution of Equations

The student is already familiar with the graphical methods used to solve quadratic equations. The following examples show how this method can be extended to solve other types of algebraic equations.

Ex. 9. Plot on the same axes the graphs of $y = x^2$ and $y = x + \dfrac{1}{x}$ between $x = 3$ and $x = -3$. Hence find one root of the equation $x^3 - x^2 - 1 = 0$.

Graph $y = x^2$.

x	-3	-2	-1	0	1	2	3
y	9	4	1	0	1	4	0

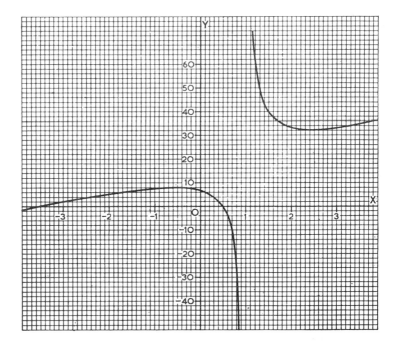

Fig. 76

Graph $y = x + \dfrac{1}{x}$.

x	± 3	± 2	± 1	$\pm \frac{1}{2}$	$\pm \frac{1}{4}$	0
y	$\pm 3\frac{1}{3}$	$\pm 2\frac{1}{2}$	± 2	$\pm 2\frac{1}{2}$	$\pm 4\frac{1}{4}$	∞

At the point of intersection of the curves the equations $y = x^2$ and $y = x + \dfrac{1}{x}$ are simultaneously satisfied, *i.e.* the co-ordinates of the point of intersection are roots of the equations

$$y = x^2,$$

$$y = x + \frac{1}{x}.$$

Eliminating y, we have $x^2 = x + \dfrac{1}{x}$,

or $$x^3 - x^2 - 1 = 0.$$

∴ The x co-ordinate of the point of intersection is a root of the equation

$$x^3 - x^2 - 1 = 0.$$

∴ From the graphs, one root of this equation is $x = 1·47$.

Ex. 10. *Find graphically the real roots of the equation*

$$x^3 + 4x - 8 = 0.$$

The solution can be effected in two ways

(a) by drawing the graph of $y = x^3 + 4x - 8$ and finding the values of x for which y is zero;

(b) by drawing the graphs of $y = x^3$ and $y = 8 - 4x$ and finding the abscissae of the common points. For at the common points,

$$\begin{matrix} y = x^3 \\ \text{and} \quad y = 8 - 4x \end{matrix} \right\} \text{ and hence,} \quad \begin{matrix} x^3 = 8 - 4x, \\ x^3 + 4x - 8 = 0. \end{matrix}$$

The second method will be used owing to the ease with which the curve $y = x^3$ and the straight line $y = 8 - 4x$ can be drawn.

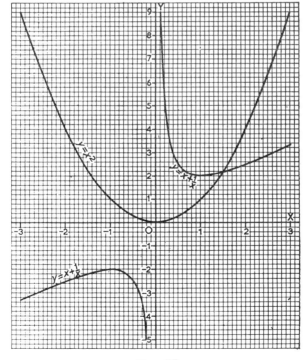

FIG. 77

From the graphs, $\qquad\qquad x^3 = 8 - 4x,$

$$i.e. \quad x^3 + 4x - 8 = 0,$$

when $\qquad\qquad\qquad\qquad x = 1{\cdot}37.$

As the straight line only meets the curve in one point, the equation has only one real root.

(B) EXAMPLES 11c

1. Solve the equations $2x^2 + xy - y^2 + 4y - 2 = 0$, $3x + 2y - 3 = 0$.

2. If $x = \dfrac{2t}{1+t^2}$, $y = \dfrac{1-t^2}{1+t^2}$, show that $x^2 + y^2 = 1$.

3. Find the values of l, m if $9x^4 - 6x^3 + 13x^2 + lx + m$ can be expressed in the form $(3x^2 + ax + b)^2$.

4. Solve the equations $\dfrac{1}{x} + \dfrac{1}{y} = \dfrac{2}{a}$, $6x - y = 3a$.

5. Find the co-ordinates of the points of intersection of the line $y - 2x = 0$ and the curve $3x^2 - xy + y^2 - 4x + 3y - 7 = 0$.

6. Eliminate t from the equations $x = t^2 + t + 1$, $y = 2t^2 - t + 2$.

7. If $\dfrac{x^3 - 2x + 2}{x(x+1)}$ can be expressed in the form $ax + b + \dfrac{c}{x} + \dfrac{d}{x+1}$, find the values of the constants a, b, c, d.

8. Solve the equations $x^3 + y^3 = 9xy$, $x + y = 6$.

9. If $x^3 - 8ax^2 - 15bx - 26 \equiv a(x-2)^3 + b(x+1)^3$, find the values of the constants a, b.

10. Solve the equations $xy = 12$, $x^2 + y^2 = 40$.

11. By plotting the graphs of $y = 2x^3$ and $y = 3x + 1$ on the same axes, find the real roots of the equation $2x^3 = 3x + 1$.

12. On the same axes, plot the graphs of $y = \dfrac{2}{x}$ and $y = x^2$, for values of x between -3 and $+3$. Use your graphs to determine

    (a) the cube root of 2;

    (b) the range of values of x for which $x^2 > \dfrac{2}{x} + 2$.

13. Plot the graphs of $y = \dfrac{1}{x-1}$ and $y = x(4-x)$ for values of x between -2 and 5. Hence solve the equation $\dfrac{1}{x-1} = x(4-x)$.

14. By drawing the graphs $y = \dfrac{2x+1}{x-2}$ and $y = x(6-x)$, solve the equation $x(x-2)(6-x) = 2x+1$.

15. Solve graphically the equation $x^3 - \dfrac{3x}{2} = 2$.

16. Solve graphically the equation $2x^1 = \dfrac{x}{2} + 4$.

17. Draw the graph of $y = 2^x$ for values of x between -3 and $+3$. Use your graph to find approximate solutions of the equation $2^x = x + 3$.

18. By plotting the graphs of $y = \log_{10}x$ and $y = \dfrac{1}{x}$ for values of x between 1 and 10, find a root of the equation $x \log_{10}x = 1$.

N.B. Before attempting to draw the graphs of the fractional functions in ex. 12, 13 and 14 above, students should refer to the work on asymptotes in Ch. 23.

CHAPTER 12

SERIES—THE ARITHMETIC AND GEOMETRIC SERIES. THE BINOMIAL SERIES

Series

A *series* is a sequence of terms each of which can be obtained by some definite law.

E.g. 1, 3, 5, 7, 9 . . . is a series. Each term is obtained by adding 2 to the preceding one.

1, 4, 16, 64 . . . is a series, each term being obtained by multiplying the preceding one by 4.

Other examples of series are:

(i) $1^2, 2^2, 3^2, 4^2 \ldots$

(ii) $\dfrac{1}{1 \cdot 2} + \dfrac{1}{2 \cdot 3} + \dfrac{1}{3 \cdot 4} + \ldots$

(iii) $x - \dfrac{x^2}{2} + \dfrac{x^3}{3} - \dfrac{x^4}{4} + \dfrac{x^5}{5} - \ldots$

The Arithmetic Series or Arithmetical Progression

An arithmetic series is a sequence of terms in which each term is obtained from the preceding one by adding some constant.

Examples of arithmetic series are

(i) $1, 2, 3, 4, 5, \ldots$

(ii) $2, 6, 10, 14, 18, \ldots$

(iii) $14, 12, 10, 8, 6, \ldots$

The difference between consecutive terms is called the *common difference*.

In ex.(i) the common difference is 1.

In ex.(ii) the common difference is 4.

In ex.(iii) the common difference is −2.

The general term and sum of an Arithmetic Series

Let the first term of the series be a, and the common difference d.

Then the series is

$$a, a+d, a+2d, a+3d, a+4d, \text{ etc.}$$

The 10th term will be $a+9d$; $i.e.$ $a+(10-1)d$.

The 21st term will be $a+20d$; $i.e.$ $a + (21-1)d$.

And generally,

the nth term will be $a+(n-1)d$.

Let $S_n = a + (a+d) + (a+2d) + (a+3d) + \ldots (a + \overline{n-1} \,.\, d)$.

Rewriting this with the R.H.S. reversed, we have,

$$S_n = (a + \overline{n-1} \,.\, d) + (a + \overline{n-2} \,.\, d) + (a + \overline{n-3} \,.\, d) + \ldots + a.$$

Adding,

$$2S_n = (2a + \overline{n-1} \,.\, d) + (2a + \overline{n-1} \,.\, d) + (2a + \overline{n-1} \,.\, d) \ldots n \text{ terms,}$$
$$= n\{2a + \overline{n-1} \,.\, d\}.$$

$$\therefore \ S_n = \frac{n}{2}\left\{2a + \overline{n-1} \,.\, d\right\}.$$

Ex. 1. *Find the* 21*st term and the sum to* 24 *terms of the series* 2, 4, 6, 8 . . .

The series is an arithmetical progression.

First term $a = 2$.

Common difference $d = 2$.

$$\therefore \ n^{\text{th}} \text{ term} = a + \overline{n-1} \,.\, d = 2 + \overline{n-1} \,.\, 2.$$

$$\therefore \ 21\text{st term} = 2 + 20 \,.\, 2 = 42.$$

$$\text{Sum to 24 terms} = \frac{n}{2}\{2a + \overline{n-1} \,.\, d\} \text{ where } n = 24,$$

$$= 12\{4 + 23 \,.\, 2\} = 600.$$

Ex. 2. *How many terms of the series* $3a + 5a + 7a + 9a + \ldots$ *must be taken for the sum to equal* $624a$?

The series is an arithmetical progression with common difference $2a$.

$$\therefore \ S_n = \frac{n}{2}\{6a + \overline{n-1} \,.\, 2a\},$$

$$624a = \frac{n}{2}\{4a + 2an\},$$

$$i.e. \quad 624 = 2n + n^2,$$

or $$n^2 + 2n - 624 = 0,$$
$$(n + 26)(n - 24) = 0,$$
$$\therefore n = 24 \text{ and } -26.$$

The negative result is clearly inadmissible. Hence required number of terms $= 24$.

Ex. 3. An arithmetical progression has 13 *terms whose sum is* 143. *The* 3rd *term is* 5. *Find the* 1st *term.*

Let the series be a, $a + d$, $a + 2d$, etc.

$$\therefore 3\text{rd term} = a + 2d = 5 \quad \ldots \ldots \ldots \ldots (1)$$

and $$S_{13} = \tfrac{13}{2}\{2a + 12d\} = 143,$$

i.e. $$a + 6d = 11. \quad \ldots \ldots \ldots \ldots (2)$$

From equations (1) and (2), $4d = 6$,

$$d = 1\tfrac{1}{2},$$

and $$a = 2.$$

$$\therefore \text{ the 1st term} = 2.$$

Ex. 4. A contractor employs 150 *men on a certain Monday and added* 12 *men to his staff every successive working day. He continued to do so for* 8 *weeks, Saturday counting as a full working day and no work being done on Sundays. How many employees had he at the end of the* 8th *week? If the wages of each man were £*3 *per day, what was the total wage bill for the* 8 *weeks?*

Total number of men after 8 weeks

$$= 150 + 47 \cdot 12 = 714.$$

Wage bill $= 3\{150 + 162 + 174 + \ldots \text{ to 48 terms}\}$ pounds,

$$= 3 \cdot \tfrac{48}{2}\{300 + 47 \cdot 12\} \text{ pounds,}$$

$$= £72\,(300 + 564),$$

$$= £62,208.$$

Arithmetic Means between two numbers

Consider the two numbers 3 and 15. We can find any number of arithmetical progressions of which 3 is the first term and 15 is the last.

e.g.
    (*a*) 3, 4, 5, ... 13, 14, 15.
    (*b*) 3, 5, 7, ... 13, 15.
    (*c*) 3, 6, 9, 12, 15.
    (*d*) 3, 9, 15.

In example (*a*), the numbers 4, 5, ... 13, 14 are called the *eleven* arithmetic means which can be inserted between 3 and 15.

Similarly in (*b*), the numbers 5, 7, . . . 13 are the *five* arithmetic means between 3 and 15.

In (*c*), we see that the three arithmetic means between 3 and 15 are 6, 9 and 12.

In case (*d*), we have one arithmetic mean, 9, between 3 and 15. This is usually spoken of as *the arithmetic mean* of the two numbers.

Ex. 5. *What is the arithmetic mean of the numbers a and b?*

Let *x* be the arithmetic mean.

Then *a*, *x*, *b* are consecutive terms of an arithmetical progression.

$$\therefore x - a = b - x = \text{common difference.}$$

$$\therefore 2x = a + b,$$

$$i.e. \quad x = \frac{a+b}{2}.$$

i.e. the arithmetic mean of two numbers is equal to half the sum of the two numbers.

Ex. 6. *Insert four arithmetic means between the numbers* 9 *and* 29.

In this case 9 will be the first term and 29 the last term of an arithmetical progression of 6 terms.

$$i.e. \quad a = 9,$$

and $\quad a + 5d = 29$, where *d* is the common difference.

$$\therefore 5d = 20,$$

$$i.e. \quad d = 4.$$

$$\therefore \text{the arithmetic means are 13, 17, 21, 25.}$$

(A) EXAMPLES 12a

1. Determine which of the following series are arithmetic series

$$\text{(i)} \ 3, 7, 11, 15, \ldots$$

$$\text{(ii)} \ 4, 2, 0, -2, \ldots$$

$$\text{(iii)} \ \tfrac{1}{2}, \tfrac{3}{2}, \tfrac{5}{2}, \tfrac{7}{2}, \ldots$$

$$\text{(iv)} \ 1, -1, 1, -1, \ldots$$

$$\text{(v)} \ 1, 2, 4, 8, 16, \ldots$$

$$\text{(vi)} \ 20, 19, 18, 17, \ldots$$

$$\text{(vii)} \ 14, 17, 20, 23, \ldots$$

$$\text{(viii)} \ 1, \tfrac{1}{2}, \tfrac{1}{3}, \tfrac{1}{4}, \ldots$$

2. For the arithmetic series in Question 1, find (*a*) the common difference, (*b*) the 8th term, (*c*) the nth term.

3. Fill in the missing terms in the following arithmetic series

$$\text{(i) } 14 — 18;$$
$$\text{(ii) } 9 — — 15;$$
$$\text{(iii) } 22 — — — 13;$$
$$\text{(iv) } 11 — — — — 5.$$

4. How many terms are there in the arithmetic series

(i) $10, 13, 16, \ldots 40$;
(ii) $22, 25, 28, \ldots 58$;
(iii) $10, 8, 6, \ldots -20$;
(iv) $a+6x, a+4x, a+2x, \ldots a-40x$?

5. Find the sums to 12 terms of the following series

(i) $1, 3, 5, \ldots$ (ii) $2, 6, 10, \ldots$
(iii) $\frac{1}{4}, \frac{3}{4}, \frac{5}{4}, \ldots$ (iv) $1, -1, -3, \ldots$
(v) $a, a+x, a+2x, \ldots$

6. Find the sums of the arithmetic series

(i) $12+16+20+\ldots$ to 16 terms;
(ii) $26+22+18+\ldots$ to 10 terms;
(iii) $22+25+28+\ldots +58$;
(iv) $38+31+24+\ldots -11-18.$

7. Find the sum of all the odd numbers between 50 and 100.

8. Find the sum of all the even numbers between 10 and 90.

9. Find (i) the sum of all numbers lying between 1 and 100; (ii) the sum of all multiples of 3 between 1 and 100; and deduce (iii) the sum of all numbers between 1 and 100 which are not multiples of 3.

10. What is the arithmetic mean of (i) 7 and 39; (ii) −6 and 12; (iii) x and $7x$?

11. Insert three arithmetic means between (i) 4 and 12; (ii) 11 and 31; (iii) $2x$ and $14x$.

12. Find the sum of the eleven arithmetic means between 4 and 44.

13. How many terms of the arithmetical progression

$$10+13+16+\ldots$$

must be taken for the sum to equal 465?

14. The first terms of an arithmetical progression is 6 and the sum to 10 terms is equal to 150. Find the common difference and the sum to 20 terms.

15. The first term of an arithmetic series is 2 and the sum to 76 terms is 4484. Find the last term.

16. The first and last terms of an arithmetical progression are -3 and 25 and the sum of all the terms is 1837. Find (i) the number of terms; (ii) the common difference; and (iii) the middle term. (J.M.B.)

17. Prove that the sum of the odd numbers from 1 to 55 inclusive is equal to the sum of the odd numbers from 91 to 105 inclusive.

18. In an arithmetical progression the first term is 4 and the common difference is 7. How many terms are required in order that the sum may exceed 500?

19. For what value of n is the nth term of the series 20, 19·6, 19·2,... equal to the 16th term of the series 1, 1·6, 2·2,...?

20. Find five numbers in arithmetical progression whose sum is 155 and whose last term is 47.

The Geometric Series or Geometrical Progression

The geometrical progression is a series of terms in which each term is obtained from the preceding one by multiplying by a constant quantity, called the *common* (or *constant*) *ratio*.

e.g. The following series are geometrical progressions:

$$(i) \quad 1, 2, 4, 8, 16, \ldots \qquad \text{Common ratio} = 2.$$
$$(ii) \quad \tfrac{1}{2}, \tfrac{1}{6}, \tfrac{1}{18}, \tfrac{1}{54}, \ldots \qquad \text{Common ratio} = \tfrac{1}{3}.$$
$$(iii) \quad 2, -8, 32, -128, \ldots \qquad \text{Common ratio} = -4.$$

The general term and sum to n terms of a Geometric Series

The most general geometric series is of the form

$$a, ar, ar^2, ar^3, ar^4, \ldots$$

i.e. 1st term a and common ratio r.

We see that the
$$3\text{rd term} = ar^2 = ar^{3-1},$$
$$4\text{th term} = ar^3 = ar^{4-1},$$
$$10\text{th term} = ar^9 = ar^{10-1},$$

and so the $\quad$ **n^{th} term** $= ar^{n-1}$.

Let $\quad S_n = a + ar + ar^2 + ar^3 + \ldots + ar^{n-1}$.

Multiplying throughout by r,

$$r \cdot S_n = ar + ar^2 + ar^3 + \ldots + ar^{n-1} + ar^n.$$

Subtracting, $S_n(1 - r) = a - ar^n$.

i.e. $\quad S_n = \dfrac{a(1 - r^n)}{1 - r}$.

Ex. 7. Find the 8th term and sum to 8 terms of the geometric series: $\frac{1}{2}$, 1, 2, 4, ...

$$\text{1st term} = a = \tfrac{1}{2}; \text{ common ratio, } r = 2.$$

$$\therefore \text{ 8th term} = ar^7 = \tfrac{1}{2} \cdot 2^7 = 2^6 = 64.$$

$$\text{Sum to 8 terms} = \frac{a(1-r^8)}{1-r} = \tfrac{1}{2}\{2^8 - 1\},$$

$$= \tfrac{255}{2} = 127\tfrac{1}{2}.$$

Ex. 8. From a sufficiently long piece of string, ten portions are cut off successively, their lengths forming a geometrical progression. The first and second portions are respectively 10 m and $9\frac{1}{2}$ m in length. What is the total length cut off to the nearest metre? If the process is continued as long as the portion cut off is not less than $\frac{1}{3}$ m how many portions are obtained?

1st term of series $a = 10$.
2nd term of series $ar = 9\frac{1}{2}$, where r is the common ratio.

$$\therefore \ r = \tfrac{19}{20}.$$

$$\therefore \text{ Sum to 10 terms} = \frac{a(1-r^{10})}{1-r} = \frac{10(1-(\tfrac{19}{20})^{10})}{\tfrac{1}{20}},$$

$$= 200 - 200 \cdot (\cdot 95)^{10},$$

$$= 200(1 - \cdot 5984) = 80 \text{ m to the nearest}$$

metre.

$$\text{Length of the } n^{\text{th}} \text{ portion} = ar^{n-1}$$

$$= 10 \cdot (\tfrac{19}{20})^{n-1} \text{ m}.$$

We require the greatest integral value of n for which

$$10 \cdot (\tfrac{19}{20})^{n-1} > \tfrac{1}{3},$$

$$\textit{i.e. } (\tfrac{19}{20})^{n-1} > \tfrac{1}{30}.$$

Solving the equation $(\tfrac{19}{20})^{n-1} = \tfrac{1}{30}$ by taking logarithms,

we have

$$n - 1 = \frac{\log \cdot 03333}{\log \cdot 95}$$

$$= \frac{\bar{2} \cdot 5228}{1 \cdot 9777}$$

$$= \frac{-1 \cdot 4772}{- \cdot 0223}$$

$$= 66 \cdot 22,$$

$$\therefore \ n = 67 \cdot 22.$$

$\therefore$ Greatest integral value of $n = 67$.

i.e. 67 portions can be obtained before the length cut off is less than $\frac{1}{3}$ m.

Ex. 9. *A man pays a premium of £30 per annum for* 18 *years. How much should he receive as a lump sum just after paying the* 18*th premium, reckoning compound interest at* 5 *per cent. per annum.*

When the 18th premium is paid, the 1st premium has been accumulating at 5 per cent. compound interest for 17 years.

∴ Amount due on account of 1st premium

$$= £30[1 + \tfrac{1}{20}]^{17} = £30(1 \cdot 05)^{17}.$$

Similarly, amount due on account of 2nd premium

$$= £30(1 \cdot 05)^{16},$$

and amount due on account of 3rd premium

$$= £30(1 \cdot 05)^{15}, \text{ etc.,}$$

and amount due on account of 18th premium

$$= £30.$$

∴ Total Amount $= £\{30 + 30(1 \cdot 05) + 30(1 \cdot 05)^2 + \ldots + 30(1 \cdot 05)^{17}\}.$

This is a geometrical progression with common ratio $1 \cdot 05$ and 18 terms.

$$\therefore \text{ Amount} = \frac{a(1 - r^n)}{1 - r},$$

$$= \frac{£30[1 - (1 \cdot 05)^{18}]}{1 - 1 \cdot 05},$$

$$= £600[(1 \cdot 05)^{18} - 1],$$

$$= £600[2 \cdot 407 - 1] \text{ (using 4-fig. tables),}$$

$$= £844 \cdot 2,$$

Geometric Mean

The *geometric mean* of two numbers a and b is x, if a, x, b are consecutive terms in a geometrical progression.

We have, $\dfrac{x}{a} = \dfrac{b}{x} = $ common ratio.

$$\therefore \ x^2 = ab. \qquad i.e. \quad x = \sqrt{ab}.$$

In the same way as we have any number of arithmetic means between two numbers so we have any number of geometric means.

Ex. 10. *Obtain* 5 *geometric means between* 3 *and* 192.

We require a geometrical progression with 1st term 3, and 7th term equal to 192.

We have $a = 3,$

and 7th term $= ar^6 = 3r^6 = 192$, where r is the common ratio.

$$\therefore r^6 = 64.$$

$$\therefore r = 2.$$

$\therefore$ The 5 geometric means are 6, 12, 24, 48, 96.

Ex. 11. *Prove that the arithmetic mean of two unequal numbers is greater than the geometric mean.*

We have, $(\sqrt{x} - \sqrt{y})^2 > 0$ if $x \neq y$ and if x and y are real.

$$\therefore x - 2\sqrt{xy} + y > 0,$$

or $\dfrac{x + y}{2} > \sqrt{xy}.$

i.e. the arithmetic mean > the geometric mean.

(A) EXAMPLES 12b

(Leave all answers in the index form)

1. Determine which of the following series are geometric series

(i) $2, 6, 18, 54, \ldots$

(ii) $\frac{1}{2}, \frac{1}{4}, \frac{1}{8}, \frac{1}{16}, \ldots$

(iii) $1, 4, 9, 16, \ldots$

(iv) $1, 3x, 6x^2, 9x^3, \ldots$

(v) $27, 18, 12, 8, \ldots$

2. For the geometric series in Ex. 1, find (i) the common ratio; (ii) the 10th term; (iii) the nth term.

3. Fill in the missing terms in the following geometric series

(i) $6, —, 24;$ (ii) $27, —, 48, —;$

(iii) $250, —, —, 54;$ (iv) $2, —, —, —, 1250.$

4. How many terms are there in the following geometric series?

(i) $2, 4, 8, \ldots 128.$ (ii) $\frac{1}{2}, \frac{1}{4}, \frac{1}{8}, \ldots \frac{1}{256}.$

(iii) $32, 48, \ldots 162.$ (iv) $ar, ar^3, ar^5, \ldots ar^{2n+1}.$

5. Find the sums of the following series

(i) $1 + 2 + 4 + 8 + \quad \ldots \ldots$ to 10 terms;

(ii) $4 - 8 + 16 - 32 + \quad \ldots \ldots$ to 8 terms;

(iii) $6 + 2 + \frac{2}{3} + \frac{2}{9} + \quad \ldots \ldots$ to 7 terms;

(iv) $1 + 3a + 9a^2 + 27a^3 + \ldots$ to 10 terms;

(v) $1 + 1 \cdot 05 + (1 \cdot 05)^2 + \quad \ldots \quad + (1 \cdot 05)^9.$

6. Find the geometric means of

      (i) 6 and 24;   (ii) 36, $20\frac{1}{4}$;   (iii) x^2 and x^4.

7. Insert three geometric means between (i) 2 and 32; (ii) 16 and $\frac{1}{16}$.

8. How many terms of the series 2, 4, 8, 16, ... must be taken for the sum to equal $2^n - 2$?

9. Find the sum of the 5 geometric means between 1 and 64.

10. Find (i) the middle term; (ii) the sum of the series $x + x^3 + x^5 + \ldots + x^{37}$.

11. In the series $36 - 12 + 4 - \frac{4}{3} + \ldots$, find (i) the 20th term; (ii) the sum of 10 terms.

12. The 1st term of a geometrical progression is $\frac{3}{11}$ and the 4th term is $2\frac{2}{11}$. Find the common ratio and the sum to 10 terms.

13. The 1st term of a geometrical progression is $\frac{3}{5}$ and the 4th term is $9\frac{3}{8}$. Find the 12th term and the sum to 12 terms.

14. The 3rd term of a geometric series exceeds the 2nd by $\frac{9}{14}$ and the 2nd exceeds the 1st by $\frac{3}{7}$. Find the common ratio.

15. Find the difference between the sum of the first 6, and the sum of the first 12 terms of the series 60, -30, 15, $-7\frac{1}{2}$, ...

16. What is the value of the 1st term of the series, 3, 12, 48, ... which exceeds 10,000?

17. The 2nd term of a geometric series is -6 and the 5th term is $20\frac{1}{4}$. Find the common ratio and the nth term.

18. The 1st and 2nd terms of a geometrical progression are a and b respectively. Find the 4th term and the nth term. Also write down the sum of n terms.

19. Prove that the square roots of the terms of a geometric series form another geometric series.

20. The 2nd term of a geometrical progression is $\frac{4}{3}$ and the 5th term is $\frac{32}{81}$. Find the 1st term and the common ratio.

Infinite Series. Convergence. Sum of an Infinite Series

Consider the series $u_1, u_2, u_3, u_4, \ldots$

If the number of terms is unlimited, we can go on adding terms endlessly and get what is called *an infinite series*.

i.e. an infinite series is a series with an unlimited number of terms.

Examples of infinite series are

$$1, 2, 3, 4, \ldots \text{to } \infty;$$

$$1, \tfrac{1}{2}, \tfrac{1}{4}, \tfrac{1}{8}, \ldots \text{to } \infty;$$

$$1 + x + \frac{x^2}{1 \cdot 2} + \frac{x^3}{1 \cdot 2 \cdot 3} + \frac{x^4}{1 \cdot 2 \cdot 3 \cdot 4} + \ldots \text{to } \infty.$$

The question arises as to the possibility of assigning a sum to an infinite series.
Consider the following examples.

Ex. 12. *Find the sum to n terms of the geometric series* $1, \tfrac{1}{2}, \tfrac{1}{4}, \ldots By$ *letting n increase without limit, determine the value to which the sum approaches as n increases.*

The sum to n terms of the geometrical progression

$$S_n = \frac{1(1 - (\tfrac{1}{2})^n)}{1 - \tfrac{1}{2}} = 2[1 - (\tfrac{1}{2})^n].$$

Now let n increase without limit, or as we say, let $n \to$ infinity.
Then $(\tfrac{1}{2})^n$ will decrease and approach zero.
Hence, as n increases, S_n approaches the value 2, without actually being equal to 2 for a finite value of n.
We say in this case that the infinite series $1 + \tfrac{1}{2} + (\tfrac{1}{2})^2 + \ldots$ to ∞ is *convergent* and converges to a sum 2.
The value 2 is often spoken of as the *sum to infinity* of the series.

Ex. 13. *Is it possible to determine a sum to infinity for the following series?*

(*a*) $1 + 2 + 3 + 4 + \ldots \infty$;
(*b*) $1 + 2 + 4 + 8 + \ldots \infty$;
(*c*) $1 + \tfrac{1}{3} + \tfrac{1}{9} + \tfrac{1}{27} + \ldots \infty$.

(*a*) It is obvious that the more terms we take the larger each term becomes, and hence the sum will go on increasing without any limit. Hence there is no sum to infinity, and the series is said *to be divergent.*

(*b*) This series is also *divergent.*

(*c*) The sum to n terms of this series

$$= \frac{1 - (\tfrac{1}{3})^n}{1 - \tfrac{1}{3}} = \tfrac{3}{2}[1 - (\tfrac{1}{3})^n].$$

Now as n increases, $(\tfrac{1}{3})^n \to 0$.
Hence the sum approaches the definite value $\tfrac{3}{2}$ as $n \to \infty$.
Thus the series is convergent, *i.e.* a sum to infinity exists and is equal to $\tfrac{3}{2}$.
At this stage we will only consider the simplest infinite series—the geometrical progression with an infinite number of terms—and obtain the

conditions necessary for the series to be convergent, *i.e.* for the series to have a sum to infinity.

Consider the geometrical progression $a, ar, ar^2, ar^3, \ldots$

The sum to n terms,

$$S_n = \frac{a(1-r^n)}{1-r} = \frac{a}{1-r} - \frac{a}{1-r} \cdot r^n.$$

Now let n increase and $\to \infty$.

The term $\dfrac{a}{1-r}$ remains unchanged and the term r^n will alter according to the value of r.

If r *is numerically less than* 1, *i.e.* $-1 < r < 1$, $r^n \to 0$, as $n \to \infty$.

Hence in this case, S_n approaches a definite value $\dfrac{a}{1-r}$ as $n \to \infty$.

Thus if r *is numerically less than unity*, the geometrical progression $(a + ar + ar^2 + \ldots \text{ to } \infty)$ is convergent and has a sum to infinity equal to $\dfrac{a}{1-r}$.

For other values of r, the series is not convergent and no sum to infinity exists.

Ex. 14. *Determine whether a sum to infinity exists for the following geometrical progressions*

$$(a) \ 4, 1, \tfrac{1}{4}, \ldots; \qquad (b) \ \tfrac{1}{2}, \tfrac{3}{4}, \tfrac{9}{8}, \ldots$$

(a) The common ratio $= \tfrac{1}{4}$.

Hence as the common ratio is numerically less than 1, the sum to infinity exists and equals $\dfrac{4}{1-\frac{1}{4}} \left(i.e. \ \dfrac{a}{1-r} \right)$,

$$i.e. \quad S_\infty = \tfrac{16}{3}.$$

(b) The common ratio $= \tfrac{3}{2}$, *i.e.* greater than 1.

Hence the series is not convergent and there is no sum to infinity.

Ex. 15. *Prove that the following series are convergent and have the same sum to infinity*

$$(1) \ 1, \tfrac{4}{5}, \left(\tfrac{4}{5}\right)^2, \ldots; \qquad (2) \ \tfrac{5}{2}, \tfrac{5}{4}, \tfrac{5}{8}, \tfrac{5}{16}, \ldots$$

Series (1) is a geometrical progression with common ratio $\tfrac{4}{5}$; hence the series is convergent.

$$\text{Sum to infinity} = \frac{1}{1-\frac{4}{5}} = 5.$$

Series (2) is a geometrical progression with common ratio $\tfrac{1}{2}$; hence the series is convergent.

$$\text{Sum to infinity} = \frac{\frac{5}{2}}{1-\frac{1}{2}} = 5.$$

(A) EXAMPLES 12c

1. Find the values to which the following expressions tend as n increases without limit (*i.e.* as n tends to infinity)

(i) $\dfrac{1}{n}$;　　(ii) $1 - \dfrac{1}{n^2}$;　　(iii) $n - \dfrac{1}{n}$;　　(iv) $(\tfrac{1}{3})^n$;

(v) $(1 \cdot 1)^n$;　　(vi) $(-\cdot 99)^n$;　　(vii) $1 - (\tfrac{1}{4})^n$;　　(viii) $2(1 - (\cdot 5)^n)$;

(ix) $\dfrac{10(1 - (\cdot 6)^n)}{1 - \cdot 6}$.

2. Write down the sums to n terms of the following series

(i) $2 + 1 + \tfrac{1}{2} + \tfrac{1}{4} + \ldots$

(ii) $\tfrac{1}{2} + 1 + 2 + 4 + \ldots$

(iii) $1 - \tfrac{1}{4} + \tfrac{1}{16} - \tfrac{1}{32} + \ldots$

(iv) $1 + 2 + 2^2 + 2^3 + \ldots$

Hence determine which of the series are convergent (*i.e.* have sums to infinity).

3. Determine which of the following series are convergent

(i) $\tfrac{1}{20} + \tfrac{1}{10} + \tfrac{1}{5} + \ldots$;

(ii) $100 + 25 + 6\tfrac{1}{4} + \ldots$;

(iii) $1 + \tfrac{1}{10} + \tfrac{1}{100} + \ldots$;

(iv) $1 + 2 + 3 + 4 + \ldots$;

(v) $2 - 1 + \tfrac{1}{2} - \tfrac{1}{4} + \ldots$;

(vi) $1 + x + x^2 + \ldots$　where x is positive and < 1;

(vii) $1 + \dfrac{1}{n} + \dfrac{1}{n^2} + \ldots$　where $n > 1$.

4. Sum to infinity the following series

(i) $1 - \tfrac{1}{4} + \tfrac{1}{16} - \tfrac{1}{64} + \ldots$;

(ii) $2 + \tfrac{1}{2} + \tfrac{1}{8} + \tfrac{1}{32} + \ldots$;

(iii) $1 + x^2 + x^4 + x^6 + \ldots$　where x is numerically < 1;

(iv) $1 + \sin \theta + \sin^2 \theta + \sin^3 \theta + \ldots$　where $\theta = 45°$.

5. The 1st term of a geometric series is 3 and the 3rd term is $\tfrac{1}{3}$. Find the sum to infinity of the series.

6. The sum to infinity of a geometric series is 10. If the 1st term of the series is 5, find the 10th term.

7. Find the sum to infinity of the series $2 + \tfrac{2}{3} + \tfrac{2}{9} + \ldots$ Find the 1st term of a 2nd geometric series with common ratio $\tfrac{5}{6}$ which has the same sum to infinity.

8. The 3rd term of a geometric series is 2 and the 6th term is $-\tfrac{1}{4}$. Find the sum to infinity.

9. Find the sums to infinity of the following series

(i) $1+2+\frac{1}{2}+\frac{1}{5}+\frac{1}{4}+\frac{1}{50}+\ldots$;

(ii) $a+b+a^2+b^2+a^3+b^3+\ldots$ where a and b are each numerically <1.

(B) EXAMPLES 12d

1. The sum of 11 terms of an arithmetical progression is 22 and the common difference is $\frac{3}{5}$. Find the nth term and the sum to 44 terms.

2. Find (i) the 20th term of the arithmetical progression, 12, 9, 6, 3,...; (ii) the sum of all the integers from 1000 to 2000 inclusive which are not multiples of 7. (J.M.B.)

3. In an arithmetical progression the 1st term is a and the common difference is $2a$. Show that the sum to $2n$ terms is always equal to four times the sum to n terms. (L.M.)

4. The first and last terms of an arithmetical progression are 7 and 127 and the sum of all the terms is 1675. Find the number of terms and the middle term of the progression. (J.M.B.)

5. In an arithmetical progression the ratio of the 2nd term to the 4th is $11:13$ and the sum of the first five terms is 30. Find the sum of thirty terms. (O.S.C.)

6. The 2nd and 3rd terms of a geometrical progression are 24 and $12(b+1)$ respectively. What is the 1st term? If the sum of these three terms is 76, find the possible values of b. (C.W.B.)

7. The two sets of numbers, 1, 4, 16, 64,... and 3, 12, 48,... are arranged in the order 1, 3, 4, 12, 16, 48,... One of the numbers of this series is 1048576. Find the numbers immediately preceding and following this. (J.M.B.)

8. How many terms of the geometric series 7, $3\frac{1}{2}$, $1\frac{3}{4}$, $\frac{7}{8}$,... must be taken in order that the sum may differ from the sum to infinity by less than 0·01?

9. The 1st, 4th and 8th terms of an arithmetic series are in geometrical progression, and the 1st term is 9. Find the common difference of the arithmetic series.

10. If a is the 1st term of an arithmetical progression, d the common difference, and l the last term, prove that the sum is equal to $\frac{1}{2}(a+l)\left(1+\dfrac{l-a}{d}\right)$. Find the sum of all multiples of 11 between 550 and 1000.

11. Sum the following series (a) to 20 terms; (b) to $2n$ terms.

(i) $1+3+2+4+3+5+\ldots$;

(ii) $2+x+3+x^2+4+x^3+\ldots$;

(iii) $a-b+a^2-b^2+a^3-b^3+\ldots$

12. A contractor undertakes to bore a well at £1 for the first 5 m, £1·25 for the next 5 m, £1·50 for the next 5 m, and so on.
(i) What is the cost of sinking a well 200 m deep?
(ii) What is the depth of a well which cost £100 to bore?

13. A ball dropped from a height h m rebounds to a height $\frac{2}{5}h$ m. If the ball is dropped from a height of 10 m, find (i) the height of rebound after the 10th bounce, and (ii) the total distance covered before the ball comes to rest.

14. On the ground are placed 20 stones in line at intervals of 5 m. How far will a person travel who, starting from the position of the first stone, brings them one by one to a basket placed at the first stone?

15. A man saves £100 each year, and invests it at the end of the year at 4 per cent. compound interest. How much will the combined savings and interest amount to at the end of 15 years?

16. £500 is invested at 5 per cent. compound interest. After how many years will the amount first exceed £750?

17. A pendulum is set swinging; its first oscillation is through 30° and each succeeding oscillation is $\frac{2}{3}$ of the one before it. What is the total angle described before the pendulum stops?

18. A man deposits £150 annually to accumulate at 5 per cent. compound interest. How much money will he have to his credit immediately after making the tenth deposit?

19. A firm borrows £5000 to be repaid by 20 equal annual instalments, the first being made one year after the loan. Allowing compound interest at 5 per cent. calculate the amount of each annual payment.

20. A company borrows £25,000 at 5 per cent. compound interest (added yearly), repaying £2000 at the end of each year. Find how much will remain unpaid after 15 years.

21. If S_n denotes the sum of the series

$$1 + 2x + 3x^2 + \ldots + nx^{n-1},$$

write down the value of xS_n and prove that

$$S_n(1-x) = \{1 + x + x^2 + \ldots + x^{n-1}\} - nx^n.$$

Hence deduce the value of S_n. If x is numerically less than unity, obtain the sum to infinity of the series

$$1 + 2x + 3x^2 + 4x^3 + \ldots$$

22. Using the method of Question 21, obtain the sums of the following series

(i) $1 + 3x + 5x^2 + \ldots + (2n-1)x^{n-1}$;
(ii) $1 + 5x + 9x^2 + \ldots + (4n-3)x^{n-1}$;
(iii) $a + 2ax + 3ax^2 + \ldots + nax^{n-1}$.

23. Sum to n terms the series

$$a + (a+d)x + (a+2d)x^2 + (a+3d)x^3 + \ldots$$

If x is numerically less than unity, deduce that the series is convergent and find the sum to infinity.

The Binomial Theorem

The Binomial Theorem enables us to obtain the expansion of a binomial term (*i.e.* a term of the form $\overline{a+x}$) raised to any power; *e.g.* $(1-3x)^{16}$, $(2-x)^{-5}$, $(1-x)^{\frac{3}{2}}$.

The proof of the theorem is beyond the scope of this book, but the general form of the result for the case when n is a positive integer can be verified by considering special cases. We have by ordinary multiplication

$$(1+x)^2 = 1 + 2x + x^2;$$

$$(1+x)^3 = 1 + 3x + 3x^2 + x^3 = 1 + 3x + \frac{3 \cdot 2}{1 \cdot 2}x^2 + x^3;$$

$$(1+x)^4 = 1 + 4x + 6x^2 + 4x^3 + x^4 = 1 + 4x + \frac{4 \cdot 3}{1 \cdot 2}x^2 + \frac{4 \cdot 3 \cdot 2}{1 \cdot 2 \cdot 3}x^3 + x^4;$$

$$(1+x)^5 = 1 + 5x + 10x^2 + 10x^3 + 5x^4 + x^5,$$

$$= 1 + 5x + \frac{5 \cdot 4}{1 \cdot 2}x^2 + \frac{5 \cdot 4 \cdot 3}{1 \cdot 2 \cdot 3}x^3 + \frac{5 \cdot 4 \cdot 3 \cdot 2}{1 \cdot 2 \cdot 3 \cdot 4}x^4 + x^5.$$

Similarly,

$$(1+x)^6 = 1 + 6x + \frac{6 \cdot 5}{1 \cdot 2}x^2 + \frac{6 \cdot 5 \cdot 4}{1 \cdot 2 \cdot 3}x^3 + \frac{6 \cdot 5 \cdot 4 \cdot 3}{1 \cdot 2 \cdot 3 \cdot 4}x^4$$

$$+ \frac{6 \cdot 5 \cdot 4 \cdot 3 \cdot 2}{1 \cdot 2 \cdot 3 \cdot 4 \cdot 5}x^5 + x^6.$$

Generally, if n is a positive integer,

$$\mathbf{(1+x)^n = 1 + nx + \frac{n(n-1)}{1 \cdot 2}x^2 + \frac{n(n-1)(n-2)}{1 \cdot 2 \cdot 3}x^3}$$

$$\mathbf{+ \frac{n(n-1)(n-2)(n-3)}{1 \cdot 2 \cdot 3 \cdot 4}x^4}$$

$$\mathbf{+ \frac{n(n-1)(n-2)(n-3)(n-4)}{1 \cdot 2 \cdot 3 \cdot 4 \cdot 5}x^5 + \ldots + x^n.}$$

Notation. The repeated product $3 \cdot 2 \cdot 1$ is called '*factorial* 3' and is written as 3!

Similarly, $4 \cdot 3 \cdot 2 \cdot 1 = 4!$ and $7 \cdot 6 \cdot 5 \cdot 4 \cdot 3 \cdot 2 \cdot 1 = 7!$ and generally, when n is a positive whole number,

$$n(n-1)(n-2)(n-3) \ldots 3 \cdot 2 \cdot 1 = n!$$

So we have, *when n is a positive integer,*

$$(1+x)^n = 1 + nx + \frac{n(n-1)}{2!}x^2 + \frac{n(n-1)(n-2)}{3!}x^3$$

$$+ \frac{n(n-1)(n-2)(n-3)}{4!}x^4 + \ldots + x^n.$$

This expansion holds for every value of x.

Ex. 16. Expand $(1+x)^7$, $(1-2x)^4$, $\left(1-\dfrac{x}{2}\right)^3$.

Using the above result,

$$(1+x)^7 = 1 + 7x + \frac{7.6}{1.2}x^2 + \frac{7.6.5}{1.2.3}x^3 + \frac{7.6.5.4}{1.2.3.4}x^4$$

$$+ \frac{7.6.5.4.3}{1.2.3.4.5}x^5 + \frac{7.6.5.4.3.2}{1.2.3.4.5.6}x^6 + x^7,$$

$$= 1 + 7x + 21x^2 + 35x^3 + 35x^4 + 21x^5 + 7x^6 + x^7.$$

$$(1+2x)^4 = 1 + 4(2x) + \frac{4.3}{1.2}(2x)^2 + \frac{4.3.2}{1.2.3}(2x)^3 + (2x)^4,$$

$$= 1 + 8x + 24x^2 + 32x^3 + 16x^4.$$

$$\left(1-\frac{x}{2}\right)^3 = 1 + 3\left(-\frac{x}{2}\right) + \frac{3.2}{1.2}\left(-\frac{x}{2}\right)^2 + \left(-\frac{x}{2}\right)^3,$$

$$= 1 - \frac{3x}{2} + \frac{3}{4}x^2 - \frac{x^3}{8}.$$

General Binomial Theorem. The expansion of $(a+b)^n$, where *n* is a positive integer.

$$(a+b)^n = a^n\left(1+\frac{b}{a}\right)^n,$$

$$= a^n\left(1 + n\left(\frac{b}{a}\right) + \frac{n(n-1)}{2!}\left(\frac{b}{a}\right)^2 + \ldots + \left(\frac{b}{a}\right)^n\right),$$

$$= a^n + na^{n-1}b + \frac{n(n-1)}{2!}a^{n-2}b^2 + \ldots + b^n.$$

So $(a+b)^n = a^n + na^{n-1}b^1 + \dfrac{n(n-1)}{2!}a^{n-2}b^2 + \dfrac{n(n-1)(n-2)}{3!}a^{n-3}b^3$

$$+ \ldots + b^n.$$

Ex. 17. *Expand* $(2+3x)^5$.

$$(2+3x)^5 = 2^5 + 5 \cdot 2^4(3x) + \frac{5 \cdot 4}{1 \cdot 2} 2^3(3x)^2 + \frac{5 \cdot 4 \cdot 3}{1 \cdot 2 \cdot 3} 2^2(3x)^3$$

$$+ \frac{5 \cdot 4 \cdot 3 \cdot 2}{1 \cdot 2 \cdot 3 \cdot 4} 2(3x)^4 + (3x)^5,$$

$$= 32 + 240x + 720x^2 + 1080x^3 + 810x^4 + 243x^5.$$

The Expansion of $(1+x)^n$ for any value of n

It has been stated that when *n* is a *positive integer*,

$$(1+x)^n = 1 + nx + \frac{n(n-1)}{2!} x^2 + \frac{n(n-1)(n-2)}{3!} x^3 + \ldots + x^n.$$

The expansion being true for *all values of x.*

It is found that an expansion of this form holds good when *n* is other than a positive integer, but with two important differences.

(i) The series never terminates, *i.e.* we get an infinite series.

(ii) The expansion is only true when *x* lies between +1 and −1,

$$\textit{i.e.} \quad (1+x)^n = 1 + nx + \frac{n(n-1)}{2!} x^2 + \frac{n(n-1)(n-2)}{3!} x^3 + \ldots \infty$$

when x is numerically less than 1.

The restriction on *x* is due to the fact that the infinite series is only convergent (*i.e.* has a definite sum) for values of *x* between +1 and −1.

Ex. 18. *Obtain the first four terms in the expansions of* (a) $\sqrt{(1+x)}$, (b) $\frac{1}{(1-2x)^2}$, *stating the values of x for which the expansions are true.*

$$(a) \quad \sqrt{(1+x)} = (1+x)^{\frac{1}{2}} = 1 + \frac{1}{2} \cdot x + \frac{(\frac{1}{2})(-\frac{1}{2})}{1 \cdot 2} x^2 + \frac{(\frac{1}{2})(-\frac{1}{2})(-\frac{3}{2})}{1 \cdot 2 \cdot 3} x^3 + \ldots$$

$$= 1 + \frac{x}{2} - \frac{x^2}{8} + \frac{x^3}{16} \ldots$$

The expansion is true for values of *x* between +1 and −1.

$$(b) \quad \frac{1}{(1-2x)^2} = (1-2x)^{-2} = 1 + (-2)(-2x) + \frac{(-2)(-3)}{1 \cdot 2} (-2x)^2$$

$$+ \frac{(-2)(-3)(-4)}{1 \cdot 2 \cdot 3} (-2x)^3 + \ldots,$$

$$= 1 + 4x + 12x^2 + 32x^3 + \ldots$$

This expansion is true for values of $(-2x)$ between +1 and −1, *i.e.* for values of *x* between $+\frac{1}{2}$ and $-\frac{1}{2}$.

Ex. 19. *Find the coefficient of* x^4 *in the expansion of* $(1+3x)^{-\frac{1}{3}}$.

In the general case, *i.e.* the expansion of $(1+x)^n$, the term in x^4 is

$$\frac{n(n-1)(n-2)(n-3)}{4!}\, x^4.$$

$\therefore$ Term in x^4 in the expansion of $(1+3x)^{-\frac{1}{3}}$

$$= \frac{(-\frac{1}{3})(-\frac{4}{3})(-\frac{7}{3})(-\frac{10}{3})}{4!}\,(3x)^4,$$

$$= \frac{4\,.\,7\,.\,10}{4\,.\,3\,.\,2}\,x^4 = \frac{35}{3}\,x^4.$$

I.e. the coefficient of $x^4 = \dfrac{35}{3}$.

Ex. 20. *If* $\dfrac{3-4x}{(1-x)(1-2x)} \equiv \dfrac{A}{1-x} + \dfrac{B}{1-2x}$ *find the values of* A *and* B *and hence obtain the first four terms in the expansion of* $\dfrac{3-4x}{(1-x)(1-2x)}$, *assuming x is sufficiently small for the expansion to exist.*

We have

$$\frac{3-4x}{(1-x)(1-2x)} \equiv \frac{A}{1-x} + \frac{B}{1-2x}.$$

$$\therefore\ 3-4x \equiv A(1-2x) + B(1-x).$$

Let $x = 1,\ -1 = -A$; *i.e.* $A = 1$.

Let $x = \frac{1}{2},\ 1 = \frac{1}{2}B$; *i.e.* $B = 2$.

$$\therefore\ \frac{3-4x}{(1-x)(1-2x)} \equiv \frac{1}{1-x} + \frac{2}{1-2x},$$

$$= (1-x)^{-1} + 2(1-2x)^{-1},$$

$$= \left\{ 1 + (-1)(-x) + \frac{(-1)(-2)}{1\,.\,2}\,(-x)^2 \right.$$

$$\left. + \frac{(-1)(-2)(-3)}{1\,.\,2\,.\,3}\,(-x)^3 + \ldots \right\},$$

$$+ 2\left\{ 1 + (-1)(-2x) + \frac{(-1)(-2)}{1\,.\,2}\,(-2x)^2 \right.$$

$$\left. + \frac{(-1)(-2)(-3)}{1\,.\,2\,.\,3}\,(-2x)^3 + \ldots \right\},$$

$$= 1 + x + x^2 + x^3 + \ldots + 2 + 4x + 8x^2 + 16x^3 + \ldots,$$

$$= 3 + 5x + 9x^2 + 17x^3 + \ldots$$

$$\therefore\ \frac{3-4x}{(1-x)(1-2x)} = 3 + 5x + 9x^2 + 17x^3 + \ldots$$

Application of the Binomial Theorem to Approximations

The following examples show how the Binomial Theorem can be used to find approximate values of given expressions.

Ex. 21. Find the values of $(1·02)^6$, $\dfrac{1}{(·998)^2}$, $\sqrt{4·004}$ to 4 places of decimals.

(i) $(1·02)^6 = (1+·02)^6 = 1+6(·02)+\dfrac{6.5}{1.2}(·02)^2+\dfrac{6.5.4}{1.2.3}(·02)^3$

$\qquad +\dfrac{6.5.4.3}{4.3.2.1}(·02)^4+\ldots,$

$\qquad = 1+·12+·006+·00016+·0000024\ldots,$

$\qquad = 1·12616 = 1·1262$ to 4 decimal places.

$\therefore (1·02)^6 = 1·1262$ to 4 places of decimals.

(ii) $\dfrac{1}{(·998)^2} = (1-·002)^{-2} = 1+(-2)(-·002)+\dfrac{(-2)(-3)}{1.2}(-·002)^2$

$\qquad +\dfrac{(-2)(-3)(-4)}{1.2.3}(-·002)^3+\ldots,$

$\qquad = 1+·004+·000012+\ldots,$

$\qquad = 1·0040$ to 4 places of decimals.

(iii) $\sqrt{4·004} = (4·004)^{\frac{1}{2}}$.

In this case, care must be taken not to try to expand in the form $(1+3·004)^{\frac{1}{2}}$ as the expansion of $(1+x)^n$, where n is not a positive integer, only exists for values of x between $+1$ and -1.

To get over the difficulty proceed as follows.

$\qquad (4·004)^{\frac{1}{2}} = (4+·004)^{\frac{1}{2}} = 4^{\frac{1}{2}}(1+·001)^{\frac{1}{2}},$

$\qquad = \sqrt{4}\left\{1+\frac{1}{2}(·001)+\dfrac{\frac{1}{2}(-\frac{1}{2})}{1.2}(·001)^2\ldots\right\},$

$\qquad = 2\{1+·0005+\ldots\},$

$\qquad = 2·0010$ to 4 places of decimals.

Ex. 22. When x is so small that its square and higher powers can be neglected, find the approximate value of the expression

$$\dfrac{\sqrt{1+2x}-\sqrt[3]{8+x}}{(1-x)^2}.$$

$\sqrt{1+2x} = (1+2x)^{\frac{1}{2}} = 1+\frac{1}{2}.(2x)+$ higher powers of x,

$\qquad = 1+x+\ldots$

$$\sqrt[3]{8+x} = (8+x)^{\frac{1}{3}} = 8^{\frac{1}{3}}\left\{1+\frac{x}{8}\right\}^{\frac{1}{3}}.$$

$$= 2\left\{1+\frac{1}{3}\cdot\frac{x}{8}+\text{higher powers of } x\right\},$$

$$= 2+\frac{x}{12}+\dots$$

$$\therefore \text{ Expression} = \left\{1+x-\left(2+\frac{x}{12}\right)\right\}(1-x)^{-2},$$

$$= \left(-1+\frac{11x}{12}\right)(1+2x)+\text{higher powers of } x,$$

$$= -1-2x+\frac{11x}{12} = -1-\frac{13x}{12} \text{ approximately.}$$

(A) EXAMPLES 12e

1. Expand $(1+a)^4$, $(1-x)^5$, $(1+2x)^6$, $\left(1-\dfrac{x}{3}\right)^3$, $(2+x)^7$, $(3-2a)^5$, $(a+b)^6$.

2. In the expansion of $(1-2x)^9$ write down (i) the coefficient of x^4; (ii) the 6th term; (iii) the number of terms.

3. Obtain the expansion of $(3x-4)^5$ in descending powers of x; and the expression of $(2+3x)^4$ in ascending powers of x.

4. Find (i) the 4th term in the expansion of $(1-4x)^8$;
(ii) the 8th term in the expansion of $(2-x)^{14}$;
(iii) the 6th term in the expansion of $(x-2y)^9$.

5. Find the coefficient of x^4 in each of the following expansions

(i) $(1-x)^{21}$;   (ii) $(1-3x)^{11}$;   (iii) $\left(1+\dfrac{x}{3}\right)^{15}$;

(iv) $(2+3x)^9$;   (v) $(1+2x^2)^6$;   (vi) $(a+x)^n$.

6. Write down, and simplify, the middle terms of the expansions of

(i) $(1+3x)^8$;   (ii) $(2-x)^{10}$;   (iii) $\left(x-\dfrac{1}{x}\right)^6$;   (iv) $\left(2a+\dfrac{1}{a}\right)^{10}$.

7. What is the coefficient of x^3 in the expansion of $(1-x)(1+2x)^6$?

8. Write down the first four terms of the following expansions, stating, in each case, the values of x for which the expansions are valid

(i) $(1+x)^{\frac{1}{4}}$;   (ii) $(1+x)^{-3}$;   (iii) $\dfrac{1}{\sqrt{(1+x)}}$;   (iv) $\dfrac{1}{(1+x)^4}$;

(v) $\sqrt[3]{(1-3x)}$;   (vi) $\left(1+\dfrac{x}{2}\right)^{\frac{2}{3}}$;   (vii) $\dfrac{1}{2x+1}$;   (viii) $(2-x)^{\frac{3}{4}}$;

(ix) $\dfrac{1}{(3+2x)^3}$;   (x) $\dfrac{1}{(2x-1)^2}$.

9. Find the coefficients of x^3 in the following expansions

(i) $(1+3x)^{\frac{1}{4}}$;   (ii) $\dfrac{1}{\left(1-\dfrac{x}{2}\right)^{\frac{2}{3}}}$;   (iii) $(3-2x)^{-2}$;   (iv) $\sqrt{(a+x)}$.

10. Find   (i) the 4th term in the expansion of $(1+3x)^{-\frac{1}{2}}$;
(ii) the 5th term in the expansion of $(2-5x)^{-6}$;

(iii) the 6th term in the expansion of $\dfrac{1}{1-5x}$.

11. Find the first four terms in the expansions of

(i) $(1+x)\sqrt{1-x}$;   (ii) $\dfrac{1+x}{1-x}$;   (iii) $\dfrac{2+x}{(1+x)^2}$.

12. By writing $(x+x^2)=z$, obtain the expansions of

(i) $(1+x+x^2)^{10}$;   (ii) $\sqrt{(1+x+x^2)}$,

as far as the terms in x^3.

13. Factorise $(1-3x-4x^2)$ and use the result to expand

$$\frac{1}{\sqrt{(1-3x-4x^2)}}$$

as far as the 4th term.

14. Find the two middle terms of the expansion of $\left(x-\dfrac{1}{x}\right)^7$.

15. Find the term independent of x in the expansion of $\left(x+\dfrac{1}{2x}\right)^8$.

16. Expand $(1-x)^{-1}+(1+2x)^{-1}$ as a series of ascending powers of x as far as the term in x^4.

17. Expand $(1-2x)^{\frac{1}{2}}-(1-3x)^{\frac{2}{3}}$ as far as the 4th term.

18. Expand $(1+x)^6$ and in the result substitute $x=\cdot01$. Hence obtain the value of $(1\cdot01)^6$ correct to 5 places of decimals.

19. Obtain the first four terms in the expansion of $(1-x)^{\frac{1}{2}}$. By putting $x = \cdot 02$ obtain the value of $\sqrt{\cdot 98}$ correct to 4 places of decimals.

20. Use the Binomial Theorem to find the values of the following expressions correct to 4 places of decimals

$$(1 \cdot 002)^4; \quad \sqrt{1 \cdot 02}; \quad (\cdot 997)^5; \quad \frac{1}{(\cdot 97)^4}; \quad \sqrt{4 \cdot 03}; \quad \frac{1}{(2 \cdot 002)^2}.$$

21. Neglecting powers of x above the first, obtain the approximate values of

(i) $\sqrt{(1+x)}$;   (ii) $\dfrac{1}{1-x}$;   (iii) $(1+2x)^{\frac{2}{3}}$;

(iv) $\dfrac{1}{\sqrt[5]{(1-3x)}}$;   (v) $(3-x)^{11}$;   (vi) $\dfrac{1}{(x-2)^3}$.

22. Expand $\dfrac{2}{1-3x^2}$ as far as the term in x^6.

(B) EXAMPLES 12f

1. Obtain the expansion of the expression $\sqrt{\left(\dfrac{1+x}{1-2x}\right)}$ up to, and including the term in x^4. By putting $x = \cdot 01$, obtain the value of $\sqrt{\dfrac{1 \cdot 01}{0 \cdot 98}}$ to four significant figures.   (J.M.B.)

2. Neglecting powers of x above the second, obtain the expansion of

$$\frac{(1+3x)^{\frac{2}{3}}(1-x)^{-\frac{1}{2}}}{(1-2x)^{\frac{1}{3}}}.$$

3. Find the coefficient of x^4 in the expansion of $(1+x+2x^2)^{\frac{1}{2}}$.

4. What is the coefficient of x^2 in the expansion of $\dfrac{(1-x)^6}{(1+x)^7}$?

5. If x, y, z are so small that the second and higher powers may be neglected, prove that

$$\frac{(1+x)^{\frac{1}{2}}(1+y)^{\frac{1}{3}}}{(1+z)^{\frac{1}{4}}} = 1 + \frac{x}{2} + \frac{y}{3} - \frac{z}{4}.$$

6. Find the coefficients of x and $\dfrac{1}{x}$ in the expansions of $\left(x + \dfrac{1}{x}\right)^5$.

7. If the first three terms of the expansion of $(a+3x)^n$ are $b + \dfrac{21}{2}bx + \dfrac{189}{4}bx^2$, find the values of a, b and n.

8. Obtain the expansion of $(a-x)^5$. If

$$(\sqrt{5}-\sqrt{3})^5 = A\sqrt{5} - B\sqrt{3},$$

find the values of **A** and **B**.

9. Expand $\dfrac{x-3}{(1-x)^2(2+x^2)}$ in ascending powers of x as far as the 4th term.

10. Find the first four terms in the expansion of $\dfrac{\sqrt{(1+3x-4x^2)}}{(1-2x)^2}$ in ascending powers of x.

11. Neglecting all powers of x above the second, find the approximate value of $\dfrac{3x}{(x-2)(x+3)^2}$.

12. Neglecting all powers of x above the first, obtain the approximate value of $\dfrac{\sqrt{(1-2x)} \cdot \sqrt[3]{(8-x)}}{(x-1)^3}$.

13. Find the values of the following expressions to 4 places of decimals

(i) $\dfrac{\sqrt{\cdot998}}{\sqrt[3]{1\cdot001}}$;   (ii) $\dfrac{(1\cdot003)^{\frac{5}{2}}}{(\cdot99)^{\frac{3}{2}}}$;   (iii) $\dfrac{(2\cdot004)^3}{\sqrt{4\cdot02}}$.

14. The formula $T = 2\pi\sqrt{\dfrac{l}{g}}$ gives the periodic time, **T** sec, of a simple pendulum of length l m.
A pendulum used to beat seconds; find the new periodic time when the length is decreased by 3 per cent., giving the answer correct to 4 decimal places.
[One beat $= \frac{1}{2}$ a period.]

15. (i) Write out the expansion of $(1+3x+2x^2)^4$ in ascending powers of x.
(ii) Find the term independent of a in the expansion of

$$\left(2a^2 - \dfrac{1}{2a^3}\right)^{10}.$$

16. If $\dfrac{1}{(1-x)(1+2x)} \equiv \dfrac{A}{1-x} + \dfrac{B}{1+2x}$, find **A** and **B**. Hence obtain the expansion of $\dfrac{1}{(1-x)(1+2x)}$ in ascending powers of x as far as the 4th term. For what values of x is the expansion valid?

17. Express the fraction $\dfrac{3}{(x+2)(x+1)}$ as the sum of two fractions of the form $\dfrac{A}{x+2}$, $\dfrac{B}{x+1}$, and hence obtain the first four terms in the expansion of the function in ascending powers of x.

18. Assuming that $\dfrac{5x-3}{(x+2)(3-x)^2} \equiv \dfrac{A}{x+2} + \dfrac{B}{3-x} + \dfrac{C}{(3-x)^2}$, find the constants A, B and C. Deduce the expansion of the expression as far as the term in x^2. State the values of x for which the expansion is valid.

19. Given that when n is a positive integer,

$$(1+x)^n = c_0 + c_1 x + c_2 x^2 + \ldots + c_n x^n,$$

what are the values of (i) c_0; (ii) c_6; (iii) c_{n-1}?

By substituting $x = 1$ and $x = -1$ in this expansion, prove that

    (i) $c_0 + c_1 + c_2 + \ldots + c_n = 2^n$;

    (ii) $c_0 + c_2 + c_4 + \ldots = c_1 + c_3 + c_5 + \ldots = 2^{n-1}$.

CHAPTER 13

ANGLES OF ANY MAGNITUDE. PROJECTIONS
SINE, COSINE AND TANGENT OF ANY ANGLE
EASY IDENTITIES AND EQUATIONS

Representation of angles of any magnitude

Take two rectangular axes **OX, OY** and draw any circle with **O** as centre. Now consider a radius of this circle, **OP**, free to rotate about **O**. Then by varying the position of **OP** it is possible to make the angle between **OP** and the positive direction of **OX** (*i.e.* $\widehat{POA}$) of any required magnitude. (Fig. 78.)

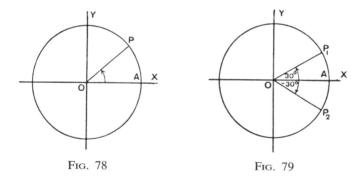

FIG. 78   FIG. 79

For *positive angles* we measure *from* **OA** in a *counter-clockwise* direction.

For *negative angles* we measure *from* **OA** in a *clockwise* direction. In the diagram (Fig. 79)

$$\widehat{P_1OA} = +30° \quad \text{and} \quad \widehat{P_2OA} = -30°.$$

The following simple examples show how it is possible to represent angles of any magnitude.

Ex. 1. *Represent angles of* 117°, 231°, 313°.

As we require positive angles the arm **OP** is rotated from the position **OA** in a counter-clockwise direction.

For all positions of **OP** in the first quadrant, the angle **POA** will be acute.

In the second quadrant we have angles between 90° and 180°, and on the diagram the angle P_1OA represents an angle of 117°.

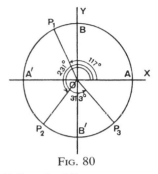

Similarly, for positions of **OP** between **OA'** and **OB'** we have angles between 180° and 270°, and for positions between **OB'** and **OA**, angles between 270° and 360°. By continuing to rotate **OP** we can represent angles of any magnitude.

Fig. 80

Ex. 2. Represent on diagrams angles of 621° *and* −71°.

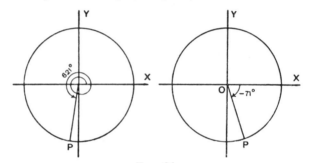

Fig. 81

Projection of a Straight Line

Definition. The projection of **AB** on **OX** is **A'B'**, where **A'**, **B'** are the feet of the perpendiculars from **A**, **B** on **OX**.

Similarly, the projection of **QP** on **OX** is **Q'P'**.

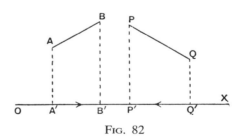

Fig. 82

In considering the projection of one line upon another we not only consider the magnitude, *i.e.* the length, of the projection but *also its direction*.

The projection **A'B'** is in the direction from **A'** to **B'**.

The projection **Q'P'** is in the direction from **Q'** to **P'**.

The projection **B′A′**, *i.e.* the projection of **BA**, is in the direction from **B′** to **A′**, and so on.

Keeping to the usual sign convention, *i.e.* distances measured to the right along **OX** are positive and distances measured to the left are negative, we see that the projection of **AB** on **OX** is a positive length whilst the projection of **QP** is a negative length.

Ex. 3. *With reference to rectangular axes, the co-ordinates of* **A** *are*
(2, 3) *and the co-ordinates of* **B** *are* (−1, 1).

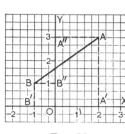

What is the projection of **AB** (i) *on* **OX**; (ii) *on* **OY**?

The projection of

$$\text{AB on OX} = \text{A′B′} = -3 \text{ units.}$$

The projection of

$$\text{AB on OY} = \text{A″B″} = -2 \text{ units.}$$

FIG. 83

N.B. Taking the usual convention for lengths along the y-axis, *i.e.* distances measured upwards are positive.

Ex. 4. *Find the projections of* **OP** *on axes* **OX, OY** *when* **P** *is the point*
(a) (3, 4); (b) (−2, 2); (c) (2, −3).

On **OX**, projection of $\text{OP}_1 = \text{OP}_1' = +3$ units;
            projection of $\text{OP}_2' = \text{OP}_2 = -2$ units;
            projection of $\text{OP}_3 = \text{OP}_3' = +2$ units.

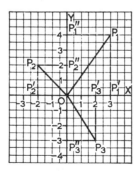

FIG. 84

On **OY**, projection of $\text{OP}_1 = \text{OP}_1'' = +4$ units;
            projection of $\text{OP}_2 = \text{OP}_2'' = +2$ units;
            projection of $\text{OP}_3 = \text{OP}_3'' = -3$ units.

Sine, Cosine, Tangent of any angle

Definition. For *all* positions of the rotating arm **OP**, the sine of an angle between the arm **OP** and **OX**,

i.e. θ_1, θ_2, θ_3, etc.,

$$= \frac{\text{projection of } \mathbf{OP} \text{ on } \mathbf{OY}}{\text{length } \mathbf{OP}},$$

and the cosine of any angle represented by the position **OP**

$$= \frac{\text{projection of } \mathbf{OP} \text{ on } \mathbf{OX}}{\text{length } \mathbf{OP}}.$$

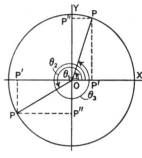

Fig. 85

The fact that the denominator is always taken as positive is stressed by writing it as the *length* **OP**. The projections of **OP** will have positive or negative signs and hence the sine and cosine of angle **POX** for various positions of **OP** will have positive or negative signs.

It follows that the tangent of angle **POX** which equals the sine divided by the cosine

$$= \frac{\text{projection of } \mathbf{OP} \text{ on } \mathbf{OY}}{\text{projection of } \mathbf{OP} \text{ on } \mathbf{OX}}.$$

Expressions for cosec θ, sec θ and cot θ where θ is the angle **POX**, follow from the definitions

$$\operatorname{cosec} \theta = \frac{1}{\sin \theta}; \quad \sec \theta = \frac{1}{\cos \theta}; \quad \cot \theta = \frac{1}{\tan \theta}.$$

To find the sine, cosine, tangent of any angle

Ex. 5. Find sin 212°, cos 114°, tan 313°.
From the diagram (Fig. 86),

$$\sin 212° = \frac{\text{projection of } \mathbf{OP} \text{ on } \mathbf{OY}}{\text{length } \mathbf{OP}},$$

$$= -\frac{\text{length } \mathbf{OP}''}{\text{length } \mathbf{OP}}.$$

Considering the right-angled $\triangle \mathbf{OPP}''$, in which $\widehat{\mathbf{OPP}''} = 32°$, we have $\dfrac{\text{length } \mathbf{OP}''}{\text{length } \mathbf{OP}} = \sin 32°$.

$$\therefore \sin 212° = -\sin 32° = -\cdot5299.$$

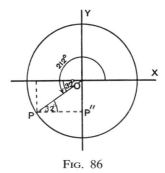

FIG. 86

From the diagram (Fig. 87),

$$\cos 114° = \frac{\text{projection } OP'}{\text{length } OP},$$

$$= -\frac{\text{length } OP'}{\text{length } OP} = -\cos 66°.$$

$$\therefore \cos 114° = -\cdot 4067.$$

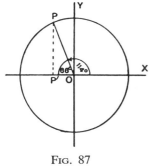

FIG. 87

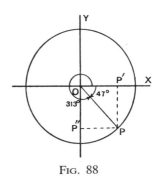

FIG. 88

From the diagram (Fig. 88),

$$\tan 313° = \frac{\text{projection } OP''}{\text{projection } OP'},$$

$$= -\frac{\text{length } OP''}{\text{length } OP'},$$

$$= -\tan 47°,$$

$$= -1\cdot 0724.$$

Negative Angles

Let the arm **OP** rotate from **OX** through an angle θ

      (*a*) in the counter-clockwise direction;

      (*b*) in the clockwise direction.

Then according to the sign convention, in Fig. 89

$$\text{angle } P_1OX = +\theta$$
$$\text{and angle } P_2OX = -\theta.$$

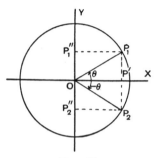

FIG. 89

From the diagram (Fig. 89), we see that

$$\sin \theta = \frac{\text{projection } OP_1''}{\text{length } OP_1} = \frac{\text{length } P'P_1}{\text{length } OP_1},$$

and

$$\sin (-\theta) = \frac{\text{projection } OP_2''}{\text{length } OP_2} = -\frac{\text{length } P'P_2}{\text{length } OP_2},$$

$$= -\sin \theta, \text{ as length } P'P_1$$

$$= \text{length } P'P_2.$$

$$\therefore \ \mathbf{sin \ (-\theta) = -sin \ \theta.}$$

$$\cos \theta = \frac{\text{projection } OP'}{\text{length } OP_1} = \frac{\text{length } OP'}{\text{length } OP_1},$$

and

$$\cos (-\theta) = \frac{\text{projection } OP'}{\text{length } OP_2} = \frac{\text{length } OP'}{\text{length } OP_2}.$$

$$\therefore \ \mathbf{cos \ (-\theta) = cos \ \theta.}$$

Hence

$$\mathbf{tan \ (-\theta) = \frac{sin \ (-\theta)}{cos \ (-\theta)} = -tan \ (\theta).}$$

Ex. 6. *Find sin* $(-71°)$ *and cos* $(-110°)$.

$$\sin (-71°) = -\sin 71° = -\cdot9455.$$

$$\cos (-110°) = \cos (110°) = -\cos 70° = -\cdot3420.$$

Graphs of sin θ, cos θ and tan θ

A study of Fig. 85 on page 160 shows that the values of sin θ, cos θ and tan θ from $\theta = 0°$ to $\theta = 360°$ are repeated for values between 360° and 720°, and so on. By means of this figure and mathematical tables we can easily obtain the graphs of sin θ, cos θ and tan θ. The shapes of these graphs are shown below.

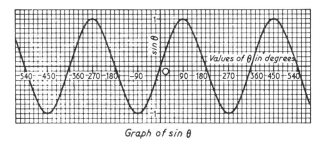

Graph of sin θ

FIG. 90

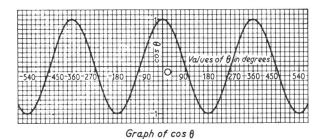

Graph of cos θ

FIG. 91

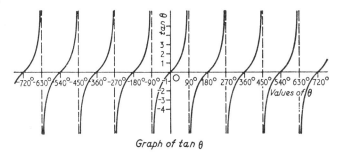

Graph of tan θ

FIG. 92

To find all angles with a given sine, cosine or tangent

Ex. 7. If sin θ = ·7636, find all values of θ between 0° and 720°.
From tables, we obtain the value θ = 49° 47'.
From the diagram (Fig. 93), we see that the only possible positions of the arm OP to give angles whose sine is the same as that of 49° 47' are OP₁ and OP₂.

Hence   θ = 49° 47', 130° 13', 360° + 49° 47', 360° + 130° 13',

i.e.   θ = 49° 47', 130° 13', 409° 47', 490° 13'.

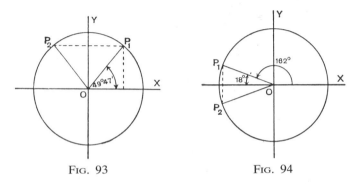

FIG. 93   FIG. 94

Ex. 8. If cos 2θ = − ·9511, find all values of θ between 0° and 360°.

From the tables, the acute angle whose cosine is ·9511 is 18°.
Hence the first positive angle whose cosine is − ·9511 is 162°.
The possible positions of the arm OP are OP₁ and OP₂. (Fig. 94.)
Hence   2θ = 162°, 198°, 360° + 162°, 360° + 198°.
N.B. We require values of 2θ between 0° and 720°.

∴ θ = 81°, 99°, 261°, 279°.

(A) EXAMPLES 13a

1. Find the projections of the line **AB** on the *x*- and *y*-axes in the following cases

(i) **A**(1, 2), **B**(3, 4);   (ii) **A**(−3, 1), **B**(6, 2);
(iii) **A**(−2, 4), **B**(0, 1);   (iv) **A**(4, 3), **B**(1, −2).

2. On a diagram represent the angles 171°, 311°, 452°, −104°, 510°, 382°.

3. Using tables, find the values of the sine, cosine and tangent of the angles in Example 2.

4. Find all the angles between 0° and 360° whose sine is +·5.

5. Find all the angles between 0° and 360° whose cosine is +·7660.

6. Find all the angles between $0°$ and $360°$ whose tangent is $1\cdot5$.

7. Solve the following equations, giving values of θ between $0°$ and $360°$

(i) $\cos\theta = -\cdot6636$;   (ii) $\tan\theta = -1\cdot7230$;
(iii) $\sin\theta = 2\sin 11° 5'$.

8. Solve the equation $3\sin\theta = 2\cos\theta$, giving all solutions between $0°$ and $360°$. $\left[\textit{Note.}\quad \tan\theta = \dfrac{\sin\theta}{\cos\theta}\right]$.

9. In the following equations find all values of θ between $0°$ and $360°$

(i) $\cos 2\theta = \cdot7765$;   (ii) $\sin 2\theta = -\cdot3636$;

(iii) $\tan 2\theta = 1\cdot9878$;   (iv) $\cos\dfrac{\theta}{2} = -\cdot3787$;

(v) $\tan\dfrac{\theta}{2} = -1\cdot0301$;   (vi) $\sin\dfrac{3\theta}{2} = \cdot6636$.

10. Using a diagram and taking x as acute, express

(i) $\sin(180°-x)$ and $\sin(180°+x)$ in terms of $\sin x$;
(ii) $\cos(180°-x)$ and $\cos(180°+x)$ in terms of $\cos x$;
(iii) $\sin(90°-x)$ and $\sin(90°+x)$ in terms of $\cos x$;
(iv) $\cos(90°-x)$ and $\cos(90°+x)$ in terms of $\sin x$.

11. On the same diagram, sketch the graphs of $\sin 2x$, $\sin\dfrac{x}{2}$, $\sin x$ for values of x between $\pm360°$.

12. On the same diagram, sketch the graphs of $\cos 2x$, $\cos\dfrac{x}{2}$, $\cos x$ for values of x between $\pm360°$.

Fundamental identities

Consider the right-angled $\triangle ABC$ (Fig. 95).
By Pythagoras' theorem, we have $b^2 = a^2 + c^2$.

FIG. 95

Dividing by b^2, $1 = \left(\dfrac{a}{b}\right)^2 + \left(\dfrac{c}{b}\right)^2$.

But $\dfrac{a}{b} = \cos\theta$ and $\dfrac{c}{b} = \sin\theta$.

$\therefore \cos^2\theta + \sin^2\theta = 1$.

Dividing by $\cos^2\theta$, $1 + \left(\dfrac{\sin\theta}{\cos\theta}\right)^2 = \left(\dfrac{1}{\cos\theta}\right)^2$,

i.e. $1 + \tan^2\theta = \sec^2\theta$.

Dividing by $\sin^2 \theta$,
$$\left(\frac{\cos \theta}{\sin \theta}\right)^2 + 1 = \left(\frac{1}{\sin \theta}\right)^2.$$
$$\therefore \ \cot^2 \theta + 1 = \operatorname{cosec}^2 \theta.$$

The three identities
$$\left\{ \begin{array}{l} \cos^2 \theta + \sin^2 \theta = 1; \\ 1 + \tan^2 \theta = \sec^2 \theta; \\ 1 + \cot^2 \theta = \operatorname{cosec}^2 \theta, \end{array} \right.$$

have been established for the case where θ is acute. It may be shown, by drawing the diagram for *any* angle, that the results hold good for all values of θ.

Ex. 9. *Express* $3 \cos^2 x - \sin^2 x$ *in terms of* (a) $\cos^2 x$; (b) $\sin^2 x$.

(a) $3 \cos^2 x - \sin^2 x = 3 \cos^2 x - (1 - \cos^2 x)$,
$$= 4 \cos^2 x - 1.$$

(b) $3 \cos^2 x - \sin^2 x = 3(1 - \sin^2 x) - \sin^2 x = 3 - 4 \sin^2 x.$

Ex. 10. *Prove that* $\tan \theta + \cot \theta = \sec \theta \operatorname{cosec} \theta.$

We have
$$\tan \theta + \cot \theta = \frac{\sin \theta}{\cos \theta} + \frac{\cos \theta}{\sin \theta},$$
$$= \frac{\sin^2 \theta + \cos^2 \theta}{\sin \theta \cos \theta},$$
$$= \frac{1}{\sin \theta \cos \theta} = \sec \theta \operatorname{cosec} \theta.$$

Ex. 11. *Solve the equations*

(a) $\sec^2 \theta = 2 \tan \theta;$

(b) $2 \cos^2 \theta = 1 - \sin \theta,$

giving values of θ *between* $0°$ *and* $360°$.

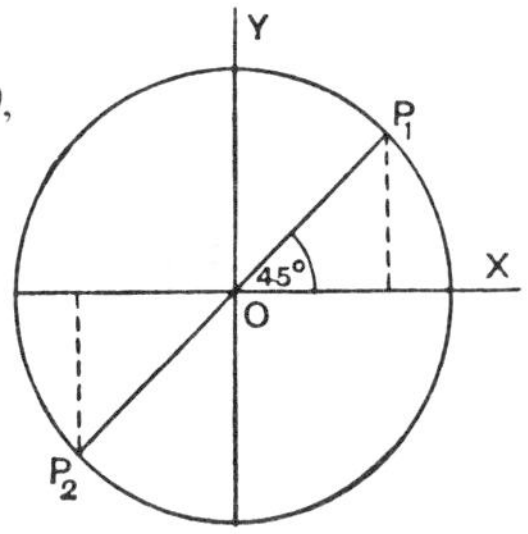

(a) $\sec^2 \theta = 2 \tan \theta.$
$$\therefore \ 1 + \tan^2 \theta = 2 \tan \theta.$$
$$\therefore \ \tan^2 \theta - 2 \tan \theta + 1 = 0.$$
$$(\tan \theta - 1)^2 = 0,$$
i.e. $\tan \theta = 1.$
$$\therefore \ \theta = 45°, 225°.$$

Fig. 96.

(Fig. 96)

(b) $2 \cos^2 \theta = 1 - \sin \theta.$

Expressing $\cos^2 \theta$ in terms of $\sin^2 \theta$, we obtain a quadratic equation in $\sin \theta$.

i.e. $2(1 - \sin^2 \theta) = 1 - \sin \theta.$
$$\therefore \ 2 \sin^2 \theta - \sin \theta - 1 = 0.$$
$$\therefore \ (2 \sin \theta + 1)(\sin \theta - 1) = 0.$$
$$\therefore \ \sin \theta = 1, -\tfrac{1}{2}.$$

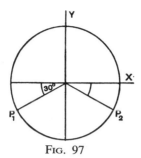

FIG. 97

When $\sin \theta = 1$,   When $\sin \theta = -\frac{1}{2}$,
$\quad \theta = 90°$. $\theta = 210°$ and $330°$.   (Fig. 97.)

$\therefore \ \theta = 90°, 210°, 330°$.

(A) EXAMPLES 13b

1. Express in terms of $\sin \theta$ the expressions

(i) $\sin^2 \theta - \cos^2 \theta$;   (ii) $2 \cos^2 \theta - 4 \sin \theta$.

2. If $\tan \theta = t$, express the following in terms of t

(i) $3 \sec^2 \theta - \tan^2 \theta$;   (ii) $(1 + 2 \sin^2 \theta)/\cos^2 \theta$.

3. Simplify (i) $\dfrac{\sin^2 \theta + \cos^2 \theta}{\cos^2 \theta}$;   (ii) $\dfrac{\sin^3 \theta + \sin \theta \cos^2 \theta}{\cos \theta}$.

4. If $2 \sin^2 \theta - \cos^2 \theta = 1$, show that $\sin^2 \theta = 2 \cos^2 \theta$, and hence find the possible values of $\tan \theta$.

5. If $3 \sec^2 \theta + \tan^2 \theta = 5$, find the possible values of (i) $\tan \theta$; (ii) $\cos \theta$; (iii) $\sin \theta$.

Solve the following equations, giving values of θ between $0°$ and $180°$.

6. $\cos^2 \theta - \sin^2 \theta = 0$.

7. $3 \tan^2 \theta = 2 \sec^2 \theta$.

8. $2 \cos^2 \theta + \sin \theta = 1$.

9. $2 \operatorname{cosec}^2 \theta = 3 \cot \theta + 1$.

10. $\cos^2 \theta + \cos \theta = \sin^2 \theta$.

11. $\tan^2 \theta = \sec \theta + 1$.

12. $1 + \sin \theta \cos^2 \theta = \sin \theta$.

CHAPTER 14

ADDITION THEOREMS

FACTORS. APPLICATIONS

To obtain expressions for sin (A+B) and cos (A+B) in terms of sin A, cos A, sin B, cos B

For our purpose it will be sufficient to consider the case when A, B, and (A+B) are acute (Fig. 98).

Let   angle AOX = A

and   angle BOA = B.

Take any point P on OB, and draw PQ perpendicular to OA. PM, QN are perpendiculars from P, Q, to OX. QR is drawn perpendicular to PM.

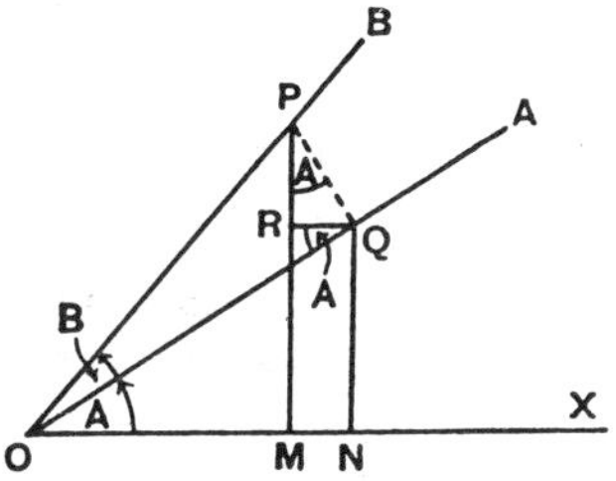

Then $\widehat{RPQ} = \widehat{RQO} = A$.

From the right-angled △ POM,

$$\sin (A+B) = \frac{PM}{OP},$$

$$= \frac{PR+RM}{OP} = \frac{PR+QN}{OP},$$

$$= \frac{PQ \cos A}{OP} + \frac{OQ \sin A}{OP}.$$

But in △ OPQ,

$$\frac{PQ}{OP} = \sin B; \quad \frac{OQ}{OP} = \cos B.$$

$$\therefore \quad \sin (A+B) = \sin A \cos B + \cos A \sin B.$$

$$\cos (A+B) = \frac{OM}{OP},$$

$$= \frac{ON-MN}{OP} = \frac{ON-RQ}{OP},$$

$$= \frac{OQ \cos A}{OP} - \frac{PQ \sin A}{OP}.$$

FIG. 98

But $\dfrac{PQ}{OP} = \sin B$ and $\dfrac{OQ}{OP} = \cos B$.

$\therefore \cos (A+B) = \cos A \cos B - \sin A \sin B.$

i.e.   **sin (A + B) = sin A cos B + cos A sin B.**   (1)

        cos (A + B) = cos A cos B − sin A sin B.   (2)

These results which we have obtained in the particular case taken, can be shown to be true for *all* values of **A** and **B**.

Alternative Proof of sin (A + B) = sin A cos B + cos A sin B when A and B are acute angles.

In the diagram (Fig. 99) **PS** is perpendicular to **QR**.

The areas of triangles **QPR, PQS, PRS** are respectively $\frac{1}{2}rq \sin (A+B)$, $\frac{1}{2}rh \sin A$ and $\frac{1}{2}qh \sin B$.

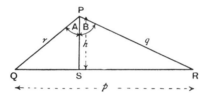

FIG. 99

But   area $\triangle$ QPR = area $\triangle$ PQS + area $\triangle$ PRS.

$\therefore \frac{1}{2}rq \sin (A+B) = \frac{1}{2}rh \sin A + \frac{1}{2}qh \sin B.$

$\therefore \sin (A+B) = \dfrac{h}{q} \sin A + \dfrac{h}{r} \sin B.$

i.e. $\sin (A+B) = \sin A \cos B + \cos A \sin B.$

To obtain the expansions of sin (A − B) and cos (A − B)

Replacing **B** by (−**B**) in (1) and (2) above, we have

$\sin (A - B) = \sin A \cos (-B) + \cos A \sin (-B).$

$\cos (A - B) = \cos A \cos (-B) - \sin A \sin (-B).$

But $\cos (-A) = \cos A$ and $\sin (-A) = -\sin A.$

$\therefore$ **sin (A − B) = sin A cos B − cos A sin B.**   (3)

    cos (A − B) = cos A cos B + sin A sin B.   (4)

Further important results

From (1) and (2) we have,

$$\tan(A+B) = \frac{\sin(A+B)}{\cos(A+B)} = \frac{\sin A \cos B + \cos A \sin B}{\cos A \cos B - \sin A \sin B},$$

$$= \frac{\tan A + \tan B}{1 - \tan A \tan B}. \quad \text{[dividing every term by } \cos A \cos B \text{]}$$

Similarly from (3) and (4) we get

$$\tan(A-B) = \frac{\tan A - \tan B}{1 + \tan A \tan B}.$$

Ex. 1. *Given* $\sin x = \frac{3}{5}$, $\cos y = \frac{5}{13}$ *where* x, y *are acute. Without using tables, find the values of* $\cos x$, $\sin y$, $\sin(x+y)$, $\cos(x-y)$, $\tan(x+y)$.

Using the right-angled triangles in Fig. 100, it follows that

$$\cos x = \tfrac{4}{5},$$

$$\sin y = \tfrac{12}{13}.$$

$$\therefore \ \sin(x+y) = \sin x \cos y + \cos x \sin y$$
$$= \tfrac{3}{5} \cdot \tfrac{5}{13} + \tfrac{4}{5} \cdot \tfrac{12}{13} = \tfrac{63}{65};$$

$$\cos(x-y) = \cos x \cos y + \sin x \sin y$$
$$= \tfrac{4}{5} \cdot \tfrac{5}{13} + \tfrac{3}{5} \cdot \tfrac{12}{13} = \tfrac{56}{65};$$

$$\tan(x+y) = \frac{\tan x + \tan y}{1 - \tan x \tan y} = \frac{\tfrac{3}{4} + \tfrac{12}{5}}{1 - \tfrac{3}{4} \cdot \tfrac{12}{5}}$$
$$= -\tfrac{63}{16}.$$

Fig. 100

Ex. 2. *Expand* $\sin(A+B+C)$.

$$\sin(A+B+C) = \sin(\overline{A+B}+C) = \sin(A+B)\cos C + \cos(A+B)\sin C,$$
$$= (\sin A \cos B + \cos A \sin B)\cos C + (\cos A \cos B -$$
$$\sin A \sin B)\sin C,$$
$$= \sin A \cos B \cos C + \sin B \cos A \cos C + \sin C \cos A$$
$$\cos B - \sin A \sin B \sin C.$$

Ex. 3. *Prove* $\tan A + \tan B = \sin(A+B)\sec A \sec B$.

We have
$$\tan A + \tan B = \frac{\sin A}{\cos A} + \frac{\sin B}{\cos B},$$

$$= \frac{\sin A \cos B + \cos A \sin B}{\cos A \cos B},$$

$$= \sin(A+B)\sec A \sec B.$$

Ex. 4. *Show that* $\cos(\alpha-\beta)+\sin(\alpha+\beta)=(\cos\beta+\sin\beta)(\cos\alpha+\sin\alpha)$ *and hence prove that*

$$\frac{\cos 4\theta+\sin 6\theta}{\cos 5\theta+\sin 5\theta}=\cos\theta+\sin\theta=\frac{\cos 2\theta+\sin 4\theta}{\cos 3\theta+\sin 3\theta}.$$

$$\cos(\alpha-\beta)+\sin(\alpha+\beta)=\cos\alpha\cos\beta+\sin\alpha\sin\beta+\sin\alpha\cos\beta$$
$$+\cos\alpha\sin\beta,$$

$$=(\cos\beta+\sin\beta)(\cos\alpha+\sin\alpha).$$

Using this result with $\alpha=5\theta$, $\beta=\theta$, we have

$$\cos 4\theta+\sin 6\theta=(\cos\theta+\sin\theta)(\cos 5\theta+\sin 5\theta).$$

So, $\quad\dfrac{\cos 4\theta+\sin 6\theta}{\cos 5\theta+\sin 5\theta}=\cos\theta+\sin\theta.$

Now take $\alpha=3\theta$, $\beta=\theta$ and get

$$\cos 2\theta+\sin 4\theta=(\cos\theta+\sin\theta)(\cos 3\theta+\sin 3\theta).$$

or, $\quad\dfrac{\cos 2\theta+\sin 4\theta}{\cos 3\theta+\sin 3\theta}=\cos\theta+\sin\theta.$

Hence the required result,

$$\frac{\cos 4\theta+\sin 6\theta}{\cos 5\theta+\sin 5\theta}=\cos\theta+\sin\theta=\frac{\cos 2\theta+\sin 4\theta}{\cos 3\theta+\sin 3\theta}.$$

(A) EXAMPLES 14a

1. If $\sin A=\frac{4}{5}$ and $\tan B=\frac{5}{12}$, and A and B are acute angles, find (i) $\cos A$, (ii) $\tan A$, (iii) $\sin B$, (iv) $\cos B$, (v) $\sin(A+B)$, (vi) $\sin(A-B)$, (vii) $\cos(A+B)$, (viii) $\cos(A-B)$, (ix) $\tan(A+B)$, (x) $\tan(A-B)$.

2. If $\cos x=\frac{3}{5}$, $\sin y=\frac{12}{13}$ and x and y are acute, find $\sin(x+y)$ and $\cos(x+y)$. Hence find $\sin(2x+y)$, $\cos(2x+y)$, $\sin(x+2y)$ and $\cos(x+2y)$.

[*Note.* $(2x+y)=(x+y)+x.$]

3. By means of a right-angled isoceles $\triangle$, deduce that

$$\sin 45°=\cos 45°=\frac{1}{\sqrt{2}}.$$

Prove that $\cos(45°+A)=\dfrac{1}{\sqrt{2}}(\cos A-\sin A)$

and $\quad\sin(45°+A)=\dfrac{1}{\sqrt{2}}(\cos A+\sin A).$

Deduce $\cos A = \dfrac{1}{\sqrt{2}}\{\cos (45° + A) + \sin (45° + A)\}$

and state the corresponding result for $\sin A$.

4. ABC is an equilateral $\triangle$ of side 2 units. AD is perpendicular to BC. Find the length AD, and hence deduce

$$\cos 60° = \sin 30° = \tfrac{1}{2},$$

and $$\cos 30° = \sin 60° = \dfrac{\sqrt{3}}{2}.$$

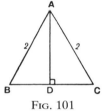

Using these results and the addition theorems, prove that

(i) $\sin (x + 60°) = \cos (x - 30°)$;

(ii) $\sin (x + 30°) = \sin (150° - x)$.

Fig. 101

5. Express as single sines or cosines

(i) $\sin 43° \cos 61° + \cos 43° \sin 61°$;

(ii) $\sin 22° \cos 18° - \cos 22° \sin 18°$;

(iii) $\cos 63° \cos 11° + \sin 63° \sin 11°$;

(iv) $\sin 41° \sin 22° - \cos 41° \cos 22°$;

(v) $\sin 2x \cos x + \cos 2x \sin x$;

(vi) $\cos^2 x - \sin^2 x$ [*i.e.* $\cos x . \cos x - \sin x . \sin x$].

6. Use the addition theorems to simplify the following

$\sin (180° - A)$, $\sin (180° + A)$, $\cos (180° - A)$, $\cos (180° + A)$, $\sin (90° + A)$, $\cos (90° + A)$.

7. By putting $A = B = \theta$ in the formula for $\cos (A - B)$, deduce $\cos^2 \theta + \sin^2 \theta = 1$.

8. Expand $\sin (3x + y)$, $\cos (x - 2y)$, $\tan (2x + y)$, $\tan (3x - y)$, $\cot (x + y)$, $\cot (x - y)$.

9. If $\tan x = 2 \tan y$, find $\tan (x + y)$ and $\tan (x - y)$ in terms of

(i) $\tan x$;   (ii) $\tan y$.

10. Expand $\cos (A + B + C)$ in terms of the sines and cosines of A, B, C.

11. Expand $\tan (A + B + C)$ in terms of $\tan A$, $\tan B$, $\tan C$.

12. If $\tan \alpha = \tfrac{1}{2}$ and $\tan \beta = \tfrac{1}{3}$ where α and β are acute, prove that $\alpha + \beta = 45°$. [Find the value of $\tan (\alpha + \beta)$.]

13. Without using tables, write down the values of $\tan 45°$ and $\tan 30°$ and deduce that $\tan 75° = 2 + \sqrt{3}$.

14. In triangle **ABC**, $\cos B = \frac{5}{13}$ and $\cos C = \frac{4}{5}$. Without using tables calculate the value of $\sin (B+C)$ and deduce the value of $\sin A$.

15. If **A, B** and **C** are the angles of a triangle, prove that

(i) $\sin (B+C) = \sin A$;   (ii) $\cos (B+C) = -\cos A$.

16. If **A, B** and **C** are the angles of a triangle, prove that

(i) $\sin B \cos C + \cos B \sin C = \sin A$;

(ii) $\cos C + \cos B \cos A = \sin A \sin B$;

(iii) $\sin A - \cos B \sin C = \sin B \cos C$.

Double and Half Angles

If in the formulae,

(1) $\sin (A+B) = \sin A \cos B + \cos A \sin B$;

(2) $\cos (A+B) = \cos A \cos B - \sin A \sin B$;

(3) $\tan (A+B) = \dfrac{\tan A + \tan B}{1 - \tan A \tan B}$,

we put $B = A$, the following important results are obtained.

(1) $\sin (A+A) = \sin 2A = \sin A \cos A + \cos A \sin A$,

i.e. $\sin 2A = 2 \sin A \cos A$.

(2) $\cos (A+A) = \cos 2A = \cos A \cos A - \sin A \sin A$,

$$i.e.\ \ \cos 2A = \cos^2 A - \sin^2 A, \left.\begin{array}{l} \\ = 2\cos^2 A - 1, \\ \\ = 1 - 2\sin^2 A. \end{array}\right\} \begin{array}{c} using \\ \cos^2 A + \sin^2 A = 1. \end{array}$$

(3) $\tan (A+A) = \tan 2A = \dfrac{\tan A + \tan A}{1 - \tan A \tan A}$,

i.e. $\tan 2A = \dfrac{2 \tan A}{1 - \tan^2 A}$.

If in these results we write $A = \dfrac{\theta}{2}$, we have,

$\sin \theta = 2 \sin \frac{1}{2}\theta \cos \frac{1}{2}\theta$.

$\cos \theta = \cos^2 \frac{1}{2}\theta - \sin^2 \frac{1}{2}\theta = 2 \cos^2 \frac{1}{2}\theta - 1 = 1 - 2 \sin^2 \frac{1}{2}\theta$.

$\tan \theta = \dfrac{2 \tan \frac{1}{2}\theta}{1 - \tan^2 \frac{1}{2}\theta}$.

Ex. 5. Express sin 3x in terms of sin x.

We have $\sin 3x = \sin(2x + x) = \sin 2x \cos x + \cos 2x \sin x$,

$$= \{2 \sin x \cos x\} \cos x + (1 - 2 \sin^2 x) \sin x.$$

{*Note.* As we require the result in terms of sin x, we use the result $\cos 2x = 1 - 2 \sin^2 x$.}

$$= 2 \sin x \cos^2 x + \sin x - 2 \sin^3 x,$$

$$= 2 \sin x\{1 - \sin^2 x\} + \sin x - 2 \sin^3 x.$$

$\therefore$ $\sin 3x = 3 \sin x - 4 \sin^3 x.$

Ex. 6. Given $\tan 30° = \dfrac{1}{\sqrt{3}}$, *find without using tables, the value of* $\tan 15°.$

Using the result $\tan 2A = \dfrac{2 \tan A}{1 - \tan^2 A}$ with $A = 15°$,

we have $\tan 30° = \dfrac{1}{\sqrt{3}} = \dfrac{2 \tan 15°}{1 - \tan^2 15°}.$

$\therefore$ $1 - \tan^2 15° = 2\sqrt{3} \tan 15°.$

i.e. $\tan^2 15° + 2\sqrt{3} \tan 15° - 1 = 0,$

$\therefore$ $\tan 15° = \dfrac{-2\sqrt{3} \pm \sqrt{12 + 4}}{2},$

$$= -\sqrt{3} \pm 2.$$

As tan 15° is positive, we must have

$$\tan 15° = 2 - \sqrt{3}.$$

Ex. 7. Solve the equation $\sin 2\theta = \sin \theta$, *giving all solutions between* 0° *and* 360°.

We have, $\sin 2\theta = 2 \sin \theta \cos \theta.$

$\therefore$ $2 \sin \theta \cos \theta = \sin \theta,$

i.e. $\sin \theta \{2 \cos \theta - 1\} = 0.$

Note. Care must be taken not to divide out the factor sin θ, as by dividing out this factor the root sin $\theta = 0$ would be omitted.

$\therefore$ $\sin \theta = 0$ or $2 \cos \theta - 1 = 0,$

i.e. $\theta = 0°, 180°, 360°$ and $\theta = 60°, 300°.$

Hence $\theta = 0°, 60°, 180°, 300°, 360°.$

Ex. 8. *Prove that* $\tan x + \cot x = \dfrac{2}{\sin 2x}$.

We have, $\tan x + \cot x = \dfrac{\sin x}{\cos x} + \dfrac{\cos x}{\sin x}$,

$$= \frac{\sin^2 x + \cos^2 x}{\sin x \cos x},$$

$$= \frac{1}{\sin x \cos x} = \frac{2}{\sin 2x}.$$

To express tan x, sin x and cos x in terms of tan $\frac{1}{2}$x

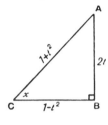

Fig. 102

We have, $\tan x = \dfrac{2 \tan \frac{1}{2}x}{1 - \tan^2 \frac{1}{2}x}$.

To simplify this expression, let

$$\tan \tfrac{1}{2}x = t.$$

$$\therefore \ \tan x = \frac{2t}{1 - t^2}.$$

In the right-angled triangle **ABC**, let $\hat{C} = x$. As $\tan x = \dfrac{2t}{1-t^2}$, we can take $\mathbf{AB} = 2t$ and $\mathbf{BC} = 1 - t^2$.

By Pythagoras, $\mathbf{AC}^2 = \mathbf{AB}^2 + \mathbf{BC}^2$,

$$= (2t)^2 + (1 - t^2)^2 = 4t^2 + 1 - 2t^2 + t^4,$$

$$= 1 + 2t^2 + t^4 = (1 + t^2)^2.$$

$$\therefore \ \mathbf{AC} = 1 + t^2.$$

$\therefore$ From the diagram, $\sin x = \dfrac{2t}{1+t^2}$, $\cos x = \dfrac{1-t^2}{1+t^2}$.

So, if $\tan \frac{1}{2}x = t$, $\tan x = \dfrac{2t}{1-t^2}$, $\sin x = \dfrac{2t}{1+t^2}$, $\cos x = \dfrac{1-t^2}{1+t^2}$.

These results are very useful in solving equations of the type $a \cos x + b \sin x = c$, where a, b and c are constants. The method adopted is illustrated by the following example.

Ex. 9. *Solve the equation* $3 \cos \theta - 8 \sin \theta = -2$, *giving solutions between* $0°$ *and* $360°$.

Let $\tan \frac{1}{2}\theta = t$; then $\cos \theta = \dfrac{1-t^2}{1+t^2}$ and $\sin \theta = \dfrac{2t}{1+t^2}$.

Substituting for $\sin\theta$ and $\cos\theta$ in the equation, we have,

$$3\frac{(1-t^2)}{1+t^2} - 8 \cdot \frac{2t}{1+t^2} = -2.$$

$$\therefore\ 3 - 3t^2 - 16t = -2 - 2t^2,$$

$i.e.\quad t^2 + 16t - 5 = 0.$

$$\therefore\ t = \frac{-16 \pm \sqrt{256+20}}{2},$$

$$= \frac{-16 \pm 16 \cdot 62}{2} = \cdot31 \text{ and } -16 \cdot 31.$$

$i.e.\quad \tan\dfrac{\theta}{2} = \cdot31, \qquad \text{and } \tan\dfrac{\theta}{2} = -16 \cdot 31.$

$$\therefore\ \frac{\theta}{2} = 17° \, 13', \quad \text{and } \frac{\theta}{2} = 180° - 86° \, 30' = 93° \, 30'.$$

(the only solutions less than $180°$)

$$\therefore\ \theta = 34° \, 26'. \qquad \theta = 187°.$$

Hence $\theta = 34° \, 26'$ and $187°$.

(A) EXAMPLES 14b

1. If $\sin\theta = \dfrac{20}{29}$ and θ is acute, find, without the use of tables, the values of $\sin 2\theta$, $\cos 2\theta$, $\sin\dfrac{\theta}{2}$, $\cos\dfrac{\theta}{2}$.

2. Given that $\tan 45° = 1$, find the value of $\tan 22\frac{1}{2}°$ in surd form.

3. Obtain the values of $\sin 15°$ and $\cos 15°$ in surd form. $\left[\text{Use } \cos 30° = \dfrac{\sqrt{3}}{2}.\right]$

4. If $\sin x = \frac{3}{5}$ and x is acute, find the values of $\tan 2x$ and $\tan\dfrac{x}{2}$.

5. If $\tan\theta = 1\cdot5$ and θ is acute, without using tables, obtain the values of $\tan 2\theta$, $\sin 2\theta$ and $\cos 2\theta$.

6. Given that $\tan x = -2$, and x lies between $0°$ and $180°$, calculate, without using tables, the value of $\sin\dfrac{x}{2}$.

7. Find, with a minimum of calculation, the values of

(i) $2\cos^2 32°$;   (ii) $\sin^2 11° 3'$;

(iii) $\sin 21° 2' \cos 21° 2'$;   (iv) $\dfrac{2 \tan 15°}{1-\tan^2 15°}$.

8. If $\tan \frac{1}{2}x = t$, express the following in terms of t

(i) $\sin x + \cos x$;   (ii) $3 \sin x - 4 \cos x$;

(iii) $3 \tan x - \cos x$;   (iv) $\dfrac{1}{1+\cos x}$;

(v) $\dfrac{\sin x}{1-2\cos x}$;   (vi) $\sec x \cot x$.

9. Using the result $\tan 2A = \dfrac{2 \tan A}{1-\tan^2 A}$, obtain an expression for $\cot 2A$ in terms of $\cot A$.

10. If $\cot A = 1\cdot2$, without using tables, find the value of $\cot \dfrac{A}{2}$.

11. Express $\cos 3x$ in terms of $\cos x$. [Write $\cos 3x = \cos(2x+x)$.]

12. Express $\tan 3x$ in terms of $\tan x$.

13. Prove that $4 \cos 2A \sin A \cos A = \sin 4A$.

14. Show that   (a) $\dfrac{1-\cos x}{\sin x} = \tan \dfrac{x}{2}$;

(b) $\dfrac{1-\cos x}{1+\cos x} = \tan^2 \dfrac{x}{2}$.

15. Express $\cos^2 2A$ in terms of $\cos 4A$.

16. Use the previous result to obtain an expression for $\cos 4A$ in terms of $\cos A$.

17. Solve the following equations, giving values of θ between $0°$ and $180°$

(i) $\sin \theta (2 \sin \theta - 1) = 0$;   (ii) $2 \sin \theta \cos \theta = \cos \theta$;

(iii) $4 \sin 2\theta = \sin \theta$;   (iv) $\cos 2\theta = \cos \theta$.

18. Draw the graph of $y = \cos^2 x$ for values of x between $0°$ and $90°$. [Use $\cos^2 x = \frac{1}{2}(\cos 2x + 1)$.]

To express $a \cos x + b \sin x$ in the form $R \cos(x-\alpha)$, where a, b, R and α are constants

Let $a \cos x + b \sin x \equiv R \cos(x-\alpha)$,

$\equiv R\{\cos x \cos \alpha + \sin x \sin \alpha\}$,

$\equiv R \cos \alpha \cos x + R \sin \alpha \sin x$.

As the identity is true for all values of x, we can equate the coefficients of $\cos x$ and $\sin x$.

$$\therefore \ a = \mathsf{R} \cos \alpha,$$
$$b = \mathsf{R} \sin \alpha.$$

Squaring and adding,

$$a^2 + b^2 = \mathsf{R}^2 (\cos^2 \alpha + \sin^2 \alpha) = \mathsf{R}^2.$$

i.e. $\mathsf{R} = \sqrt{a^2 + b^2}$, taking the positive sign for the root.

Dividing,

$$\frac{b}{a} = \frac{\sin \alpha}{\cos \alpha} = \tan \alpha.$$

$$\therefore \ \alpha = \tan^{-1} \frac{b}{a}.$$

Hence $a \cos x + b \sin x \equiv \sqrt{a^2 + b^2} \cos (x - \alpha)$ where $\tan \alpha = \dfrac{b}{a}$.

In the same way we can show that

(i) **$a \cos x - b \sin x \equiv \sqrt{a^2 + b^2} \cos (x + \alpha)$ where $\tan \alpha = \dfrac{b}{a}$.**

(ii) **$b \sin x - a \cos x \equiv \sqrt{a^2 + b^2} \sin (x - \beta)$ where $\tan \beta = \dfrac{a}{b}$.**

The previous results have two important applications

(a) to find the maximum and minimum values of expressions of the form $a \cos x \pm b \sin x$;

(b) to solve equations of the form $a \cos x \pm b \sin x = c$, where a, b, c are constants.

[An alternative method of solving these equations has already been discussed on page 175.]

Ex. 10. *Find the maximum and minimum values of the expression*

$$2 \cos x - 3 \sin x.$$

Let
$$2 \cos x - 3 \sin x \equiv \mathsf{R} \cos (x + \alpha),$$
$$\equiv \mathsf{R} \cos x \cos \alpha - \mathsf{R} \sin x \sin \alpha.$$

Equating coefficients of $\cos x$ and $\sin x$,

$$2 = \mathsf{R} \cos \alpha,$$
$$3 = \mathsf{R} \sin \alpha.$$

Dividing,
$$\tan \alpha = \tfrac{3}{2}; \quad \textit{i.e.} \quad \alpha = 56° \ 19'.$$

Squaring and adding, $R^2 = 2^2 + 3^2 = 13.$

$$\therefore R = \sqrt{13}.$$

$$\therefore 2 \cos x - 3 \sin x \equiv \sqrt{13} \cos (x + 56° 19').$$

But the maximum and minimum values of $\cos (x + 56° 19')$ are $+1$ and -1 respectively.

$\therefore$ Maximum and minimum values of $2 \cos x - 3 \sin x$ are $+\sqrt{13}$ and $-\sqrt{13}$ respectively.

Note. $\cos (x + 56° 19')$ is a maximum when

$$x + 56° 19' = 0°, 360°, \text{etc.}$$

$$i.e. \quad x = -56° 19', 303° 41', \text{etc.}$$

I.e.   the expression is a maximum when

$$x = -56° 19', 303° 41', \text{etc.}$$

Similarly the expression is a minimum when

$$x + 56° 19' = 180°, 540°, \text{etc.}$$

$$I.e. \quad \text{when } x = 123° 41', 483° 41', \text{etc.}$$

Ex. 11. *Solve the equation* $\cos x + 2 \sin x = 1.$ [*x between* 0° *and* 360°.]

We first express $\cos x + 2 \sin x$ in the form $R \cos (x - \alpha)$.

Let $\cos x + 2 \sin x \equiv R \cos (x - \alpha),$

$$= R \cos \alpha \cos x + R \sin \alpha \sin x.$$

and $\left.\begin{array}{l} \therefore \ 1 = R \cos \alpha \\ \quad 2 = R \sin \alpha \end{array}\right\}.$

Dividing, $\tan \alpha = 2; \quad \alpha = 63° 26'.$

[*Note.*   It is only necessary to take one solution for α.]

Squaring and adding, $R^2 = 2^2 + 1^2 = 5.$

$$R = \sqrt{5}.$$

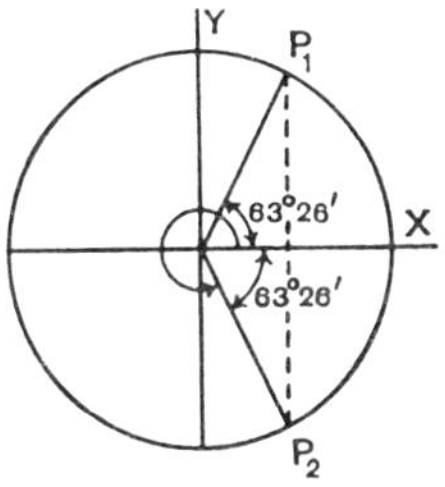

FIG. 103

Hence $\sqrt{5} \cos (x - 63° 26') = 1$,

$$\therefore \cos (x - 63° 26') = \frac{1}{\sqrt{5}} = \frac{\sqrt{5}}{5} = \frac{2·236}{5} = ·4472.$$

$$\therefore (x - 63° 26') = 63° 26', 296° 34', \text{ and } -63° 26'.$$

[These results are obtained by considering the diagram, Fig. 103.]

$$\therefore x = 126° 52', 0°.$$

Values of x between $0°$ and $360°$ are $0°$ and $126° 52'$.

(B) EXAMPLES 14c

1. Express each of the following in the form $R \cos (x - \alpha)$

    (i) $\sin x + \cos x$;   (ii) $4 \cos x + 3 \sin x$;

    (iii) $\sqrt{2} \cos x + \sin x$;   (iv) $3 \cos x + \sqrt{3} \sin x$.

2. Express the following in the form $R \cos (x + \alpha)$

    (i) $\cos x - \sin x$;   (ii) $4 \cos x - 3 \sin x$;

    (iii) $2 \cos x - \sqrt{2} \sin x$;   (iv) $5 \cos x - 12 \sin x$.

3. Express the following in the form $R \sin (x - \alpha)$

    (i) $2 \sin x - \cos x$;   (ii) $\sqrt{3} \sin x - 4 \cos x$;

    (iii) $3 \sin x - 4 \cos x$;   (iv) $a \sin x - b \cos x$.

4. Find the maximum and minimum values of the expressions in examples (1), (2) and (3).

5. Find the values of x for which the expression $3 \cos x - 5 \sin x$ is a maximum.

6. Find the values of x for which the expression $2 \sin x - 3 \cos x$ is a minimum.

7. Solve the following equations, giving values of x between $0°$ and $360°$

    (i) $3 \cos x + 4 \sin x = 2$;   (ii) $\sqrt{2} \cos x - \sin x = 1$;

    (iii) $5 \cos x - 12 \sin x = 7·5$;   (iv) $\cos x + \sin x = 1$.

8. Express $2 \sin 2x + \cos 2x$ in the form $R \sin (2x + \alpha)$.

Hence solve the equation,

    $2 \sin 2x + \cos 2x = 1$, giving values of x between $0°$ and $360°$.

Factors

We have,
$$\sin (A + B) = \sin A \cos B + \cos A \sin B \qquad (1)$$
$$\sin (A - B) = \sin A \cos B - \cos A \sin B \qquad (2)$$
$$\cos (A + B) = \cos A \cos B - \sin A \sin B \qquad (3)$$
$$\cos (A - B) = \cos A \cos B + \sin A \sin B \qquad (4)$$

Adding (1) and (2)
$$\sin (A + B) + \sin (A - B) = 2 \sin A \cos B. \qquad (a)$$

Subtracting (2) from (1),
$$\sin (A + B) - \sin (A - B) = 2 \cos A \sin B. \qquad (b)$$

Adding (3) and (4),
$$\cos (A + B) + \cos (A - B) = 2 \cos A \cos B. \qquad (c)$$

Subtracting (4) from (3),
$$\cos (A + B) - \cos (A - B) = -2 \sin A \sin B. \qquad (d)$$

The results (a), (b), (c), (d) can be simplified by putting $A + B = S$ and $A - B = T$.

Then
$$A = \frac{S + T}{2}, \qquad B = \frac{S - T}{2}.$$

We have,
$$\sin S + \sin T = 2 \sin \frac{S + T}{2} \cos \frac{S - T}{2}.$$

$$\sin S - \sin T = 2 \cos \frac{S + T}{2} \sin \frac{S - T}{2}.$$

$$\cos S + \cos T = 2 \cos \frac{S + T}{2} \cos \frac{S - T}{2}.$$

$$\cos S - \cos T = -2 \sin \frac{S + T}{2} \sin \frac{S - T}{2}.$$

These results should be remembered in words.

(1) The *sum* of two *sines* =
     Twice sine {half sum} cos {half difference}.

(2) The *difference* of two *sines* =
     Twice cos {half sum} sin {half difference}.

(3) The *sum* of two *cosines* =
     Twice cos {half sum} cos {half difference}.

(4) The *difference* of two *cosines* =
     Minus twice sin {half sum} sin {half difference}.

Ex. 12. *Factorise* (*a*) $\sin 3A + \sin A$;

        (*b*) $\cos 3A - \cos 2A$.

(*a*) $\sin 3A + \sin A$ $= 2 \sin \dfrac{3A+A}{2} \cos \dfrac{3A-A}{2}$,

                $= 2 \sin 2A \cos A.$

(*b*) $\cos 3A - \cos 2A = -2 \sin \dfrac{3A+2A}{2} \sin \dfrac{3A-2A}{2}$,

               $= -2 \sin \dfrac{5A}{2} \sin \dfrac{A}{2}.$

Ex. 13. *Factorise, and hence evaluate without the use of tables, the expression* $\sin 75° - \sin 15°$.

$$\sin 75° - \sin 15° = 2 \cos \left(\frac{75°+15°}{2}\right) \sin \left(\frac{75°-15°}{2}\right),$$

$$= 2 \cos 45° \sin 30°,$$

$$= 2 \cdot \frac{1}{\sqrt{2}} \cdot \frac{1}{2} = \frac{1}{\sqrt{2}} = \frac{\sqrt{2}}{2} = \cdot 7071.$$

Ex. 14. *Prove that* $\dfrac{\sin 3\theta + \sin \theta}{\cos 3\theta + \cos \theta} = \tan 2\theta.$

To simplify the fraction, factorise numerator and denominator.

We have $\dfrac{\sin 3\theta + \sin \theta}{\cos 3\theta + \cos \theta} = \dfrac{2 \sin 2\theta \cos \theta}{2 \cos 2\theta \cos \theta} = \tan 2\theta.$

Ex. 15. *Solve the equation* $\cos 5\theta - \cos \theta = \sin 3\theta$, *for values of* θ *between* 0° *and* 180°.

$$\cos 5\theta - \cos \theta = \sin 3\theta.$$

$$\therefore \ -2 \sin 3\theta \sin 2\theta = \sin 3\theta.$$

$$\therefore \ \sin 3\theta \{-2 \sin 2\theta - 1\} = 0.$$

$$\therefore \ \sin 3\theta = 0 \quad \text{or} \quad \sin 2\theta = -\tfrac{1}{2}.$$

$$\therefore \ 3\theta = 0°, 180°, 360°, 540° \quad \text{and} \quad 2\theta = 210°, 330°.$$

$\therefore$ Values of θ between 0° and 180° are

$$0°, 60°, 105°, 120°, 165°, 180°.$$

(A) EXAMPLES 14d

1. Factorise

(i) $\sin 3A - \sin A$;　　　　　　　(ii) $\cos 3A + \cos A$;

(iii) $\cos 3A - \cos A$;　　　　　　(iv) $\sin 2A + \sin 2B$;

(v) $\cos 2A - \cos 2B$;　　　　　　(vi) $\cos 5A + \cos 3A$;

(vii) $\sin 7A + \sin 3A$;　　　　　　(viii) $\cos A - \cos 7A$;

(ix) $\sin (\theta + \phi) - \sin (\theta - \phi)$;　(x) $\cos (2\alpha + \beta) + \cos (\alpha + 2\beta)$.

2. Evaluate, without the use of tables

(i) $\cos 75° + \cos 15°$;　　　　　(ii) $\cos 75° - \cos 15°$.

3. Prove that $\dfrac{\sin \alpha + \sin \beta}{\cos \alpha + \cos \beta} = \tan \dfrac{\alpha + \beta}{2}$.

4. Prove that $\sin 37° + \cos 47° = 2 \sin 40° \cos 3°$.

[*Note.* $\cos A = \sin \overline{90° - A}$.]

5. Factorise $\cos (\alpha + \beta - \gamma) + \cos (\alpha - \beta + \gamma)$.

6. Solve the following equations, giving roots between $0°$ and $180°$.

(i) $\cos 3x + \cos x = 0$.　　　　(ii) $\sin 5x - \sin x = 0$.

(iii) $\cos 7x = \cos x$.　　　　　　(iv) $\sin 3x = \sin x$.

(v) $\cos 5x + \cos x = \cos 3x$.　(vi) $\sin 9x + \sin 3x = \sin 6x$.

(vii) $\sin x + \sin 3x + \sin 5x = 0$.　(viii) $\cos 2x + \cos 4x + \cos 6x = 0$.

7. Simplify the following fractions

(i) $\dfrac{\sin 2x + \sin x}{\cos 2x + \cos x}$;

(ii) $\dfrac{\sin 2x - \sin x}{\cos 2x - \cos x}$;

(iii) $\dfrac{\sin 3\theta + \sin \theta}{\cos 3\theta - \cos \theta}$;

(iv) $\dfrac{\cos 7\theta + \cos 3\theta}{\sin 7\theta + \sin 3\theta}$;

(v) $\dfrac{\cos 3\theta + \cos 2\theta}{\cos 2\theta + \cos \theta}$;

(vi) $\dfrac{\sin \dfrac{3\theta}{2} + \sin \dfrac{\theta}{2}}{\cos \dfrac{5\theta}{2} + \cos \dfrac{3\theta}{2}}$.

(B) EXAMPLES 14e

1. Given that the sine of an obtuse angle A is equal to $\frac{20}{29}$, find the values of $\cos 2A$, $\tan 2A$ and $\sin \dfrac{A}{2}$.

2. If $\tan \theta = \dfrac{b}{a}$, find the value of $a \cos 2\theta + b \sin 2\theta$.

3. Prove that $\sin(A+B)\sin(A-B) = \sin^2 A - \sin^2 B$. Hence, without using tables, obtain the value of $\sin 75° \sin 15°$.

4. Prove that $\cos(A+B)\cos(A-B) = \cos^2 A - \sin^2 B$, and deduce the value of $\cos 75° \cos 15°$.

5. Show that

$$\sin(45°+A)\cos(45°-B) + \cos(45°+A)\sin(45°-B) = \cos(A-B).$$

6. Prove (i) $\sin 105° + \cos 105° = \cos 45°$;   [Write $105° = 60° + 45°$.]

   (ii) $\sin 75° - \sin 15° = \cos 105° + \cos 15°$.

Prove the identities in questions 7 to 13.

7. $\dfrac{\sin 2A}{1 + \cos 2A} = \tan A$.

8. $\dfrac{\sin 2A}{1 - \cos 2A} = \cot A$.

9. $\tan A + \cot A = 2 \operatorname{cosec} 2A$.

10. $\tan A - \cot A = -2 \cot 2A$.

11. $\tan\left(\dfrac{\pi}{4} + \theta\right) - \tan\left(\dfrac{\pi}{4} - \theta\right) = 2 \tan 2\theta$.

12. $\tan^2 A - \tan^2 B = \dfrac{\sin^2 A - \sin^2 B}{\cos^2 A \cos^2 B}$.

13. $\operatorname{cosec} 2A + \cot 2A = \cot A$.

14. Solve the following equations, giving all values of x between $0°$ and $180°$.

   (i) $6 \sin^2 x + 5 \cos x = 5$.   (ii) $6 \cos^2 x - 5 \cos x + 1 = 0$.

   (iii) $3 \sec^2 x - 5 \tan x = 1$.   (iv) $\sec^2 x - \dfrac{1}{\cos x} - 6 = 0$.

15. Find all roots of the following equations which lie between $0°$ and $360°$.

   (i) $\cos 2\theta = \cos \theta$.   (ii) $\cos 2\theta + \cos \theta = 0$.

   (iii) $2 \sin 2\theta - \cos \theta = 0$.   (iv) $\tan 2\theta = \tan \theta$.

16. By expressing $\sin \theta$ and $\cos \theta$ in terms of t, where $t = \tan \tfrac{1}{2}\theta$, or by any other method, solve the following equations, giving all solutions between $0°$ and $180°$.

   (i) $\sin \theta + \cos \theta = \sqrt{2}$.   (ii) $\sqrt{3} \cos \theta + \sin \theta = \sqrt{2}$.

   (iii) $5 \sin \theta + 2 \cos \theta = 5$.   (iv) $6 \cos \theta + 8 \sin \theta = 9$.

   (v) $3 \sin \theta - 4 \cos \theta = 4$.   (vi) $\operatorname{cosec} \theta = \cot \theta + \sqrt{2}$.

17. Solve the equation $\cos(\theta + 30°) = 2\sin(\theta + 60°)$, by expanding and dividing throughout by $\cos\theta$. Give one solution.

18. Using the method of Ex. 17, find roots of the following equations lying between $0°$ and $180°$

(i) $\sin(\theta - 15°) = \cos(\theta + 15°)$;

(ii) $2\cos(\theta + 30°) = \sin(\theta - 60°)$;

(iii) $2\sin(\theta + 45°) + \cos(\theta + 45°) = 0$.

19. Show that

(i) $\sec 2\theta - \tan 2\theta = \dfrac{\cos\theta - \sin\theta}{\cos\theta + \sin\theta}$;

(ii) $\dfrac{1 + \sin 2\theta - \cos 2\theta}{1 + \sin 2\theta + \cos 2\theta} = \tan\theta$;

and from (ii) obtain the value of $\tan 22\tfrac{1}{2}°$.   (L.U.)

20. Prove that $\cos\phi - \cos\theta = 2\sin\dfrac{\theta - \phi}{2}\sin\dfrac{\phi + \theta}{2}$.

What is the least value of $2\sin\theta\sin(\theta + \alpha)$ if α is constant and θ varies?

Solve the equation $\cos\theta - \cos(\theta + 60°) = 0\cdot4$, giving the values of θ which lie between $0°$ and $360°$.   (L.U.)

Prove the identities, numbers 21 to 27.

21. $\dfrac{\sin 3A + \sin A}{\cos 3A + \cos A} = \tan 2A$.

22. $\dfrac{\sin 3A - \sin A}{\cos 3A - \cos A} = -\cot 2A$.

23. $\dfrac{\sin 7\theta - \sin 5\theta}{\cos 7\theta + \cos 5\theta} = \tan\theta$.

24. $\dfrac{\sin A + \sin 2A}{\cos A - \cos 2A} = \cot\dfrac{A}{2}$.

25. $\dfrac{\cos 2B + \cos 2A}{\cos 2B - \cos 2A} = \cot(A + B)\cot(A - B)$.

26. $\dfrac{\cos 2B - \cos 2A}{\sin 2B + \sin 2A} = \tan(A - B)$.

27. $\dfrac{\sin A + 2\sin 3A + \sin 5A}{\sin 3A + 2\sin 5A + \sin 7A} = \dfrac{\sin 3A}{\sin 5A}$.

28. If $A + B + C = 180°$, prove

(i) $\sin 2A + \sin 2B = 2 \cos (A - B) \sin C$;

(ii) $\sin 2A + \sin 2B + \sin 2C = 2 \sin C \{\cos (A - B) + \cos C\}$
$$= 4 \sin A \sin B \sin C.$$

29. If $A + B + C = 180°$, prove

$$\sin 2A + \sin 2B - \sin 2C = 2 \sin C \{\cos (A - B) - \cos C\}$$
$$= 4 \cos A \cos B \sin C.$$

30. Solve the following equations, giving solutions between $0°$ and $180°$ in each case

(i) $\sin 3\theta + \sin \theta = \sin 2\theta$;

(ii) $\sin \theta + \sin 7\theta = \sin 4\theta$;

(iii) $\cos \theta + \cos 3\theta + \cos 2\theta = 0$.

CHAPTER 15

CIRCULAR MEASURE. ARC LENGTH AND AREA OF A SECTOR. SMALL ANGLES

Up to the present we have measured our angles in terms of degrees and minutes, where the definition of one degree is

one degree is $\frac{1}{360}$th of the total angle round a point.

It is convenient at this stage to introduce a new unit for fixing the magnitude of an angle. This unit is particularly useful when we come to differentiate the trigonometrical functions, as will be seen in a later chapter, and also in problems connected with a circle or number of circles.

Circular Measure

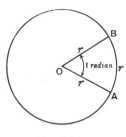

FIG. 104

Consider any circle centre **O**, radius r. Take an arc **AB** of this circle of length equal to the radius r. Then the angle subtended by this arc at the centre **O** is defined as *one unit of circular measure*.

The unit of circular measure is called *the radian*. Hence, *one radian* is the angle subtended at the centre of a circle by an arc of length equal to the radius.

Connection between radians and degrees

As the total circumference of a circle $= 2\pi \times$ radius, then the total angle subtended by the entire circumference at the centre $= 2\pi$ radians.

i.e. the total angle round a point is 2π radians.

Hence 2π radians $= 360°$,

or π radians $= 180°$.

$$\therefore \text{ radian} = \frac{180°}{\pi} = \frac{180°}{3 \cdot 142}$$

$$= 57° \ 18' \text{ (approx.).}$$

Ex. 1. *Express in degrees* $\dfrac{\pi}{2}, \dfrac{\pi}{3}, \dfrac{2\pi}{5}, \dfrac{\pi}{4}$, $\cdot 2\pi$ *radians.*

As π radians $= 180°$,

we have

$$\dfrac{\pi}{2} \text{ radians} = 90°; \qquad \dfrac{\pi}{3} \text{ radians} = 60°;$$

$$\dfrac{2\pi}{5} \text{ radians} = 72°; \qquad \dfrac{\pi}{4} \text{ radians} = 45°;$$

and $\qquad \cdot 2\pi \text{ radians} = \cdot 2 \times 180° = 36°.$

Ex. 2. *Express in radians:* (*a*) 35°; (*b*) 112° 30′.

(*a*) $\qquad 1° = \dfrac{\pi}{180}$ radians,

$$\therefore \; 35° = \dfrac{35\pi}{180} = \dfrac{7\pi}{36} \text{ radians.}$$

(*b*) $\qquad 112° 30′ = 112\tfrac{1}{2} \times \dfrac{\pi}{180}$ radians,

$$= \dfrac{225}{2} \times \dfrac{\pi}{180} = \dfrac{5\pi}{8} \text{ radians.}$$

Ex. 3. *Find, using tables, the sine and cosine of the angle* $\cdot 24$ *radians.*

As the tables do not give us sines and cosines of angles measured in radians we must first change $\cdot 24$ radians into degrees and minutes.

We have, $\qquad \cdot 24 \text{ radians} = \cdot 24 \times \dfrac{180°}{\pi},$

$$= \dfrac{\cdot 24 \times 180°}{3 \cdot 142} = 13 \cdot 75°,$$

$$= 13° 45′.$$

$\therefore \;\; \sin (\cdot 24 \text{ radians}) = \sin 13° 45′ = \cdot 2376$ ⎱
and $\quad \cos (\cdot 24 \text{ radians}) = \cos 13° 45′ = \cdot 9713.$ ⎰

Arc Length and area of a sector of a circle

Let the angle subtended by the arc **AB** at the centre of the circle **O** be θ radians.

Let the length of the arc **AB** be s.

Let the area of the sector **AOB** be **A**.

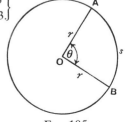

Fig. 105

Then

$$\frac{\text{length of arc } \mathbf{AB}}{\text{whole circumference}} = \frac{\text{angle subtended by } \mathbf{AB} \text{ at } \mathbf{O}}{\text{angle subtended by the circumference at } \mathbf{O}},$$

$$\therefore \quad \frac{s}{2\pi r} = \frac{\theta}{2\pi}.$$

i.e. $s = \mathbf{r\theta}.$

Also,
$$\frac{\text{area of the sector } \mathbf{AOB}}{\text{area of the circle}} = \frac{\theta}{2\pi},$$

$$\therefore \quad \frac{\mathbf{A}}{\pi r^2} = \frac{\theta}{2\pi},$$

$$\therefore \quad \mathbf{A} = \tfrac{1}{2}\mathbf{r^2\theta}.$$

Area of a segment of a circle

Consider the two segments **ACB** and **ADB** formed by the chord **AB**.

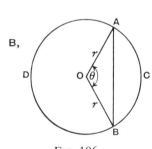

FIG. 106

Area of the segment **ACB**

$= $ area of sector **AOB** $-$ area of $\triangle$ **AOB**,

$= \tfrac{1}{2}r^2\theta - \tfrac{1}{2}r^2 \sin \theta,$

$= \tfrac{1}{2}r^2(\theta - \sin \theta).$

Hence,

area of the segment **ADB**

$= \pi r^2 - \tfrac{1}{2}r^2(\theta - \sin \theta),$

$= \tfrac{1}{2}r^2(\overline{2\pi - \theta} + \sin \theta).$

Ex. 4. Find the arc length and area of the sector of a circle, radius 6 cm, which contains an angle of 36°.

Arc length $s = r\theta$ where θ is measured in *radians*.

We have
$$36° = 36 \times \frac{\pi}{180} \text{ radians} = \frac{\pi}{5} \text{ radians}.$$

$\therefore$ Arc length $s = 6 \cdot \dfrac{\pi}{5} = \dfrac{6\pi}{5} \text{ cm}.$

Area of sector $= \tfrac{1}{2}r^2\theta = \tfrac{1}{2} \cdot 6^2 \cdot \dfrac{\pi}{5} \text{ cm}^2,$

$$= \frac{18\pi}{5} \text{ cm}^2.$$

Ex. 5. *A cylindrical log, radius of circular section* 24 *cm, length* 10 *m, is floating in water with its axis horizontal and its highest point* 4 *cm above the level of the water. Find the volume not immersed.*

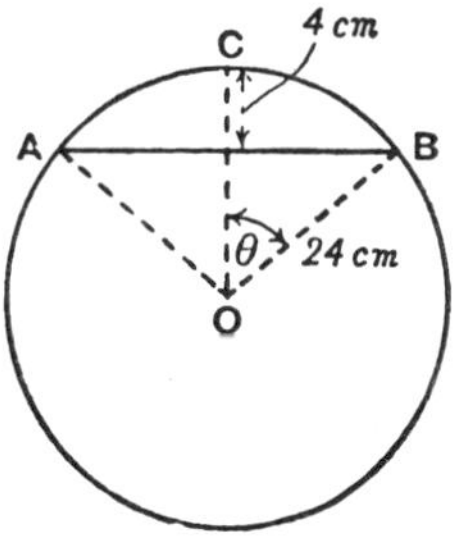

From the figure, $\cos \theta = \frac{20}{24} = \cdot 8333$.

$$\therefore \ \theta = 33° 34'.$$

$$\therefore \ \widehat{\text{AOB}} = 2\theta = 67° 8'.$$

$$\therefore \ \widehat{\text{AOB}} = 67\frac{2}{15} \times \frac{\pi}{180} \text{ radians},$$

$$= 1 \cdot 172 \text{ radians}.$$

$$\therefore \ \text{Area of sector } \text{AOB} = \tfrac{1}{2} r^2 \theta = \tfrac{1}{2} . (0 \cdot 24)^2 . 1 \cdot 172,$$

$$= 0 \cdot 03375 \text{ m}^2$$

and   area of $\triangle$ AOB $= \tfrac{1}{2} r^2 \sin \theta = \tfrac{1}{2} . (0 \cdot 24)^2 . \sin 67° 8'$,

$$= 0 \cdot 02654 \text{ m}^2.$$

$\therefore$ Area of segment **ACB** $= 0 \cdot 03375 - 0 \cdot 02654,$

$$= 0 \cdot 0072 \text{ m}^2. \qquad \text{(to 2 sig. figures)}$$

Hence

volume above water $= 0 \cdot 0072 \times 10 \text{ m}^3$,

$$= 0 \cdot 072 \text{ m}^3.$$

Fig. 107

Ex. 6. *A chord of a circle subtends an angle of* θ *radians at the centre of the circle. If the area of the minor segment cut off by the chord is one-quarter the area of the circle, prove that* $\sin \theta = \theta - \dfrac{\pi}{2}$. *Solve this equation graphically.*

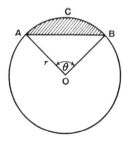

Fig. 108

Graphs of $y = \sin \theta$ and $y = \theta - \dfrac{\pi}{2}$.

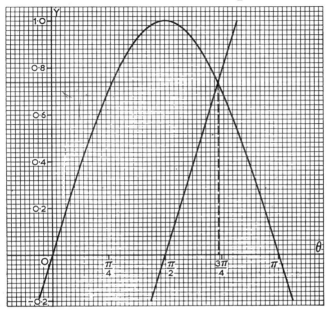

FIG. 109

Area of segment **ACB** $= \frac{1}{2}r^2\{\theta - \sin \theta\}$.   (Fig. 108)

$$\therefore \; \tfrac{1}{2}r^2\{\theta - \sin \theta\} = \tfrac{1}{4} \text{ area of circle} = \frac{\pi r^2}{4}.$$

$$\therefore \; \theta - \sin \theta = \frac{\pi}{2},$$

$$\text{i.e.} \quad \sin \theta = \theta - \frac{\pi}{2}.$$

To solve this equation we plot the graphs $y = \sin \theta$ and $y = \theta - \dfrac{\pi}{2}$, and find the points of intersection.

By considering Fig. 108, it is obvious that θ is less than π radians, and so we will plot the graphs for the range $\theta = 0$ to $\theta = \pi$.

$y = \sin \theta$.

θ	0	$\dfrac{\pi}{6}$	$\dfrac{\pi}{4}$	$\dfrac{\pi}{3}$	$\dfrac{\pi}{2}$	$\dfrac{2\pi}{3}$	$\dfrac{3\pi}{4}$	$\dfrac{5\pi}{6}$	π
y	0	·5	·7071	·866	1	·866	·7071	·5	0

$$y = \theta - \frac{\pi}{2}.$$

Note. This is an equation of the 1st degree, and so the graph is a straight line.

θ	$\dfrac{3\pi}{4}$	$\dfrac{\pi}{2}$
y	$\dfrac{\pi}{4}$	0

From the graphs, $\sin \theta = \theta - \dfrac{\pi}{2}$

when $\theta = \cdot737\pi$ radians.

(A) EXAMPLES 15a

[Take $\pi = 3\cdot142$]

1. Express the following angles in radians 18°, 72°, 162°, 300°, 270°, 85°, 63° 30′, 22° 12′, 191° 14′.

2. Express the following angles in degrees and minutes

$$\frac{\pi}{3}, \frac{2\pi}{9}, \frac{5\pi}{6}, \frac{7\pi}{3}, 3\pi, \frac{3\pi}{8}, \frac{\pi}{8}, \frac{\pi}{16}, \cdot4, \cdot65, 1\cdot32 \text{ radians.}$$

3. Use tables to find the values of the sines of the following angles measured in radians $\dfrac{5\pi}{6}, \dfrac{\pi}{8}, \dfrac{2\pi}{3}, \dfrac{7\pi}{9}, 2\pi, \cdot4, 1\cdot436.$

4. Find the arc lengths of the following sectors.

(i)	Radius	6 cm	Angle at centre	30°.
(ii)	,,	9 cm	,,	,,	120°.
(iii)	,,	4 cm	,,	,,	·35 radians.
(iv)	,,	15 cm	,,	,,	36°.
(v)	,,	10 cm	,,	,,	162°.

5. Find the areas of the sectors in Question 4.

6. Find the areas of the segments into which a circle, radius 8 cm, is divided by a chord of length (a) 8 cm, (b) 10 cm, (c) 12 cm.

7. Find the angle, in degrees, subtended by an arc length 10 cm at the centre of a circle, radius 8 cm.

8. The area of a sector is $5\cdot024 \text{ cm}^2$ and its angle is 36°. Find the radius.

9. Find the area of a sector of a circle, radius 5 cm, bounded by an arc of length 8 cm.

10. Express the following angular velocities in radians per second.

(i) 1 revolution per second.
(ii) 5 revolutions per minute.
(iii) 1 revolution in 5 seconds.

11. A wheel turns through 3 radians each second. How many revolutions is it making per minute?

12. Use tables to find the value of A cos (ωt) where the angle (ωt) is in radians, when $A = 36$, $\omega = 3\pi$, $t = \cdot 1$.

13. A string of length 12 cm is wrapped on the circumference of a circle radius 16 cm. How far are the ends apart?

14. The radius of a cycle wheel is 14 cm. How many radians does the wheel turn through in travelling 20 cm?

15. Find the area enclosed by two concentric circles of radii 10 cm and 6 cm, and two radii inclined at an angle of 50°.

16. What is the ratio of the radii of two circles at the centres of which two arcs of the same length subtend angles of 60° and 75°?

17. The radius of a circle is 3 m. Find the length of an arc of this circle if the length of the chord of the arc is 3 m.

Small Angles

To prove that when θ is acute, sin $\theta < \theta <$ tan θ, where θ is measured in radians.

In Fig. 110, θ is an acute angle, *i.e.* $0 < \theta < \frac{1}{2}\pi$.
AT is the tangent at A to the circle, centre O, radius r.

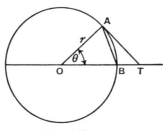

Fig. 110

Then in the $\triangle$ OAT, we have

$$AT = OA \tan \theta = r \tan \theta.$$

$$\therefore \text{ Area of } \triangle OAT = \frac{1}{2}r^2 \tan \theta.$$

Also,   area of $\triangle$ **AOB** $= \frac{1}{2}r^2 \sin\theta$,

and   area of sector **AOB** $= \frac{1}{2}r^2\theta$.

From the diagram, we see that, so long as θ is acute,

Area $\triangle$ **AOB** $<$ area of sector **AOB** $<$ area of $\triangle$ **OAT**.

i.e. $\frac{1}{2}r^2 \sin\theta < \frac{1}{2}r^2\theta < \frac{1}{2}r^2 \tan\theta$.

Hence $\sin\theta < \theta < \tan\theta$.

As θ is acute and positive, $\sin\theta$ is positive, and hence we can divide each term of the inequality by $\sin\theta$,

$$\therefore\ 1 < \frac{\theta}{\sin\theta} < \frac{\tan\theta}{\sin\theta},$$

$$\therefore\ 1 < \frac{\theta}{\sin\theta} < \frac{1}{\cos\theta}.$$

Now let θ decrease and approach zero. Then the inequality is still true, that is, the value of $\dfrac{\theta}{\sin\theta}$ lies between 1 and $\dfrac{1}{\cos\theta}$. But as $\theta \to 0$, $\dfrac{1}{\cos\theta} \to 1$, and thus $\dfrac{\theta}{\sin\theta}$ must also approach nearer and nearer to the value 1.

I.e.   as $\theta \to 0, \dfrac{\theta}{\sin\theta} \to 1$,

or, $\lim\limits_{\theta\to0} \dfrac{\theta}{\sin\theta} = 1$.

Similarly, by dividing throughout the inequality

$$\sin\theta < \theta < \tan\theta,$$

by $\tan\theta$, we obtain the result

$$\cos\theta < \frac{\theta}{\tan\theta} < 1,$$

and hence as $\theta \to 0$, $\dfrac{\theta}{\tan\theta} \to 1$,

i.e. $\lim\limits_{\theta\to0} \dfrac{\theta}{\tan\theta} = 1$.

In other words, if θ *is small*, $\dfrac{\theta}{\sin\theta}$ and $\dfrac{\theta}{\tan\theta}$ are each approximately equal to unity, the approximation getting closer as θ gets smaller.

∴ **When θ is small, we have the approximate result,**

$$\theta = \sin \theta = \tan \theta.$$

The truth of this last result is readily verified by using tables. *e.g.* take

$\theta = 2° = \dfrac{2\pi}{180}$ radians $= \dfrac{\pi}{90} = \cdot0349$ to 4 dec. places.

Referring to tables, we find sin 2° = ·0349

and   tan 2° = ·0349.

Showing that, to four places of decimals, the result

$$\theta = \sin \theta = \tan \theta$$

is true for $\theta = 2° = \cdot0349$ radians.

Ex. 7. Without using tables, find the sine of 18'.

As the angle is very small, the sine is equal to the angle when expressed in radians.

Hence $\sin 18' = \dfrac{18}{60} \times \dfrac{\pi}{180} = \dfrac{\pi}{600} = \dfrac{3\cdot1417}{600}$,

$$= \cdot005236.$$

i.e.   sin 18' = ·005236 approximately.

Ex. 8. A tower, height 100 m, stands on a horizontal plane. When

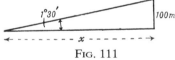

viewed from a point on the plane the angle of elevation of the top of the tower is 1° 30'. Find the distance of the foot of the tower from the point of observation.

Fig. 111

We have $\dfrac{100}{x} = \tan 1° 30'$, where x m is the distance of the foot of

tower from the point of observation.

But   tan 1° 30' = value of the angle when expressed in radians,

$$= \frac{3}{2} \times \frac{\pi}{180} = \frac{\pi}{120}.$$

Hence $x = \dfrac{100 \times 120}{\pi} = 3820$ m.

Hence distance of foot of tower from point of observation

$$= 3820 \text{ m approximately.}$$

Ex. 9. ABC is a triangle right-angled at C, AC = 1, BC = 7. Prove, without using tables, that angle ABC lies between 0·141 and 0·143 radians.

By Pythagoras' theorem,

$$AB = \sqrt{50}.$$

$$\therefore \sin \widehat{ABC} = \frac{1}{\sqrt{50}} = \frac{\sqrt{50}}{50} = \frac{7 \cdot 071}{50},$$

$$= 0 \cdot 141 \text{ (to 3 decimal places)}$$

and

$$\tan \widehat{ABC} = \frac{1}{7} = \cdot 143.$$

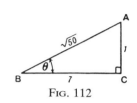

FIG. 112

But as **ABC** is an acute angle,

$$\sin \theta < \theta < \tan \theta.$$

Hence θ, *i.e.* $\widehat{ABC}$, lies between $0 \cdot 141$ and $0 \cdot 143$ radians.

Ex. 10. *A weather-vane stands on the top of a spire 45 m high. A man 200 m from the foot of the spire and on the same level, observes that the difference of the angles of elevation of the bottom and top of the vane is 9'. Find the height of the vane without using trigonometrical tables.*

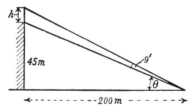

FIG. 113

Let h m be the height of the vane.

We have,

$$h + 45 = 200 \tan (\theta + 9'),$$

$$= 200 \cdot \left\{ \frac{\tan \theta + \tan 9'}{1 - \tan \theta \tan 9'} \right\}.$$

But

$$\tan 9' = \frac{\overset{3}{\cancel{9}}}{\underset{20}{\cancel{60}}} \times \frac{\pi}{\underset{60}{\cancel{180}}} = \frac{\pi}{1200}.$$

$$\therefore h + 45 = \frac{200 \cdot \left\{ \dfrac{45}{200} + \dfrac{\pi}{1200} \right\}}{1 - \dfrac{45}{200} \cdot \dfrac{\pi}{1200}}, \text{ as } \tan \theta = \frac{45}{200},$$

$$h + 45 = 200\left\{\frac{9}{40} + \frac{\pi}{1200}\right\}\left(1 - \frac{3\pi}{16000}\right)^{-1},$$

$$\simeq 200\left\{\frac{9}{40} + \frac{\pi}{1200}\right\}\left(1 + \frac{3\pi}{16000}\right),$$

$$\simeq 45 + \tfrac{1}{6}\pi + \frac{27\pi}{3200} = 45 + 0 \cdot 524 + 0 \cdot 028 = 45 \cdot 55 \text{ to 4 s.f.}$$

$$\therefore h \simeq 0 \cdot 55 \text{ to 2 s.f.}$$

$$\therefore \text{ Height of vane} = 0 \cdot 55 \text{ m to 2 s.f.}$$

(A) EXAMPLES 15b

1. Without using tables, find the approximate values of θ in degrees and minutes in the following cases

  (i) $\sin \theta = \cdot 005$; (ii) $\tan \theta = \cdot 0106$; (iii) $\sin \theta = \cdot 1096$;

  (iv) $\sin 2\theta = \cdot 0034$; (v) $\tan 3\theta = \cdot 003$; (vi) $\sin \dfrac{\theta}{2} = \cdot 0056$.

2. Find, without using tables, the sines of the following angles $30'$, $1° 5'$, $10'$, $2° 11'$, $\cdot 0035$ radians.

3. A hill, 15 km away, has an angle of elevation of $35'$. Find the approximate height of the hill in metres.

4. At what distance does a man, height $1 \cdot 80$ m, subtend an angle of $15'$?

5. Find approximately, in minutes, the inclination to the horizon of an incline which rises $1 \cdot 50$ m in 250 m.

6. A church spire, height 33 m, subtends an angle of $9'$ at the eye. Find its approximate distance.

7. Find approximately the distance at which a sphere, 7 cm in diameter, will subtend an angle of $33\frac{1'}{2}$.

8. A comet, ten million kilometres away, subtends at the earth an angle of 16 seconds. Find its approximate diameter.

9. In the triangle ABC, $\hat{B} = 90°$, AB $= 4$ cm, BC $= 3$ cm. Prove that the value of angle A lies between $\frac{3}{5}$ and $\frac{3}{4}$ radians.

(B) EXAMPLES 15c

1. A chord of a circle, radius 20 cm, divides the circumference in the ratio 1:3. Find the ratio of the areas of the segments into which the circle is divided by the chord.

2. Tangents TA, TB are drawn from a point T to a circle, centre O, radius 10 cm. If OT $= 15$ cm, find (a) the length of the minor arc AB; (b) the area of the minor segment AB.

3. Find the ratio of the areas of the segments into which a circle, radius a, is divided by a chord of length a.

4. A cylindrical log, length 1·44 m, radius 8 cm, floats with its axis horizontal with its highest points 3 cm above the water level. Find the volume of the log immersed.

5. A log of rectangular section, 8 cm by 6 cm, is to be cut from a cylindrical log, length 2·40 m, radius 5 cm. Find the volume of wood wasted.

6. Three circular discs, each of radius 6 cm, touch each other. Find the area enclosed between them.

7. If α is a small angle, expressed in radians, prove the approximate results,

$$\sin (x \pm \alpha) = \sin x \pm \alpha \cos x.$$

Hence obtain, without tables, the approximate values of (i) $\sin 30° \, 30'$; (ii) $\sin 60° \, 15'$; (iii) $\sin 120° \, 11'$.

8. If α is a small angle, expressed in radians, show that

$$\tan (x + \alpha) = \frac{\tan x + \alpha}{1 - \alpha \tan x}.$$

Deduce the approx. values of (i) $\tan 45° \, 10'$; (ii) $\cot 135° \, 32'$.

9. A weather-vane stands on the top of a spire of height 75 m. When viewed from a point 250 m from the foot of the tower the weather-vane subtends an angle of $6'$. Find the height of the weather-vane to the nearest tenth of a metre.

10. A ship, length 800 m, is sailing directly towards a light-house of height 250 m. If the ship subtends an angle of $7'$ at the top of the lighthouse, find its approximate distance from the lighthouse.

11. Using the same axes, sketch roughly the graphs of $\sin \theta$, $\sin 2\theta$, $\sin \frac{\theta}{2}$, where θ is measured in radians, for values of θ between $-\pi$ and $+\pi$.

12. Plot, on the same diagram, the graphs of $\cos \theta$, $\cos 2\theta$, $\cos \frac{\theta}{2}$ for values of θ between 0 and 2π radians.

13. Plot the graph $y = 2 \tan \frac{\theta}{2}$ for values of θ between -2π and $+2\pi$ radians.

14. Plot on the same axes the graphs $y = \sin x$ and $y = 2x - \frac{1}{2}$ for values of x between 0 and $\frac{\pi}{2}$. Hence find a value of x, in radians, which satisfies the equation $2 \sin x = 4x - 1$.

15. Plot on the same diagram the graphs of $y = \sin \theta$ and $y = 2 \cos 2\theta$ for values of θ between 0 and $\frac{\pi}{2}$ radians. Hence find an approximate solution of the equation

$$\sin \theta = 2 \cos 2\theta.$$

16. Find, graphically, a value of θ between 0 and $\frac{\pi}{2}$ radians, which satisfies the equation

$$\tan \theta = 2\theta + 1.$$

17. Plot the graph $y = 2 \sin x° - \cos x°$ for values of x between 0° and 90°. Hence find a value of x for which $2 \sin x° - \cos x° = 0.5$.

18. Draw the graph of $y = 3 \cos x° + 4 \sin x°$ from $x = 0°$ to $x = 360°$. Find, from the graph, the values of x within this range for which $3 \cos x° + 4 \sin x° = 0.8$.

19. Draw the graph of $y = \tan x°$ from $x = -90°$ to $x = 90°$. Hence find an approximate solution of the equation

$$x + 50 \tan x° = 150.$$

20. A chord **AB**, subtending an angle θ radians at the centre of a circle, divides the circle into segments whose areas are in the ratio $1:2$. Prove that $\sin \theta = \theta - \frac{2\pi}{3}$.

By drawing the graphs $y = \sin \theta$ and $y = \theta - \frac{2\pi}{3}$ for values of θ between $\frac{\pi}{2}$ and π, obtain an approximate value for θ in degrees and minutes.

REVISION PAPERS III

PAPER A (1)

1. If α, β are the roots of the equation $3x^2 - 2x - 2 = 0$, find the value of $\alpha^3 + \beta^3$.

2. (i) Solve the equations $xy = 16$, $y - 5x = 2$.

(ii) If $x \equiv A(1+x) + B(1-x)$, find the values of **A** and **B**.

3. (i) Find the sum of all the odd numbers between 30 and 90.

(ii) A geometrical progression has 5 for its third term and 28 for its eighth term. Find the common ratio correct to two decimal places.

4. Use the Bionomial Theorem to obtain the expansion of $\dfrac{1}{\sqrt[3]{1+2x}}$ up to, and including, the term in x^3.

5. (i) Write down the values of sin 325°, tan 220°, cos 160°.

(ii) Evaluate $2 \cos 2\theta$ where $\theta = \cdot 4$ radians.

PAPER A (2)

1. Prove that the expression $x^2 + 3x + 3$ is positive for all real values of x. What is the minimum value of the expression?

2. Solve the equations $3x^2 + xy - 2y^2 = -7$, $x + 2y = 3$.

3. (i) Find the fourth term in the expansion of $(\sqrt{2}+3)^7$ by the Bionomial Theorem.

(ii) Find the sum to 58 terms of the geometrical progression, $1+3+9+27+\ldots$ [Do not work out the answer.]

4. An angle **A** is known to be between 270° and 360°, and $\sin^2 \mathbf{A} = \frac{9}{25}$. Find sin **A**, cos **A**, tan **A**.

5. Three equal circles, radius 4 cm, are placed in the same plane so that each touches the other two. Find the area contained between them.

PAPER A (3)

1. Find the equation whose roots are those of the equation $x^2 + 5x - 11 = 0$ each increased by 4.

2. Given $x = \dfrac{3(\lambda - 1)}{\lambda + 1}$, $y = \dfrac{8}{\lambda + 1}$, find λ (i) in terms of x; (ii) in terms of y and hence find the equation connecting x and y.

3. Show that the sum of the first 20 terms of the geometrical progression, $7 + 2 \cdot 1 + 0 \cdot 63 + \ldots$ differs from 10 by an amount which is less than half the twentieth term.

4. (i) Write down the first 4 terms in the expansion of $(a - b)^{20}$.

(ii) Find the coefficient of x^5 in the expansion of $(1 + 2x)^{\frac{1}{2}}$.

5. (i) Express, in degrees and minutes, the angle $\cdot 34$ radians. ($\pi = 3 \cdot 142$.)

(ii) If $\sin 2\theta = \cdot 6670$, find all values of θ between $0°$ and $180°$.

PAPER A (4)

1. If the roots of the quadratic equation $x^2 + mx + n = 0$ are in the ratio $a : b$, find the relation between m, n, a and b.

2. Solve the simultaneous equations $xy = 6$, $xy + x + y = 1$.

3. Show that the sum of the first n odd numbers is a perfect square. Show also that $57^2 - 13^2$ is the sum of certain consecutive odd numbers, and find them.

4. Expand $(2 - x)^{\frac{3}{2}}$ up to, and including, the term in x^3. For what values of x is the expansion valid?

5. Find the area of the smaller segment cut off by a chord of length 1 m from a circle of radius $0 \cdot 75$ m.

PAPER A (5)

1. Find, algebraically, the maximum value of the expression $2 - x - x^2$.

2. Solve the equations $\dfrac{1}{x} + \dfrac{1}{y} = \dfrac{5}{2}$, $3x - 2y = 5$.

3. (i) An arithmetical progression has 13 terms whose sum is 143. The third term is 5. Find the first term.

(ii) Write down the ninth term and the sum to 9 terms of the series, $1\frac{1}{4}, 1, \frac{4}{5}, \frac{16}{25}, \ldots$

4. Use the Binomial Theorem to evaluate, to four significant figures

$$\text{(i) } (1 \cdot 02)^4; \qquad \text{(ii) } \frac{1}{\sqrt{\cdot 998}}.$$

5. (i) Use tables to find the value of $\cot 210°$.

(ii) Without using tables, find the angle whose sine is $\cdot 035$.

(iii) If $2 \sin \theta = \cos \theta$, θ in radians, find the value of θ lying between 0 and $\dfrac{\pi}{2}$.

PAPERS A (6)

1. If the roots of the equation $3x^2 - 2x - 5 = 0$ are α and β, find the equation whose roots are $\dfrac{\alpha^2}{\beta}$ and $\dfrac{\beta^2}{\alpha}$.

2. (i) Express x^2 in the form $A(x+1)^2 + B(x+1) + C$.

(ii) Eliminate t from the equations $x = 3 + t^2$, $y = 1 + 2t$.

3. For all values of r the sum of the first r terms of a progression is $3r^2 - 2r$. Find the first three terms.

4. (i) Find the coefficient of x^4 in the expansion of $(1 - 4x)^{\frac{1}{2}}$.

(ii) Find the middle term of the expansion of $(3x - 5y)^8$.

5. Given that $\sin 20° = 0.3420$, $\cos 20° = 0.9397$, find, without using the tables, the values of

(i) $\sin 160° + \cos 160°$;   (ii) $\sin 250° + \cos 250°$.

PAPER B (1)

1. If α and β are the roots of the equation $x^2 - 6x + k = 0$, and if $\dfrac{1-\alpha}{\alpha}$ and $\dfrac{1-\beta}{\beta}$ are the roots of the equation $8x^2 + 10x + l = 0$, find k and l.

2. (i) If x^3 is expressed in the form

$$a + b(x-1) + c(x-1)(x-2) + d(x-1)(x-2)(x-3),$$

find the values of a, b, c, d.

(ii) Show that the roots of the equation $x^2 + (1-k)x - k = 0$ are real for all real values of k.

3. State the first four terms of $(a + b)^n$.

If x is so small that its fourth and higher powers may be neglected, show that $\sqrt[4]{1+x} + \sqrt[4]{1-x} = a - bx^2$, and find the numbers a and b.

Hence by putting $x = \frac{1}{16}$ show that the sum of the fourth roots of 17 and of 15 is 3·9985 approximately.   (J.M.B.)

4. Draw the graph of $y = \tan x°$ from $x = 0°$ to $x = 180°$. Hence find, approximately, the values of x within this range for which

(a) $5 \sin x° = 3 \cos x°$;

(b) $x + 10 \tan x° = 50$.   (J.M.B.)

5. (i) Find the circular measure of the angle through which the second-hand of a watch turns in a second of time.

(ii) The angle subtended at the earth by the diameter of the moon varies between $29'\,34''$ and $33'\,32''$ owing to the alteration of the distance between the moon and the earth. When the angle is $31'\,7''$ the distance of the moon is 382,100 km. Find, to the nearest thousand kilometres, the least and greatest distances of the moon. (J.M.B.)

PAPER B (2)

1. (i) Find the sign of $2x - 5 - 4x^2$ for real values of x.

(ii) Given that $x^2 - 3x + 5 \equiv (x-a)^2 + b$, find the numerical values of a and b. Deduce the minimum value of $x^2 - 3x + 5$.

2. The first and last terms of an arithmetical progression are -3 and 25, and the sum of all the terms is 1837. Find the number of terms and the common difference. Find also the middle term of the progression. (J.M.B.)

3. Use the same axes draw the graphs of $y = x^2(x-2)$ and $y = 3 - x$ from $x = -1$ to $x = 2\cdot5$, plotting at half-unit intervals for x.

Using the graphs to find the real root of the equation $x^3 - 2x^2 + x - 3 = 0$ correct to one decimal place.

4. (i) Expand $(1-x)^{-\frac{1}{2}}$ in ascending powers of x as far as the term in x^3. For what values of x is the expansion valid?

(ii) Use the Binomial Theorem to calculate correct to four decimal places the value of $\left(\dfrac{1+x}{1-x}\right)^{\frac{1}{2}} + (4-x)^{-\frac{1}{2}}$ when $x = 0\cdot01$. (J.M.B.)

5. Explain how the length of the projection of a finite straight line on any other straight line is calculated.

A rod **AB** is suspended from a point **O** by two strings **OA** and **OB** fastened to its ends, each string making an angle of $32°$ with the vertical. If **OA** $= 13$ cm and **OB** $= 8$ cm, calculate the length of **AB** and the angle **AB** makes with the horizontal. (J.M.B.)

PAPER B (3)

1. If $\dfrac{x^2 + x}{x^4 - 16} \equiv \dfrac{A}{x-2} + \dfrac{B}{x+2} + \dfrac{Cx + D}{x^2 + 4}$, find the values of **A, B, C, D**.

2. Given that one root of the equation $x^2 + bx + c = 0$ is three times the other root, show that $3b^2 = 16c$.

If one root of the equation $x^2 + (3k+2)x + (k^2 - 2k - 5) = 0$ is three times the other root, find the possible values of k. (J.M.B.)

3. (i) Find the sum of the series $1 + x + x^2 + x^3 + \ldots$ to n terms. Under what circumstances can the series be summed when the number of terms is infinite, and what is then its sum?

(ii) Given the series $1 + 2x + 3x^2 + 4x^3 + \dots$

    (a) Find the sum of the first n terms when $x = 1$,

    (b) find, by multiplying by $1 - x$, the sum of the first n terms when x is not equal to unity.   (J.M.B.)

4. Define the sine of an angle so as to include angles of any magnitude, and prove from a figure that $\sin(270° - A) = -\cos A$, when A is an acute angle. Obtain the values of A, between $0°$ and $360°$, which satisfy the equation

$$8 \cos^2 A = 5 + 2 \sin A. \qquad \text{(J.M.B.)}$$

5. A circular cylinder of radius 12 cm floats in water with its axis horizontal. The highest point of a circular section is 6 cm above the water level. What fraction

    (a) of the volume of the cylinder is immersed?

    (b) of the curved surface area of the cylinder is immersed?

PAPER B (4)

1. Find the sum and difference of the roots of the equation

$$ax^2 + bx + c = 0.$$

Show that the ratio of the roots satisfies the equation

$$ac(x + 1)^2 - b^2 x = 0.$$

2. (i) The expression $p(x - 2) + q(x - 4)$ is equal to $x + 2$ for all values of x. Find the values of p and q.

(ii) Eliminate y from the equations $3x + 2y = 5$, $3x^2 + 2y^2 = k$ and show that the resulting equation in x has no real solutions if $k < 5$.

3. Draw the graph of $\frac{1}{4}x^3 - 3x$ for values of x between -4 and $+4$. By means of the graph, find the roots of the equation $x^3 - 12x + 12 = 0$ approximately.   (J.M.B.)

4. If $\dfrac{3}{(1-x)(2-x)} \equiv \dfrac{A}{1-x} + \dfrac{B}{2+x}$, find the values of A and B. Hence obtain the expansion of the expression $\dfrac{3}{(1-x)(2+x)}$ in ascending powers of x, as far as the term in x^4.

5. A chord AB of a circle, of radius 5 cm, is 3 cm long, and the tangents to the circle at A and B meet at T. Find the area enclosed by TA, TB and the larger arc AB.   (J.M.B.)

PAPER B (5)

1. If α, β are the roots of the equation $x^2 - 5x + 7 = 0$, prove that $\alpha^3 + \beta^3 - 5(\alpha^2 + \beta^2) + 7(\alpha + \beta) = 0$, and hence, or otherwise, prove that $\alpha^3 + \beta^3 = 20$.

2. (i) If a and b are the first and last terms of an arithmetical progression of $(r + 2)$ terms, find the second and the $(r + 1)^{\text{th}}$ terms.

(ii) At the beginning of each year a man puts by £50 to accumulate at compound interest, interest at the rate of 5 per cent. being added at the end of each year. Find, to the nearest pound, the total amount of his accumulated savings at the end of the tenth year.

(J.M.B.)

3. Write down the expansion of $(a + b)^4$.

Let $a = 1 + x$, $b = -x^2$ and deduce the expansion of $(1 + x - x^2)^4$. By putting $x = 1$, obtain the sum of the coefficients in this expansion.

4. (i) Show that $\tan x = -\cot(90° + x)$, when x is an angle in the second quadrant.

(ii) **ABCD** is a rectangle having **AB** $= 10$ cm. E is a point in **BC** such that **BE** $= 6$ cm. There is a point **O** in **CD** produced such that angle **OAD** is $25°$ and angle **OEC** is $70°$. If **EC** $= x$ cm, name the line in the figure whose length in cm is $(x + 6) \tan 25°$. Calculate the length of **EC** correct to 1 place of decimals.

(J.M.B.)

5. If x is an acute angle measured in radians, show geometrically that

$$\sin x < x < \tan x.$$

A triangle **XYZ** has a right angle at **Z**; **XZ** $= 1$, **YZ** $= 7$. Without using tables, prove that angle **XYZ** lies between $0 \cdot 1414$ and $0 \cdot 1429$ radians. Show that these two angles differ by less than $0 \cdot 1$ degrees.

(J.M.B.)

PAPER B (6)

1. (i) Find the sign of $2x^2 - 13x + 15$ for real values of x.

(ii) The roots of the equation $3x^2 - 2kx + k + 4 = 0$ are α and β. If $\alpha^2 + \beta^2 = \frac{16}{9}$, find the possible values of k.

2. Draw the graph of $x^2(x - 5)$ for values of x from -2 to $+6$.

It is required to construct a circular cylinder of volume 30 cm^3, such that the radius (x) of the base and the height are together equal to 5 cm. Use your graph to determine the possible values of x. (J.M.B.)

3. Explain what is meant by "the sum to infinity" of a geometrical progression, and state under what circumstances such a sum exists.

The first term of a geometrical progression is $\frac{3}{5}$. A new series is formed by taking the square of each term. Prove that the new series is a geometrical progression and, if its sum to infinity is nine-tenths of the sum to infinity of the first progression, find the common ratio of the first progression.

(J.M.B.)

4. (i) Expand $(x + 2x^{-1})^7$ by the Binomial Theorem.

(ii) Find the coefficient of x^4 in the expansion of

$$(1+x)^4 \times (1-x)^4.$$

(iii) Show that, when x is so small that x^3 and higher powers of x are neglected, $(1+3x)^{\frac{1}{3}}(1-2x)^{-\frac{1}{2}} = 1 + 2x + \frac{3}{2}x^2$.

(J.M.B.)

5. Two circles of radii 3 m and 4 m have their centres 5 m apart. Find the area common to the two circles.

CHAPTER 16

SOLUTION OF TRIANGLES.
PROBLEMS IN THREE DIMENSIONS

The Sine Rule

Let **ABC** be any triangle.

Let **O** be the centre, and **R** the radius of the circumcircle of the triangle.

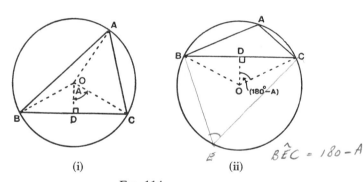

(i)   (ii) $B\hat{E}C = 180 - A$

FIG. 114

With the usual notation, $BC = a$, $CA = b$, $AB = c$.

OD, the perpendicular from **O** on **BC**, bisects **BC**, and hence $DC = \frac{1}{2}a$, and **OD** also bisects $\widehat{BOC}$.

But in Fig. (i),　　$\widehat{BOC} = 2A$ (angle at the centre).

$$\therefore \widehat{DOC} = A.$$

and in Fig. (ii),　　$\widehat{DOC} = 180° - A$.

$$= B\hat{E}C$$

207

Hence from the right-angled △ DOC,

$$\frac{DC}{OC} = \sin A \text{ in Fig. (i)},$$

and $\sin(180° - A)$ in Fig. (ii).

∴ In both figures, $\dfrac{a}{2R} = \sin A$,

or $2R = \dfrac{a}{\sin A}$.

Similarly in △ AOC, $2R = \dfrac{b}{\sin B}$

and also $2R = \dfrac{c}{\sin C}$.

Hence $\dfrac{a}{\sin A} = \dfrac{b}{\sin B} = \dfrac{c}{\sin C} = 2R$.

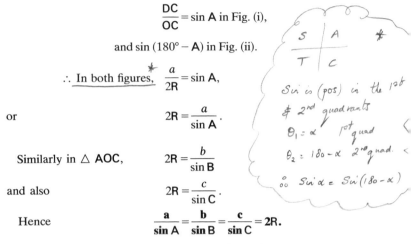

The Sine Rule can be used to solve a triangle in two cases

    (*a*) given two angles and one side;

    (*b*) given two sides and the NON-INCLUDED angle.

Solution of a triangle when two angles and one side are given

Ex. 1. Solve the △ ABC, given $\hat{A} = 102° 31'$, $\hat{B} = 51° 2'$, $c = 4·71$ *cm. Also find the radius of the circumcircle of the triangle.*

We have, $\hat{C} = 180° - (A + B)$,

$$= 180° - 153° 33',$$

$$= 26° 27'.$$

By the Sine Rule, $\dfrac{a}{\sin A} = \dfrac{b}{\sin B} = \dfrac{c}{\sin C}$.

∴ $\dfrac{a}{\sin 102° 31'} = \dfrac{4·71}{\sin 26° 27'}$

i.e. $a = \dfrac{4·71 \sin 102° 31'}{\sin 26° 27'}$

$$= \dfrac{4·71 \sin 77° 29'}{\sin 26° 27'}.$$

By logarithms,

$$a = 10·32 \text{ cm.}$$

No.	Log.	Log.
4·71	0·6730	
sin 77° 29'	$\bar{1}$·9895	
4·71 × sin 77° 29'	0·6625	0·6625
sin 26° 27'		$\bar{1}$·6488
a = 10·32		1·0137

		No.	Log.	Log.
Also		4·71	0·6730	
$\dfrac{b}{\sin 51° 2'} = \dfrac{4·71}{\sin 26° 27'}$		$\sin 51° 2'$	$\bar{1}·8907$	
$\therefore b = \dfrac{4·71 \sin 51° 2'}{\sin 26° 27'}$		$4·71 \times \sin 51° 2'$	0·5637	0·5637
$\therefore b = 8·22$ cm.		$\sin 26° 27'$		$\bar{1}·6488$
Also $2R = \dfrac{c}{\sin C}$		$b = 8·221$		0·9149
$i.e.\ \ 2R = \dfrac{4·71}{\sin 26° 27'}$		4.71	0·6730	
		$\sin 26° 27'$	$\bar{1}·6488$	
$\therefore 2R = 10·58$				
$\therefore R = 5·29$ cm.		2R	1·0242	

$$\therefore a = 10·32 \text{ cm}, \qquad b = 8·22 \text{ cm}, \qquad \hat{C} = 26° 27'.$$

Radius of circumcircle = 5·29 cm.

Solution of a triangle when two sides and the non-included angle are given

If asked to construct a triangle having been given two sides and the non-included angle, in general, we find that there are *two* possible triangles. However, in certain cases we find there is only one solution.

Ex. 2. Construct the triangle **ABC**, *in which*

$$\hat{C} = 22°,\ c = 5\ cm,\ a = 8\ cm.$$

We find there are two possible $\triangle$'s, **A₁CB** and **A₂CB** (Fig. 115).

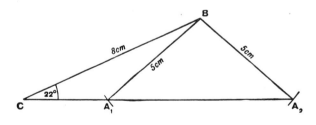

FIG. 115

Ex. 3. Construct the triangle **ABC**, *in which*

$$\hat{C} = 30°,\ c = 3\ cm,\ a = 6\ cm.$$

In this case there is only one $\triangle$ **ABC** (Fig. 116).

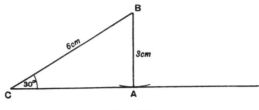

FIG. 116

Ex. 4. *Construct the triangle* **ABC** *in which*

$$\hat{C} = 33°, c = 6·5 \ cm, a = 4 \ cm.$$

There is only one △ **ABC** satisfying the given conditions (Fig. 117).

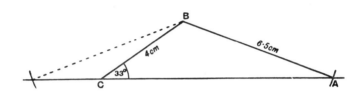

FIG. 117

[*Note.* The previous diagrams are on a reduced scale.]

Hence, in any case where we are asked to solve a triangle given two sides and the non-included angle, care will be required to determine whether there is, or is not, more than one solution. The case when two triangles satisfy the given conditions is called the *Ambiguous Case*, because the triangle is not *definitely* determined. The following examples illustrate different cases which arise.

Ex. 5. *Solve the* △ **ABC** *where*

$$\hat{B} = 21° \ 11', b = 5·63, a = 10·12.$$

By the Sine Rule, $\dfrac{a}{\sin A} = \dfrac{b}{\sin B} = \dfrac{c}{\sin C}$.

$$\therefore \ \frac{10·12}{\sin A} = \frac{5·63}{\sin 21° \ 11'}.$$

$$\therefore \ \sin A = \frac{10·12 \sin 21° \ 11'}{5·63}.$$

There are two possible values of **A** in this case, an acute angle and the supplement of this angle.

Note. The working is much simplified by first obtaining the logarithms of $\dfrac{b}{\sin B}$ and $\dfrac{\sin B}{b}$.

No.	Log.
5·63	0·7505
$\sin 21° 11'$	$\bar{1}·5579$
$\dfrac{b}{\sin B}$	1·1926
$\dfrac{\sin B}{b}$	$\bar{2}·8074$
10·12	1·0052
$\dfrac{\sin B}{b}$	$\bar{2}·8074$
$\sin A$	$\bar{1}·8126$
$\sin 118° 18'$	$\bar{1}·9447$
$\dfrac{b}{\sin B}$	1·1926
c	1·1373

We have $\sin A = 10·12 . \dfrac{\sin B}{b}$,

$\therefore$ A = 40° 31' or 139° 29'
$\therefore$ C = 118° 18' or 19° 20'

Case 1.   When $A = 40° 31', C = 118° 18'$.

By the Sine Rule, $c = \sin 118° 18' . \dfrac{b}{\sin B}$.

$\therefore c = 13·72$.

Case 2. When $A = 139° 29', C = 19° 20'$.

By the Sine Rule, $c = \sin 19° 20' . \dfrac{b}{\sin B}$.

$\therefore c = 5·16$.

No.	Log.
$\sin 19° 20'$	$\bar{1}·5199$
$\dfrac{b}{\sin B}$	1·1926
c	0·7125

Hence we have two solutions

      (i) A = 40° 31',   C = 118° 18',   c = 13·72

and   (ii) A = 139° 29',   C = 19° 20',   c = 5·16

Ex. 6. *Solve the* $\triangle$ ABC *in which*

    A = 33° 5',   a = 16·2 cm,   b = 12·1 cm.

By the Sine Rule, $\dfrac{a}{\sin A} = \dfrac{b}{\sin B} = \dfrac{c}{\sin C}$.

First obtain the logarithms of $\dfrac{a}{\sin A}$ and $\dfrac{\sin A}{a}$.

Then $\sin B = b \cdot \dfrac{\sin A}{a} - 12 \cdot 1 \cdot \dfrac{\sin A}{a}$.

$\therefore$ B $= 24° 4'$ or $155° 56'$.

Now in this case, as $a > b$, then A $>$ B, hence B must be $< 33° 5'$, and so the obtuse value for B is not admissible, and we have only one solution.

i.e. B $= 24° 4'$; C $= 122° 51'$.

By the Sine Rule, $c = \sin C \cdot \dfrac{a}{\sin A}$

$= \sin 122° 51' : \dfrac{a}{\sin A}$.

$\therefore c = 24\cdot 9$ cm.

$\therefore$ B $= 24° 4'$, C $= 122° 51'$, $c = 24.9$ cm.

No.	Log.
$\dfrac{16\cdot 2}{\sin 33° 5'}$	$\begin{array}{l}1\cdot 2095\\ \overline{1}\cdot 7371\end{array}$
$\dfrac{a}{\sin A}$	$1\cdot 4724$
$\dfrac{\sin A}{a}$	$\overline{2}\cdot 5276$
$12\cdot 1$	$1\cdot 0828$
$\dfrac{\sin A}{a}$	$\overline{2}\cdot 5276$
$\sin B$	$\overline{1}\cdot 6104$
$\sin 122° 51'$	$\overline{1}\cdot 9243$
$\dfrac{a}{\sin A}$	$1\cdot 4724$
c	$1\cdot 3967$

We notice that when solving a triangle by means of the Sine Rule, it is possible to do all the calculations by means of logarithms, *i.e.* the Sine Rule is suitable for logarithmic calculations.

The Cosine Rule

In any triangle ABC, to prove, $a^2 = b^2 + c^2 - 2bc \cos A$ with corresponding results for b^2 and c^2.

We consider the cases of an acute-angled triangle and an obtuse-angled triangle (Fig. 118).

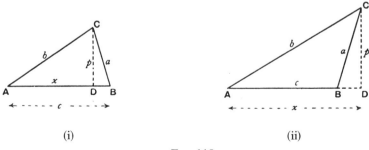

(i)   (ii)

FIG. 118

In each case **CD** is drawn perpendicular to **AB** [produced in case (ii)].

Let **AD** = x.

In case (i)

By Pythagoras' theorem from right-angled △ **CDB**,

$$BC^2 = CD^2 + DB^2$$

i.e. $a^2 = p^2 + (c - x)^2.$

In case (ii)

By Pythagoras' theorem from right-angled △ **CDB**,

$$BC^2 = CD^2 + DB^2$$

i.e. $a^2 = p^2 + (x - c)^2.$

∴ in both cases, $a^2 = p^2 + c^2 + x^2 - 2cx.$

Also from right-angled △ **ACD**,

$$b^2 = p^2 + x^2.$$

Hence $\qquad a^2 = b^2 + c^2 - 2cx.$

But $\qquad x = b \cos A.$

And ∴ in both cases, $\mathbf{a^2 = b^2 + c^2 - 2bc \cos A.}$

Similarly $\qquad \mathbf{c^2 = a^2 + b^2 - 2ab \cos C,}$

and in Fig. (i), $\qquad \mathbf{b^2 = c^2 + a^2 - 2ca \cos B.}$

For the case of the obtuse angle **B** in Fig. 118 (ii).

$$b^2 = AD^2 + CD^2 = (c + BD)^2 + CD^2,$$
$$= c^2 + 2c \cdot BD + BD^2 + CD^2,$$
$$= c^2 + 2cBD + a^2 \text{ (from rt. angled △ CBD)}.$$

Now $\qquad BD = a \cos \widehat{CBD} = a \cos (180° - B) = -a \cos B.$

Hence in this case also $b^2 = c^2 + a^2 - 2ca \cos B.$

Ex. 7. *Find the smallest angle of the* △ **ABC** *in which* $a = 11$ *cm,* $b = 7$ *cm,* $c = 8$ *cm.*

The smallest angle is the one opposite the smallest side, *i.e.* B.
We have by the Cosine Rule, $b^2 = a^2 + c^2 - 2ac \cos B,$

$$i.e. \quad \cos B = \frac{a^2 + c^2 - b^2}{2ac},$$

$$= \frac{121 + 64 - 49}{176} = \frac{136}{176} = \frac{17}{22}.$$

∴ $\cos B = \cdot 7727.$

∴ $B = 39° 24'.$

i.e. The smallest angle of the triangle $= 39° 24'.$

Ex. 8. If in the △ **ABC**, *b* = 82·7, *c* = 98·3, **A** = 41° 6′, *find a.*

No.	Log.
82·7	1·9175
	2
6839	3·8350
98·3	1·9926
	2
9665	3·9852
2	·3010
82·7	1·9175
98·3	1·9926
cos 41° 6′	$\bar{1}$·8771
12 550	4·0882
4254	3·6288
a	1·8144

By the Cosine Rule,

$$a^2 = b^2 + c^2 - 2bc \cos \mathbf{A},$$

$$= 6839 + 9665 - 12250,$$

$$= 4254 \text{ (to 3 s.f.)}.$$

$$\therefore \ a = 65 \cdot 2.$$

This example (Ex. 8) shows that, where logarithms are necessary, the cosine formula is not suitable for logarithmic calculations as each term required must be evaluated separately.

We will consider how the Cosine Rule can be adapted to a form suitable for logarithmic calculation.

We have, $\cos \mathbf{A} = \dfrac{b^2 + c^2 - a^2}{2bc}.$

But $\cos \mathbf{A} = 2 \cos^2 \dfrac{\mathbf{A}}{2} - 1.$

$$\therefore \ 2 \cos^2 \frac{\mathbf{A}}{2} - 1 = \frac{b^2 + c^2 - a^2}{2bc}.$$

$$\therefore \ 2 \cos^2 \frac{\mathbf{A}}{2} = 1 + \frac{b^2 + c^2 - a^2}{2bc} = \frac{(b^2 + c^2 + 2bc) - a^2}{2bc},$$

$$= \frac{(b+c)^2 - a^2}{2bc} = \frac{(b+c+a)(b+c-a)}{2bc}.$$

Let $a + b + c = 2s.$ *i.e.* $s = \dfrac{a+b+c}{2}.$

Then $\qquad b+c-a = 2s-2a$, etc.

$$\therefore \ 2\cos^2\frac{A}{2} = \frac{(2s)(2s-2a)}{2bc}.$$

$$\therefore \ \cos^2\frac{A}{2} = \frac{s(s-a)}{bc},$$

$$\therefore \ \cos\frac{A}{2} = \sqrt{\frac{s(s-a)}{bc}}.$$

Similarly $\qquad \cos\dfrac{B}{2} = \sqrt{\dfrac{s(s-b)}{ac}},$

and $\qquad \cos\dfrac{C}{2} = \sqrt{\dfrac{s(s-c)}{ab}}.$

These formulae are suitable for logarithmic computation.

Similarly we can obtain formulae for $\sin\dfrac{A}{2}$, $\sin\dfrac{B}{2}$, $\sin\dfrac{C}{2}$.

For we have $\quad \cos A = 1 - 2\sin^2\dfrac{A}{2}.$

$\therefore$ By the Cosine Rule

$$1 - 2\sin^2\frac{A}{2} = \frac{b^2+c^2-a^2}{2bc}.$$

$$\therefore \ 2\sin^2\frac{A}{2} = \frac{a^2-b^2-c^2+2bc}{2bc},$$

$$= \frac{a^2-(b-c)^2}{2bc} = \frac{(a-b+c)(a+b-c)}{2bc}.$$

$$\therefore \ 2\sin^2\frac{A}{2} = \frac{(2s-2b)(2s-2c)}{2bc}.$$

$$\therefore \ \sin^2\frac{A}{2} = \frac{(s-b)(s-c)}{bc}.$$

i.e. $\quad \sin\dfrac{A}{2} = \sqrt{\dfrac{(s-b)(s-c)}{bc}}$, with corresponding results

for $\sin\dfrac{B}{2}$ and $\sin\dfrac{C}{2}$.

Also $\qquad \tan\dfrac{A}{2} = \dfrac{\sin\dfrac{A}{2}}{\cos\dfrac{A}{2}} = \dfrac{\sqrt{\dfrac{(s-b)(s-c)}{bc}}}{\sqrt{\dfrac{s(s-a)}{bc}}}$

$$i.e. \quad \tan\frac{A}{2} = \sqrt{\frac{(s-b)(s-c)}{s(s-a)}} \, .$$

$\sqrt{\dfrac{(s-b)(s-c)}{s(s-a)}}$ may be written $\sqrt{\dfrac{(s-a)(s-b)(s-c)}{s}} \Big/ (s-a).$

Thus if r denotes the symmetrical expression $\sqrt{\dfrac{(s-a)(s-b)(s-c)}{s}},$

$$\tan\frac{A}{2} = \frac{r}{s-a} \, .$$

Similarly
$$\tan\frac{B}{2} = \sqrt{\frac{(s-a)(s-c)}{s(s-b)}} = \frac{r}{s-b} \, ,$$

and
$$\tan\frac{C}{2} = \sqrt{\frac{(s-a)(s-b)}{s(s-c)}} = \frac{r}{s-c} \, .$$

To solve a triangle when three sides are given

Ex. 9. Solve the $\triangle$ **ABC**, *in which* $a = 111 \cdot 1$, $b = 173 \cdot 4$, $c = 211 \cdot 3$.

$$a = 111 \cdot 1, \qquad b = 173 \cdot 4, \qquad c = 211 \cdot 3.$$
$$\therefore \ 2s = a + b + c = 495 \cdot 8.$$
$$\therefore \ s = 247 \cdot 9.$$

$\therefore \ (s-a) = 136 \cdot 8; \qquad \log(s-a) = 2 \cdot 1360.$

$(s-b) = 74 \cdot 5; \qquad \log(s-b) = 1 \cdot 8722.$

$(s-c) = 36 \cdot 6; \qquad \log(s-c) = 1 \cdot 5635.$

$\overline{\log(\text{product}) = 5 \cdot 5715}$

$s = 247 \cdot 9 \qquad\qquad \therefore \ \log s = 2 \cdot 3943$

(subtracting) $\overline{3 \cdot 1774}$

$\therefore \ \log r = 1 \cdot 5887$ (dividing by 2 for square root).

Now
$$\tan\frac{A}{2} = \frac{r}{s-a} \, .$$

$$\therefore \ \text{Log} \tan\frac{A}{2} = \log r - \log(s-a),$$

$$= 1 \cdot 5887 - 2 \cdot 1360,$$

$$= \bar{1} \cdot 4527.$$

$$\therefore \ \frac{A}{2} = 15° \, 50' \text{ and } A = 31° \, 40'.$$

$$\text{Log tan} \frac{B}{2} = \log r - \log (s - b),$$
$$= 1 \cdot 5887 - 1 \cdot 8722,$$
$$= \bar{1} \cdot 7165.$$
$$\therefore \frac{B}{2} = 27° 30' \text{ and } B = 55° 0'.$$

$$\text{Log tan} \frac{C}{2} = \log r - \log (s - c),$$
$$= 1 \cdot 5887 - 1 \cdot 5635,$$
$$= 0 \cdot 0252.$$
$$\therefore \frac{C}{2} = 46° 40' \text{ and } C = 93° 20'.$$

$$\therefore A = 31° 40', B = 55°, C = 93° 20'.$$

[Check $A + B + C = 180°$.]

N.B. (i) We obtain A, B and C, and check by showing that the sum of the resulting values is 180°.

(ii) Only four logarithms are used, and the arithmetic is reduced to a minimum.

To solve a triangle when two sides and the included angle are given

This case can be solved at once by means of the Cosine Rule, but this is laborious for all but the very simplest cases, as the rule is not suitable for logarithmic computation.

A rule suitable for this case is obtained as follows.

We have by the Sine Rule, $\dfrac{a}{\sin A} = \dfrac{b}{\sin B}$.

$$\therefore \frac{a - b}{a + b} = \frac{\sin A - \sin B}{\sin A + \sin B},$$

$$= \frac{2 \cos \dfrac{A+B}{2} \sin \dfrac{A-B}{2}}{2 \sin \dfrac{A+B}{2} \cos \dfrac{A-B}{2}} = \cot \frac{A+B}{2} \tan \frac{A-B}{2}.$$

Since $A + B = 180° - C$, $\qquad \dfrac{A+B}{2} = 90° - \dfrac{C}{2}.$

$$\therefore \cot \left(\frac{A+B}{2} \right) = \cot \left(90° - \frac{C}{2} \right) = \tan \frac{C}{2}.$$

$$\therefore \frac{a - b}{a + b} = \tan \frac{C}{2} \tan \frac{A-B}{2},$$

or $$\tan\left(\frac{A-B}{2}\right) = \frac{a-b}{a+b}\cot\frac{C}{2}.$$

With similar results for $\tan\left(\frac{A-C}{2}\right)$ and $\tan\left(\frac{B-C}{2}\right)$.

When a, b and C are given, the above result gives $A-B$, and as $(A+B)$ is known (for C is given), A and B can be found.

Ex. 10. *In the* $\triangle$ **ABC**, $a = 75\cdot31$, $b = 56\cdot17$, $\hat{C} = 51° 12'$.

To find angles B and A and the side c.

We have $a - b = 75\cdot31 - 56\cdot17 = 19\cdot14,$

           $a + b = 75\cdot31 + 56\cdot17 = 131\cdot48.$

As $\tan\left(\frac{A-B}{2}\right) = \frac{a-b}{a+b}\cot\frac{C}{2},$

$\therefore \tan\left(\frac{A-B}{2}\right) = \frac{19\cdot14}{131\cdot48}\cot 25° 36',$

$= \frac{19\cdot14}{131\cdot48}\tan 64° 24'.$

$\therefore \frac{A-B}{2} = 16° 54',$

$\therefore A-B = 33° 48'.$

No.	Log.
19·14	1·2819
tan 64° 24'	0·3196
	1·6015
131·48	2·1188
$\tan\dfrac{A-B}{2}$	$\bar{1}·4827$

But $A+B = 180° - C,$

           $= 128° 48'.$

Adding, $2A = 162° 36',$

           $\therefore A = 81° 18'.$

Subtracting, $2B = 95°,$

           $\therefore B = 47° 30'.$

By the Sine Rule,

$$\frac{c}{\sin C} = \frac{a}{\sin A}.$$

$\therefore c = \frac{75\cdot31 \cdot \sin 51° 12'}{\sin 81° 18'}.$

$\therefore c = 59\cdot37.$

No.	Log.
75·31	1·8769
sin 51° 12'	$\bar{1}$·8917
	1·7686
sin 81° 18'	$\bar{1}$·9950
c	1·7736

Area of a Triangle

Area of △ ABC, $\triangle = \frac{1}{2}bp$ (Fig. 119).

In both figures, $p = a \sin C.$

       $\therefore \ \triangle = \frac{1}{2}\textbf{ab sin C.}$

Again, in figure (i) $p = c \sin A,$

and in figure (ii) $p = c \sin(180° - A) = c \sin A.$

$\therefore$ in both figures $\triangle = \frac{1}{2}\textbf{bc sin A.}$

Similarly $\triangle = \frac{1}{2}\textbf{ca sin B.}$

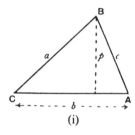 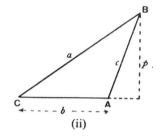

      (i)   (ii)

Fig. 119

The area of a triangle in terms of the sides.

$$\triangle = \frac{1}{2}bc \sin A = \frac{1}{2}bc \cdot 2 \sin \frac{A}{2} \cos \frac{A}{2},$$

$$= bc \sqrt{\frac{(s-b)(s-c)}{bc}} \sqrt{\frac{s(s-a)}{bc}},$$

$$\triangle = \sqrt{\textbf{s(s-a)(s-b)(s-c)}}, \text{ where } s = \tfrac{1}{2}(a+b+c).$$

(A) EXAMPLES 16a

1. In the △ ABC, $\hat{A} = 90°$, $a = 12$, $b = 3$. Find the length of the altitude through A.

2. Solve the following triangles

    (i) $a = 4,$ $b = 6,$ $c = 9;$

    (ii) $A = 51° 30',$ $B = 66° 10',$ $a = 20;$

    (iii) $a = 3,$ $b = 4,$ $C = 40°.$

3. Find the largest angle of the triangle whose sides are of lengths 2 cm, 4 cm, 5 cm.

4. Find the smallest angle of the triangle whose sides are 3 m, 4 m, 6 m.

5. Solve the $\triangle$ ABC in which $A = 31°$, $B = 14°$, $c = 7$. Also find the area of the triangle and its largest altitude.

6. In the $\triangle$ PQR, $q = 2\sqrt{3}$, $r = 6$, $P = 30°$. Prove that the triangle is isosceles.

7. Find the shortest altitude of the $\triangle$ ABC in which $a = 6$, $b = 7$, $c = 8$.

8. Find the altitudes of the triangle whose sides are 5 cm, 6 cm, 10 cm.

9. Find the area of the triangle in Question 8.

10. Find the area of the triangle in which $AB = 16\cdot1$, $BC = 11\cdot4$, $\widehat{ABC} = 21° \, 10'$.

11. Two sides of a triangle are $\sqrt{3}+1$ and $\sqrt{3}-1$, and the included angle is $60°$. Find the other side.

12. In each of the following cases show that two triangles can be drawn satisfying the given data, and solve the triangles

$$\begin{array}{llll} \text{(i)} & a = 3, & b = 4, & A = 20°; \\ \text{(ii)} & b = 6, & c = 10, & B = 15°; \\ \text{(iii)} & c = 8, & b = 12, & C = 30°; \\ \text{(iv)} & a = 100, & c = 100\sqrt{3}, & A = 30°. \end{array}$$

13. Given $B = 30°$, $b = 150$, $c = 150\sqrt{3}$, prove that of the two triangles which satisfy the data, one will be isosceles and the other right-angled.

14. Find the difference in area of the two triangles which satisfy each set of conditions in Question 12.

15. The base angles of a triangle are $22\frac{1}{2}°$ and $112\frac{1}{2}°$. Prove that the base is equal to twice the height.

16. If the angles of a triangle are in the ratios $5:10:21$ and the side opposite the smallest angle is 4 m, find the other sides.

17. Find the acute angles of a right-angled triangle whose hypotenuse is four times as long as the perpendicular drawn to it from the opposite vertex.

18. Adjacent sides of a parallelogram are of lengths 6 cm and 4 cm, and the angle included between them is $32°$. Find the lengths of the diagonals.

19. If $2b = 3a = 3c$, prove $\tan \dfrac{A}{2} = \dfrac{1}{\sqrt{7}}$.

Deduce that $\sin A = \frac{1}{4}\sqrt{7}$.

20. If $A = 45°$, $B = 75°$, $C = 60°$, prove that $a + c\sqrt{2} = 2b$.

21. When given c, a, C, show that two triangles are possible if $a > c > a \sin C$.

(B) EXAMPLES 16b

1. Solve the triangles in which

    (i) $a = 351 \cdot 2$, $B = 61° 50'$, $C = 75° 34'$;

    (ii) $c = 85 \cdot 3$, $A = 114° 16'$, $B = 34° 28'$;

    (iii) $b = 11 \cdot 1$, $A = 25° 10'$, $C = 78° 21'$.

2. Solve the following triangles

    (i) $a = 461 \cdot 2$, $b = 413 \cdot 3$, $c = 136$;

    (ii) $a = 38 \cdot 2$, $b = 19 \cdot 4$, $c = 31 \cdot 4$;

    (iii) $a = 4 \cdot 62$, $b = 6 \cdot 10$, $c = 5 \cdot 72$.

3. Solve the following triangles

    (i) $b = 8 \cdot 7$, $c = 9 \cdot 2$, $A = 59° 30'$;

    (ii) $c = 2532$, $a = 779 \cdot 7$, $B = 68° 40'$;

    (iii) $a = 21 \cdot 2$, $b = 30 \cdot 6$, $C = 111° 10'$.

4. The lengths of the sides of a triangle are 28, 41 and 59 cm. Find the greatest angle and the area of the triangle.

5. Solve the triangle ABC, in which $a = 32 \cdot 9$, $b = 50 \cdot 9$, $c = 65 \cdot 4$. Find also the longest altitude of the triangle.

6. Solve the triangle ABC if $a = 10 \cdot 82$, $c = 21 \cdot 71$, $B = 66° 39'$. Find also the area of the triangle.

7. Find the area of the trapezium ABCD, in which AB is parallel to CD, AB = 8 cm, BC = 11 cm, CD = 17·6 cm, $\hat{B} = 102° 18'$.

8. The sides of a parallelogram are 7·21 cm and 9·32 cm respectively, and one diagonal measures 8·3 cm. Calculate (i) the length of the other diagonal, (ii) the angles of the parallelogram, and (iii) the angles between the diagonals.

9. In the $\triangle$ ABC, $b = 30$, $c = 44$, $B = 38°$. Show that there are two possible triangles, and find the difference in the values of a.

10. In the $\triangle$ ABC, $c = 87 \cdot 37$ m, $a = 94 \cdot 16$ m, $B = 61° 38'$. Find C and b.

11. In the $\triangle$ ABC, $a : b : c = 3\sqrt{5} : 2\sqrt{3} + 3 : 2\sqrt{3} - 3$. Find B.

12. Find the difference in the areas of the two triangles which satisfy the following conditions, $b = 16 \cdot 12$, $c = 24 \cdot 82$, $B = 32° 14'$.

13. In the ambiguous case given a, b and A, prove that the difference between the two values of c is $2\sqrt{a^2 - b^2 \sin^2 A}$.

14. In the quadrilateral ABCD, the angles A, C are right angles, and angle D is 123°; AD is 16 cm and CD is 10 cm. Find the length of BD.

15. Find the area of the quadrilateral **ABCD**, given that $AB = 7$ cm, $BC = 4.5$ cm, $CD = 14.5$ cm, $DA = 12$ cm and angle $B = 110° 30'$.

16. B is 378 m N. 48° 12' E. of **A**; **C** is 413 m N. 21° 10' W. of **B**. Find the distance and bearing of **C** from **A**.

17. In the quadrilateral **PQRS**, $PQ = 4$ cm, $QR = 6$ cm, $RS = 8$ cm, $\widehat{QRS}$ is a right angle and $\cos\widehat{PQR} = -0.2$. Find (i) the lengths of the diagonals and (ii) the area of the quadrilateral.

18. The angle included by two sides of a triangle, whose lengths are 216.8 m and 261.3 m, is $62° 36'$. Calculate the length of the remaining side of the triangle. Also find the radius of the circumcircle of the triangle.

19. If **R** is the radius of the circumcircle of $\triangle$ **ABC**, prove that $R = \dfrac{a}{2\sin A} = \dfrac{b}{2\sin B} = \dfrac{c}{2\sin C}$. Deduce the formula $R = \dfrac{abc}{4\triangle}$, where $\triangle$ is the area of the triangle.

20. A tower, 50 m high, stands on the top of a mound. From a point on the ground the angles of elevation of the top and bottom of the tower are $74° 30'$ and $43° 30'$ respectively. Find the height of the mound.

21. A ship sailing north sees two lighthouses, which are 6 km apart, in a line due west; after an hour's sailing one of them bears S.W. and the other S.S.W. Find the speed of the ship.

22. A station **A** is 10 km west of **B**. The bearing of a rock from **A** is $72° 18'$ E. of N., and its bearing from **B** is $24° 45'$ W. of N. How far is the rock north of the line **AB**?

23. Triangle **ABC** is right-angled at **C**. The bisector of angle **C** meets **AB** in **D**. Prove that the length **CD** is $c\sqrt{2}\,\dfrac{\sin B}{1 + \tan B}$.

24. In the triangle **ABC**, $A = 56° 12'$, $B = 47° 17'$, and the length of the altitude from **C** is 16.9 cm. Find the lengths of the sides of the triangle.

25. A tower subtends an angle α at a point on the same level as the foot of the tower, and at a second point, h m above the first, the angle of depression of the foot of the tower is β. Prove that the height of the tower is $h \tan \alpha \cot \beta$.

26. A tower and a spire on the top of the tower subtend equal angles at a point whose distance from the foot of the tower is a. If h be the height of the tower, prove that the height of the spire is $h\,\dfrac{a^2 + h^2}{a^2 - h^2}$.

Problems in Three Dimensions

Angle between two planes

Let **AB** be the line of intersection of two planes.

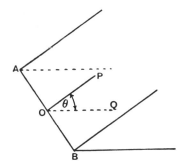

Fig. 120

Take any point **O** on **AB**, and draw lines **OP, OQ** in the two planes each perpendicular to **AB**. Then the angle θ between these two lines is called *the angle between the two planes.*

Angle between a line and a plane

Let **A'B'** be the projection of **AB** on the plane—*i.e.* **A', B'** are the feet of the perpendiculars from **A** and **B** on to the plane.

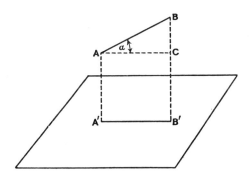

Fig. 121

Then the angle between **AB** and its projection **A'B'**, *i.e.* $\widehat{BAC}$, is defined as the angle between the line and the plane.

Ex. 11. *A rectangular solid, length* 10 *cm, width* 8 *cm, height* 6 *cm, rests on a horizontal plane. Find the length of a diagonal, and the inclination of a diagonal to the horizontal.*

BS is a diagonal of the solid.

By Pythagoras, in △ **DAB**,

$$DB^2 - 8^2 + 10^2,$$

$$= 164.$$

$$\therefore DB = \sqrt{164}.$$

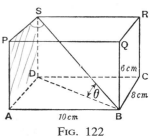

Fig. 122

By Pythagoras, in △ **SDB**,

$$BS^2 = DB^2 + DS^2,$$

$$= 164 + 36 = 200.$$

$$\therefore BS = \sqrt{200} = 10\sqrt{2} = 14 \cdot 14 \text{ cm}.$$

Also **DB** is the projection of **BS** on the horizontal plane.

∴ Angle between diagonal and the horizontal plane = θ,

and

$$\tan \theta = \frac{DS}{DB} = \frac{6}{\sqrt{164}} = \frac{3}{\sqrt{41}}.$$

$$\therefore \theta = 25° 6'.$$

∴ Length of diagonal = 14·14 cm. ⎫
Inclination to horizontal = 25° 6'. ⎬

Ex. 12. *The lid of a desk is a board* 1 *m long and* 50 *cm wide. It slopes at* 40° *to the horizontal with its length horizontal. Find the inclination of its diagonal to the horizon.*

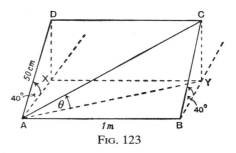

Fig. 123

Let **ABCD** be the lid of the desk. **DX** and **CY** are drawn perpendicular to the horizontal plane through **AB**.

Then $\widehat{DAX} = \widehat{CBY} = 40°$.

We have, $\dfrac{CY}{CB} = \sin 40°.$

$$\therefore \ CY = 50 \ . \ \sin 40° \ \text{cm}.$$

Also by Pythagoras, $AC^2 = AB^2 + BC^2 = 100^2 + 50^2 = 12500.$

$$\therefore \ \sin \theta = \dfrac{CY}{AC} = \dfrac{50 \sin 40°}{\sqrt{12500}} = \dfrac{\sin 40°}{\sqrt{5}}.$$

$$\therefore \ \theta = 16° \ 42'.$$

$\therefore$ Inclination of diagonal to the horizontal $= 16° \ 42'$.

Ex. 13. An aeroplane is sighted at the same instant from two points A
and B *on a horizontal line,* A *being due east of* B *and* AB $= 6760$ m.
From A *the bearing of the aeroplane is* 53° N. *of* W., *and from* B *the
bearing is* 33° N. *of* E. *The angle of elevation of the aeroplane from* A *is*
29°. *Find (a) the vertical height of the aeroplane above the ground, and
(b) its angle of elevation from* B. (J.M.B.)

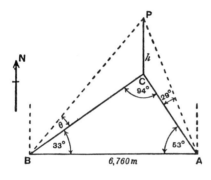

Fig. 124

Let P be the position of the aeroplane.
Let h m be the height of the aeroplane.
Using the Sine Rule to solve the triangle ABC, we have

$$\frac{a}{\sin A} = \frac{b}{\sin B} = \frac{c}{\sin C} \ .$$

$$\therefore \ \frac{b}{\sin 33°} = \frac{6760}{\sin 94°}$$

i.e. $b = \dfrac{6760 \sin 33°}{\sin 86°} \ .$

$$\therefore \ b = 3691 \ \text{m}.$$

	No.	Log.
	6760	3·8299
	sin 33°	$\bar{1}$·7361
		3·5660
	sin 86°	$\bar{1}$·9989
	b	3·5671

Also $\dfrac{a}{\sin 53°} = \dfrac{6760}{\sin 94°}$,

$\therefore a = 5412 \text{ m.}$

From triangle PCA, $\dfrac{h}{AC} = \tan 29°$,

$\therefore h = 3691 \tan 29°.$

i.e. $h = 2046 \text{ m.}$

From triangle BCP, $\tan \theta = \dfrac{h}{BC}$,

$\therefore \theta = 20° 43'.$

No.	Log.
6760	3·8299
sin 53°	$\bar{1}$·9023
	3·7322
sin 86°	$\bar{1}$·9989
a	3·7333
3691	3·5671
tan 29°	$\bar{1}$·7438
h	3·3109
a	3.7333
tan θ	$\bar{1}$·5776

$\therefore$ height of aeroplane = 2046 m.⎫
Angle of elevation from **B** = 20° 43' ⎬

Ex. 14. *Two lines at right angles to one another in a plane, are inclined at angles* α, β *respectively to the horizon. If* θ *be the inclination of the plane to the horizon, prove*

$$\sin^2 \theta = \sin^2 \alpha + \sin^2 \beta.$$

Let **XY** be the line of intersection of the given plane with the horizontal (Fig. 125).

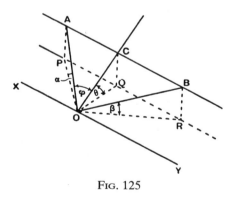

FIG. 125

OA, OB are lines in the plane perpendicular to each other.

OC is the line of greatest slope of the plane through **O**, *i.e.* **OC** is perpendicular to **XY**.

ACB is drawn parallel to **XY**.

AP, CQ, BR are perpendiculars from A, C, B respectively on to the horizontal plane.

Then the inclination of OA to the horizon $= \widehat{AOP} = \alpha$.
The inclination of OC to the horizon = inclination of the plane to the horizon,

$$= \widehat{COQ} = \theta,$$

and the inclination of OB to the horizon $= \widehat{BOR} = \beta$.

As AB is parallel to XY, AB is horizontal and thus

$$AP = CQ = BR.$$

Let $\qquad\qquad \widehat{AOC} = \phi$.

Then $\qquad\qquad \widehat{BOC} = 90° - \phi$.

We have $\qquad\qquad \sin \alpha = \dfrac{AP}{AO} = \dfrac{CQ}{AO}$.

But $\qquad\qquad AO = \dfrac{OC}{\cos \phi}$,

$$\therefore \ \sin \alpha = \dfrac{CQ}{OC} \cos \phi = \sin \theta \cos \phi,$$

i.e. $\quad \sin \alpha = \sin \theta \cos \phi$.

Similarly $\qquad\qquad \sin \beta = \sin \theta \cos (90° - \phi) = \sin \theta \sin \phi$.

$$\therefore \ \sin^2 \alpha + \sin^2 \beta = \sin^2 \theta \cos^2 \phi + \sin^2 \theta \sin^2 \phi,$$

$$= \sin^2 \theta \{\cos^2 \phi + \sin^2 \phi\},$$

$$= \sin^2 \theta.$$

$$\therefore \ \sin^2 \alpha + \sin^2 \beta = \sin^2 \theta.$$

(B) EXAMPLES 16c

1. Find the length of the diagonal of a cube of side 12 cm. Find also (a) the angle made by this diagonal with an adjacent face, and (b) the angle between this diagonal and an adjacent edge.

2. ABCDPQRS is a rectangular block, fixed with the base ABCD horizontal. AB = CD = PQ = RS = 12 cm, AD = BC = PS = QR = 8 cm, and AP = BQ = CR = DS = 6 cm. X is the middle point of PS. Find the lengths BX, CX and the angle between them. Also find the inclination of BX and CX to the horizontal.

3. The lid of a desk is 96 cm long and 54 cm wide. It slopes at 32° to the horizon with its length horizontal. Find the inclination of its diagonal to the horizon.

4. A rectangular door ABCD, AB = 2·40 m and AD = 1·20 m, is free to rotate about AB, which is vertical. If the door is opened through an angle of 44° from the position ABCD to the position ABC'D', find (a) the length CD' and (b) its inclination to the horizontal.

5. Find the angle between the two diagonals of a rectangular solid of length 20 cm, breadth 16 cm, height 12 cm, which pass through the two upper corners of one of the smaller faces.

6. A plane is inclined at 30° to the horizon. Find the inclination to the horizon of a line on the plane, which makes an angle of 20° with a line of greatest slope.

7. A pyramid OABCD has a square base ABCD. O is vertically above the centre of the square ABCD. If AB = 8 cm and OA = 12 cm, find
   (i) the height of the pyramid;
   (ii) the inclination of OA to the base ABCD;
   (iii) the angle between a sloping face and the base;
   (iv) the angle between adjacent sloping faces.

8. The legs of a tripod are 1 m long. The upper ends of the legs meet in a point and the feet are at the vertices of an equilateral triangle of side 50 cm. Find the height of the tripod and the inclination of a leg to the ground.

9. The legs of a tripod OP, OQ, OR are each 25 cm in length, and they are opened so that P, Q, R are on a horizontal plane with PQ = PR = 12 cm and QR = 9 cm. Find the height of the tripod and the inclinations of the legs to the horizon.

10. A weight is supported by three chains attached to points in a ceiling at the corners of an equilateral triangle of side 4 m. The chains are of lengths 6 m, 6 m, 5 m. Find the distance of the weight below the ceiling and the inclinations of the chains to the horizontal.

11. From a point A due north of a tower of height 100 m, the elevation of the top of the tower is 32° 30'. Find the elevation of the top of the tower from a point B, 120 m S. 60° E. of A.

12. From the top of a cliff 100 m high two buoys are observed in directions N. 32° W. and N. 25° W., their respective angles of depression being 3° 12' and 2° 30'. Find the distance between the buoys to the nearest ten metres.

13. A and B are two points on a horizontal plane and C is a point 250 m above the plane. A is due south, and B is south-west of C; the angles of depression of A and B from C are 22° and 12° respectively. Find the distance and bearing of A from B.

14. A point B is due west of A and AB = 80 m. A tower which stands on the same horizontal plane as **A** and **B** is due south of **A**. The angles of elevation of the top of the tower from **A, B** are 32° 30′ and 19° 22′ respectively. Find the height of the tower.

15. P, Q, R are three points on a horizontal plane, with QR = 6000 m and $\widehat{QPR}$ = 110° 20′. From each of the points P, Q, R the angle of elevation of the summit of a mountain is 45°. Find the height of the mountain summit above the plane.

16. A tower **AB** is observed from two points P, Q in the same horizontal plane as **B**, the foot of the tower. The length PQ = 375 m, and the angles BPQ, BQP, APB are ⁻58°, 44° 30′, 26° 34′ respectively. Find the height of the tower to the nearest metre.

17. An observer sees an aeroplane bearing N. 34° W. at an elevation of 22°, and one minute later its bearing is N. 75° W. and its elevation is 25° 30′. Assuming that the aeroplane is travelling at a uniform speed in a straight line at a height of 5000 m, find its speed and direction.

18. At two points, the distance between which subtends an angle 32° 28′ at the base of the flagstaff, the angles of elevation of its top are 34° 24′ and 25° 14′ respectively. If the distance between the two points is 56·5 m, find the height of the flagstaff, assuming the two points and the foot of the flagstaff are in the same horizontal plane.

19. At a point **A** the angle of elevation of the top of a hill, 1000 m high, is 12° 30′ and the bearing from **A** is N. 16° 22′ E. At a point **B** on the same horizontal plane as **A**, the elevation is 19° and the bearing is S. 43° 12′ E. Find the distance and bearing of **B** from **A**.

20. A man in a balloon observes that the angle of depression of an object on the ground bearing due north is 37°. The balloon drifts 5·28 km due east and the angle of depression is now found to be 23° 34′. Find the height of the balloon.

21. A tower is observed from two stations **A** and **B**. It is due north of **A** and north-west of **B**. **B** is due east of **A** and distant 100 m from it. The elevation of the tower as seen from **A** is the complement of the elevation as seen from **B**. Find the height of the tower.

22. A flagstaff, 20 m high, stands at the centre of a plot of land in the shape of an equilateral triangle. From the top of the flagstaff each side of the plot subtends an angle of 40°. Find the length of a side of the plot.

23. From a point **A** on a level plane the angle of elevation of a kite is α, the kite being due south of **A**. From a point **B**, which is a distance of d m south of **A**, the kite is seen northwards at an elevation β. Find the height of the kite.

CHAPTER 17

DIFFERENTIATION AND INTEGRATION OF TRIGONOMETRICAL FUNCTIONS

To differentiate sin x, where the angle x is expressed in radians

Let $\qquad y = \sin x.$

Let x increase by a small amount δx and let the corresponding increase in y be δy.

Then $\qquad y + \delta y = \sin(x + \delta x).$

$$\therefore \ \delta y = \sin(x + \delta x) - \sin x,$$

$$= 2 \cos\left(\frac{2x + \delta x}{2}\right) \sin\left(\frac{\delta x}{2}\right).$$

[Using the results of Ch. 14.]

$$\therefore \ \frac{\delta y}{\delta x} = \frac{2 \cos\left(x + \dfrac{\delta x}{2}\right) \sin\left(\dfrac{\delta x}{2}\right)}{\delta x},$$

$$= \cos\left(x + \frac{\delta x}{2}\right) \frac{\sin\left(\dfrac{\delta x}{2}\right)}{\dfrac{\delta x}{2}}.$$

Now let $\delta x \to 0$.

Then $\dfrac{\delta y}{\delta x} \to \dfrac{dy}{dx}$, $\cos\left(x + \dfrac{\delta x}{2}\right) \to \cos x$, and $\dfrac{\sin\left(\dfrac{\delta x}{2}\right)}{\dfrac{\delta x}{2}} \to 1.$

$\left[\text{Using the result } \lim\limits_{\theta \to 0} \dfrac{\sin \theta}{\theta} = 1, \text{ with } \theta = \dfrac{\delta x}{2}.\right]$

The student will see at this stage the reason for taking x in radians, for the result $\lim\limits_{\theta \to 0} \dfrac{\sin \theta}{\theta} = 1$ is only true for θ in radians.

Hence
$$\frac{dy}{dx} = \cos x.$$

i.e.
$$\frac{d}{dx}(\sin x) = \cos x.$$

To differentiate cos x, where x is in radians

Let
$$y = \cos x.$$

Let x increase by a small amount δx and let the corresponding increase in y be δy.

Then
$$y + \delta y = \cos(x + \delta x).$$

$$\therefore \ \delta y = \cos(x + \delta x) - \cos x,$$

$$= -2\sin\left(\frac{2x + \delta x}{2}\right)\sin\left(\frac{\delta x}{2}\right).$$

[Using the results of Ch. 14.]

$$\therefore \ \frac{\delta y}{\delta x} = \frac{-2\sin\left(x + \frac{\delta x}{2}\right)\sin\left(\frac{\delta x}{2}\right)}{\delta x},$$

$$= -\sin\left(x + \frac{\delta x}{2}\right)\frac{\sin\left(\frac{\delta x}{2}\right)}{\frac{\delta x}{2}}.$$

Now let $\delta x \to 0$.

Then
$$\frac{\delta y}{\delta x} \to \frac{dy}{dx}, \ \sin\left(x + \frac{\delta x}{2}\right) \to \sin x \text{ and } \frac{\sin\left(\frac{\delta x}{2}\right)}{\frac{\delta x}{2}} \to 1.$$

Hence
$$\frac{dy}{dx} = -\sin x.$$

i.e.
$$\frac{d}{dx}(\cos x) = -\sin x.$$

To differentiate tan x, where x is in radians

Let
$$y = \tan x.$$

Let x increase by a small amount δx and let the corresponding increase in y be δy.

Then $y + \delta y = \tan(x + \delta x)$.

$$\therefore \; \delta y = \tan(x + \delta x) - \tan x,$$

$$= \frac{\sin(x + \delta x)}{\cos(x + \delta x)} - \frac{\sin x}{\cos x},$$

$$= \frac{\sin(x + \delta x)\cos x - \cos(x + \delta x)\sin x}{\cos(x + \delta x)\cos x},$$

$$= \frac{\sin \delta x}{\cos(x + \delta x)\cos x}.$$

[Using $\sin(A - B) = \sin A \cos B - \cos A \sin B$.]

$$\therefore \; \frac{\delta y}{\delta x} = \frac{1}{\cos(x + \delta x)\cos x} \cdot \frac{\sin \delta x}{\delta x}.$$

Let $\delta x \to 0$.

Then $\dfrac{\delta y}{\delta x} \to \dfrac{dy}{dx}$, $\cos(x + \delta x) \to \cos x$ and $\dfrac{\sin \delta x}{\delta x} \to 1$.

Hence $\dfrac{dy}{dx} = \dfrac{1}{\cos^2 x} = \sec^2 x.$

i.e. $\dfrac{d}{dx}(\tan x) = \sec^2 x.$

The previous results are very important, and we will collect them together for reference.

$$\frac{d}{dx}(\sin x) = \cos x;$$

$$\frac{d}{dx}(\cos x) = -\sin x; \quad \left.\begin{array}{l} \text{In each case, } x \text{ being} \\ \text{measured in radians.} \end{array}\right.$$

$$\frac{d}{dx}(\tan x) = \sec^2 x.$$

These results can be written in the form of integrals as follows

$$\int \cos x\, dx = \sin x + c;$$

$$\int \sin x\, dx = -\cos x + c; \quad \left.\begin{array}{l} \text{where } c \text{ is an arbitrary} \\ \text{constant.} \end{array}\right.$$

$$\int \sec^2 x\, dx = \tan x + c.$$

Ex. 1. Differentiate $3 \sin x - 4 \cos x$ *with respect to x.*

Let $\qquad y = 3 \sin x - 4 \cos x.$

Then $\qquad \dfrac{dy}{dx} = 3 \cos x - 4(-\sin x),$

$$= 3 \cos x + 4 \sin x.$$

Ex. 2. If $s = \sin t + 2 \cos t,$ *find* $\dfrac{d^2 s}{dt^2}.$

$$\dfrac{ds}{dt} = \cos t - 2 \sin t.$$

$$\therefore \dfrac{d^2 s}{dt^2} = -\sin t - 2 \cos t.$$

Ex. 3. Find the maximum and minimum values of the function $2 \sin x + \cos x$ *[x in radians].*

Let $\qquad y = 2 \sin x + \cos x.$

$$\therefore \dfrac{dy}{dx} = 2 \cos x - \sin x.$$

For maximum or minimum values of y, $\dfrac{dy}{dx} = 0.$

$$\therefore 2 \cos x - \sin x = 0.$$

$$i.e. \quad \sin x = 2 \cos x.$$

$$\therefore \dfrac{\sin x}{\cos x} = 2, \text{ or } \tan x = 2.$$

The equation $\tan x = 2$ is satisfied by (i) a value of x between 0 and $\frac{1}{2}\pi$, and (ii) a value of x between π and $\frac{3}{2}\pi$—the tangent being positive for angles in the first and third quadrants (Fig. 126).

In case (i) both $\sin x$ and $\cos x$ will be positive.
In case (ii) both $\sin x$ and $\cos x$ will be negative.

We have, $\quad \tan^2 x = \sec^2 x - 1.$

$$\therefore \sec^2 x = 5, \sec x = \pm\sqrt{5}.$$

And $\qquad \cot^2 x = \operatorname{cosec}^2 x - 1.$

$$\therefore \operatorname{cosec} x = \pm\dfrac{\sqrt{5}}{2}.$$

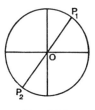

FIG. 126

Hence y is a maximum or minimum when

$$(i) \ \sin x = \dfrac{2}{\sqrt{5}}, \qquad \cos x = \dfrac{1}{\sqrt{5}};$$

$$(ii) \ \sin x = -\dfrac{2}{\sqrt{5}}, \qquad \cos x = -\dfrac{1}{\sqrt{5}}.$$

On differentiating again, $\dfrac{d^2y}{dx^2} = -2\sin x - \cos x$.

Case (i) makes $\dfrac{d^2y}{dx^2}$ *negative* and hence makes y a *maximum*.

Case (ii) makes $\dfrac{d^2y}{dx^2}$ *positive* and hence makes y a *minimum*.

On substitution.

$$\text{Maximum value of } y = \frac{4}{\sqrt{5}} + \frac{1}{\sqrt{5}} = \frac{4\sqrt{5}}{5} + \frac{1}{\sqrt{5}} = \sqrt{5}.$$

$$\text{Minimum value of } y = \frac{-4}{\sqrt{5}} - \frac{1}{\sqrt{5}} = -\sqrt{5}.$$

N.B. (a) Having obtained $\tan x = 2$, we could have used tables to find x and then proceeded in the usual way;

(b) The method employed on page 178 gives a neater solution.

Ex. 4. *Evaluate* $\displaystyle\int_0^{\pi} (3\sin x - \cos x)\,dx.$

$$\int_0^{\pi} (3\sin x - \cos x)\,dx = \left[-3\cos x - \sin x\right]_0^{\pi},$$
$$= (-3\cos\pi - \sin\pi) - (-3\cos 0 - \sin 0),$$
$$= (3) - (-3) = 6.$$
$$\therefore \int_0^{\pi} (3\sin x - \cos x)\,dx = 6.$$

Note. It is important to remember the following results

$$\left.\begin{array}{l} \sin 0 = 0 \\ \cos 0 = 1 \end{array}\right\}; \quad \left.\begin{array}{l} \sin\dfrac{\pi}{2} = \sin 90° = 1 \\ \cos\dfrac{\pi}{2} = \cos 90° = 0 \end{array}\right\}; \quad \left.\begin{array}{l} \sin\pi = \sin 180° = 0 \\ \cos\pi = \cos 180° = -1 \end{array}\right\};$$

$$\left.\begin{array}{l} \sin\dfrac{3\pi}{2} = \sin 270° = -1 \\ \cos\dfrac{3\pi}{2} = \cos 270° = 0 \end{array}\right\}; \quad \left.\begin{array}{l} \sin 2\pi = \sin 360° = 0 \\ \cos 2\pi = \cos 360° = 1 \end{array}\right\}.$$

Ex. 5. *Find the area contained between the x-axis and one loop of the curve* $y = \sin x$, *where x is measured in radians.*

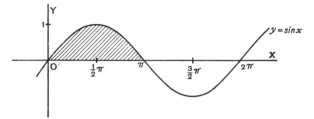

Fig. 127

Area of one loop = area between $x = 0$ and $x = \pi$,

$$= \int_0^\pi y\,dx = \int_0^\pi \sin x\,dx,$$

$$= \left[-\cos x\right]_0^\pi = (-\cos \pi) - (-\cos 0) = 1 - (-1),$$

$$= 2 \text{ sq. units.}$$

(A) EXAMPLES 17a

[*Note.* In the following exercises, the angle is expressed in radians except where otherwise stated.]

1. Differentiate with respect to θ the following functions

(i) $2 \sin \theta$;   (ii) $3 \cos \theta$;   (iii) $\dfrac{\tan \theta}{2}$;

(iv) $\sin \theta - \cos \theta$;   (v) $3 \cos \theta + 4 \sin \theta$;   (vi) $2 \sin \theta - \tan \theta$;

(vii) $1 - 3 \cos \theta$;   (viii) $\dfrac{\sin \theta + 4 \cos \theta}{3}$.

2. Evaluate the following integrals

(i) $\displaystyle\int 3 \sin x\,dx$;   (ii) $\displaystyle\int \dfrac{\cos x}{2}\,dx$;

(iii) $\displaystyle\int 4 \sec^2 x\,dx$;   (iv) $\displaystyle\int (2 \sin x - 3 \cos x)\,dx$;

(v) $\displaystyle\int \dfrac{\sin x + \cos x}{2}\,dx$;   (vi) $\displaystyle\int_0^{\frac{\pi}{2}} \cos x\,dx$;

(vii) $\displaystyle\int_0^\pi (2 \sin x - \cos x)\,dx$;   (viii) $\displaystyle\int_0^{\frac{\pi}{4}} \sec^2 x\,dx$;

(ix) $\displaystyle\int_{\frac{\pi}{2}}^\pi (2 \cos x - \sin x)\,dx$;   (x) $\displaystyle\int_{\frac{3\pi}{2}}^{2\pi} (3 \cos x + 4 \sin x)\,dx$.

3. Find $\dfrac{d^2s}{dt^2}$ in the following cases

(i) $s = 3 \sin t$;   (ii) $s = 4 \cos t - 3 \sin t$;

(iii) $s = 2 \sin t + \cos t$;   (iv) $s = 1 - 2 \sin t$.

4. Find the values of tan x for which $\dfrac{dy}{dx}$ is zero in the following cases

(i) $y = \sin x + \cos x$;   (ii) $y = 3 \sin x - \cos x$;

(iii) $y = 4 \sin x - 2 \cos x$.

5. Find the gradients of the following curves at the points (a) where $x = 0$, (b) where $x = \dfrac{\pi}{4}$.

(i) $y = \sin x$.   (ii) $y = \cos x$.   (iii) $y = \tan x$.

(iv) $y = 3 \sin x + \cos x$.   (v) $y = 4 \sin x - \tan x$.

6. Find the equations of the tangents and normals to the curves in Question 5 at the point $x = \dfrac{\pi}{4}$.

7. Find the rates of increase of y with respect to x for the following functions when $x = \dfrac{\pi}{3}$

(i) $y = 2 \sin x + 7 \cos x$;   (ii) $y = 3 \tan x - \cos x + 4 \sin x$.

8. Using the results $\cos x = 1 - 2 \sin^2 \dfrac{x}{2} = 2 \cos^2 \dfrac{x}{2} - 1$, find the following

(i) $\dfrac{d}{dx}\left(\sin^2 \dfrac{x}{2}\right)$;   (ii) $\dfrac{d}{dx}\left(\cos^2 \dfrac{x}{2}\right)$;

(iii) $\displaystyle\int \sin^2 \dfrac{x}{2}\, dx$;   (iv) $\displaystyle\int \cos^2 \dfrac{x}{2}\, dx$.

9. Find the areas contained between the x-axis and

(i) the curve $y = \sin x$ and the ordinates $x = 0$ and $x = \dfrac{\pi}{2}$;

(ii) ,, ,, $y = \cos x$,, ,, $x = 0$,, $x = \dfrac{\pi}{2}$;

(iii) ,, ,, $y = \sec^2 x$,, ,, $x = 0$,, $x = \dfrac{\pi}{4}$;

(iv) ,, ,, $y = 3 \sin x + \cos x$,, $x = 0$,, $x = \dfrac{\pi}{4}$;

(v) ,, ,, $y = 2 \cos x - \sin x$,, $x = -\dfrac{\pi}{2}$ and $x = \dfrac{\pi}{3}$.

10. Find the volumes of revolution formed when the portions of the curves (i) $y = \sqrt{\sin x}$, (ii) $y = 2\sqrt{\cos x}$, between $x = 0$ and $x = \dfrac{\pi}{4}$, are rotated about the x-axis.

Differentiation of the sine, cosine and tangent of multiple angles

Ex. 6. Differentiate sin 2x [x in radians] with respect to x.

Let $\qquad\qquad\qquad\qquad y = \sin 2x$

and let $\qquad\qquad\qquad\left. \begin{aligned} u &= 2x. \\ y &= \sin u. \end{aligned} \right\}$
Then

Using the result $\qquad\qquad \dfrac{dy}{dx} = \dfrac{dy}{du} \times \dfrac{du}{dx},$

we have $\qquad\qquad\qquad \dfrac{dy}{dx} = \cos u \times 2 = 2\cos 2x.$

$\qquad\qquad i.e. \quad \dfrac{d}{dx}(\sin 2x) = 2\cos 2x.$

Ex. 7. Differentiate cos 4x with respect to x.

Let $\qquad\qquad\qquad\qquad y = \cos 4x$

and let $\qquad\qquad\qquad\left. \begin{aligned} u &= 4x. \\ \therefore\ y &= \cos u. \end{aligned} \right\}$

Hence $\qquad\qquad \dfrac{dy}{dx} = \dfrac{dy}{du} \times \dfrac{du}{dx} = -\sin u \times 4 = -4\sin 4x.$

$\qquad\qquad i.e. \quad \dfrac{d}{dx}(\cos 4x) = -4\sin 4x.$

Ex. 8. Differentiate tan 8x with respect to x.

Let $\qquad\qquad\qquad\qquad y = \tan 8x.$

and let $\qquad\qquad\qquad\left. \begin{aligned} u &= 8x. \\ \therefore\ y &= \tan u. \end{aligned} \right\}$

Hence $\qquad\qquad \dfrac{dy}{dx} = \dfrac{dy}{du} \times \dfrac{du}{dx} = \sec^2 u \times 8 = 8\sec^2 8x.$

$\qquad\qquad \therefore\ \dfrac{d}{dx}(\tan 8x) = 8\sec^2 8x.$

General case

To differentiate $a \sin mx$, where a and m are constants.

Let $\qquad\qquad\qquad y = a \sin mx$

and let $\qquad u = mx.$ $\qquad \therefore \dfrac{du}{dx} = m.$

$$\therefore y = a \sin u. \quad \therefore \dfrac{dy}{du} = a \cos u.$$

We have $\qquad \dfrac{dy}{dx} = \dfrac{dy}{du} \times \dfrac{du}{dx} = a \cos u \times m = am \cos mx.$

$$i.e. \quad \dfrac{d}{dx}(a \sin mx) = am \cos mx.$$

Similarly $\qquad \dfrac{d}{dx}(a \cos mx) = -am \sin mx,$

and $\qquad \dfrac{d}{dx}(a \tan mx) = am \sec^2 mx.$

$$e.g. \quad \dfrac{d}{dx}(\sin 15x) = 15 \cos 15x.$$

$$\dfrac{d}{dx}(2 \cos 11x) = 2 \times (-11 \sin 11x) = -22 \sin 11x.$$

$$\dfrac{d}{dx}(6 \tan 6x) = 6 \cdot 6 \sec^2 6x = 36 \sec^2 6x.$$

Integration of sin mx, cos mx, sec² mx, where m is a constant

As $\qquad \dfrac{d}{dx}(\cos mx) = -m \sin mx,$

then $\qquad \dfrac{d}{dx}\left(-\dfrac{\cos mx}{m}\right) = \sin mx.$

Hence $\qquad \displaystyle\int \sin mx \, dx = -\dfrac{\cos mx}{m} + c.$

Also $\qquad \dfrac{d}{dx}(\sin mx) = m \cos mx,$

then $\qquad \dfrac{d}{dx}\left(\dfrac{\sin mx}{m}\right) = \cos mx.$

Hence $\qquad \displaystyle\int \cos mx \, dx = \dfrac{\sin mx}{m} + c.$

Similarly $\qquad \displaystyle\int \sec^2 mx \, dx = \dfrac{\tan mx}{m} + c.$

We will collect these results for reference.

$$\int \sin mx dx = -\frac{\cos mx}{m} + c;$$

$$\int \cos mx dx = \frac{\sin mx}{m} + c;$$

where c is an arbitrary constant.

$$\int \sec^2 mx dx = \frac{\tan mx}{m} + c.$$

Ex. 9. *Evaluate* (a) $\displaystyle\int \sin 2x dx$; (b) $\displaystyle\int \cos 4x\, dx$;

(c) $\displaystyle\int_0^{\frac{\pi}{3}} 2 \sin 3x dx$; (d) $\displaystyle\int_0^{\frac{\pi}{8}} \frac{\sec^2 2x}{2}\, dx$.

(a) $\displaystyle\int \sin 2x dx = -\frac{\cos 2x}{2} + c.$

(b) $\displaystyle\int \cos 4x dx = \frac{\sin 4x}{4} + c.$

(c) $\displaystyle\int_0^{\frac{\pi}{3}} 2 \sin 3x dx = \left[-2 \cdot \frac{\cos 3x}{3}\right]_0^{\frac{\pi}{3}},$

$$= (-\tfrac{2}{3} \cos \pi) - (-\tfrac{2}{3} \cos 0),$$

$$= \tfrac{2}{3} - (-\tfrac{2}{3}) = \tfrac{4}{3}.$$

(d) $\displaystyle\int_0^{\frac{\pi}{8}} \frac{\sec^2 2x}{2}\, dx = \left[\frac{1}{2} \cdot \frac{\tan 2x}{2}\right]_0^{\frac{\pi}{8}},$

$$= \left(\frac{1}{4} \tan \frac{\pi}{4}\right) - \left(\frac{1}{4} \tan 0\right),$$

$$= \frac{1}{4}.$$

Ex. 10. *The portion of the curve $y = \sin x$ between $x = 0$ and $x = \pi$ is rotated about the x-axis. Find the volume swept out.*

$$\text{Volume of revolution} = \pi \int_a^b y^2 dx,$$

$$= \pi \int_0^\pi \sin^2 x dx.$$

To evaluate this integral, we express $\sin^2 x$ in terms of $\cos 2x$.

We have $\qquad \cos 2x = 1 - 2\sin^2 x; \quad i.e. \quad \sin^2 x = \dfrac{1 - \cos 2x}{2}.$

$$\therefore \text{ Volume} = \pi \int_0^\pi \left(\frac{1}{2} - \frac{\cos 2x}{2} \right) dx = \pi \left[\frac{x}{2} - \frac{\sin 2x}{4} \right]_0^\pi,$$

$$= \pi \left[\frac{\pi}{2} - \frac{\sin 2\pi}{4} \right] - \pi \left[0 - \frac{\sin 0}{4} \right],$$

$$= \frac{\pi^2}{2}.$$

$$\therefore \text{ Volume of revolution} = \frac{\pi^2}{2}.$$

(A) EXAMPLES 17b

1. Differentiate with respect to x: $\sin 8x$, $\cos \dfrac{x}{2}$, $\sin \dfrac{x}{2}$, $\tan 3x$, $\sin \dfrac{11x}{2}$, $\cos \cdot 3x$, $\sin \dfrac{x}{10}$, $2\tan 10x$, $2\cos 3x$, $\frac{1}{2}\sin 2x$, $4\tan \dfrac{x}{2}$, $\sin x°$, $\cos x°$, $\tan x°$.

2. Integrate with respect to x: $\sin 3x$, $\cos 3x$, $\sec^2 3x$, $\sin \dfrac{x}{2}$, $\cos \dfrac{x}{3}$, $2\sec^2 \dfrac{x}{2}$, $4\cos 4x$, $\frac{1}{5}\sin \dfrac{x}{5}$, $\frac{1}{2}\cos 2x$, $8\sec^2 4x$, $\cdot 2\cos \cdot 5x$, $1\cdot 1\sin \cdot 1x$.

3. Evaluate the following integrals

(i) $\displaystyle\int_0^{\frac{\pi}{4}} \cos 2x\, dx$; (ii) $\displaystyle\int_0^{\frac{\pi}{3}} \sin 3x\, dx$;

(iii) $\displaystyle\int_0^{\frac{\pi}{8}} \sec^2 2x\, dx$; (iv) $\displaystyle\int_{-\frac{\pi}{2}}^{\frac{\pi}{2}} (2\sin 2x + 1)\, dx$;

(v) $\displaystyle\int_0^\pi \frac{1 - \cos 2x}{2}\, dx$, (vi) $\displaystyle\int_{\frac{\pi}{2}}^\pi (3\sin 2x - 2\cos 2x)\, dx$.

4. Find $\dfrac{dy}{dx}$ in each of the following cases

(i) $y = 3\sin 3x - 2\cos 2x + x$;

(ii) $y = \frac{1}{2}\cos \dfrac{x}{2} + 3\sin \dfrac{x}{2} + 2$;

(iii) $y = 2\tan x + \sin 3x$;

(iv) $y = \cos^2 x$; $\left.\begin{array}{l}\\ \\\end{array}\right\}$ express in terms of $\cos 2x$.
(v) $y = \sin^2 x$.

5. Find the gradient of the curve $y = 2 \sin 2x$ at the point where $x = \dfrac{\pi}{6}$. Hence find the equation of the tangent at this point.

6. If $r = \sin 3t - 2 \cos t$, find the value of $\dfrac{d^2 r}{dt^2}$ when $t = \dfrac{\pi}{4}$ rads.

7. If $y = 2 \sin 2x$, find the approximate change in y when x increases from $\dfrac{\pi}{8}$ to $\dfrac{11\pi}{80}$.

8. Find values of θ, between 0 and π, for which $\dfrac{dr}{d\theta}$ vanishes where $r = \cos 2\theta - \cos \theta$.

9. For what values of θ, between 0 and π, is the function $\cos 2\theta - \cos \theta$ increasing?

10. Find the area enclosed between the curve $y = 2 \sin 2x$, the x-axis and the ordinates $x = 0$, $x = \dfrac{\pi}{4}$.

11. Find the area enclosed between the curve $y = \cos \dfrac{x}{2}$, the x-axis and the ordinates $x = -\pi$, $x = \pi$.

12. Find the volume swept out when the portion of the curve $y = \sqrt{\sin 3x}$, between $x = 0$ and $x = \dfrac{\pi}{3}$, is rotated about the axis of x.

To differentiate sin (ax + b), where a and b are constants

Let $\qquad\qquad\qquad\qquad y = \sin (ax + b)$

and let $\qquad\qquad\qquad\quad u = ax + b.$

Then $\qquad\qquad\qquad\quad y = \sin u.$

$$\therefore \frac{dy}{dx} = \frac{dy}{du} \times \frac{du}{dx}$$

$$= \cos u \times a = a \cos (ax + b).$$

i.e. $\qquad\dfrac{d}{dx} \{\sin (ax+b)\} = a \cos (ax+b).$

Similarly $\qquad\dfrac{d}{dx} \{\cos (ax+b)\} = -a \sin (ax+b);$

$\qquad\qquad\dfrac{d}{dx} \{\tan (ax+b)\} = a \sec^2 (ax+b).$

Hence
$$\int \sin (ax+b)dx = -\frac{\cos (ax+b)}{a}+c;$$
$$\int \cos (ax+b)dx = \frac{\sin (ax+b)}{a}+c;$$
$$\int \sec^2 (ax+b)dx = \frac{\tan (ax+b)}{a}+c.$$

e.g. $\dfrac{d}{dx}\{\sin (x+2)\}=\cos (x+2).$

$\dfrac{d}{dx}\{\cos (3x-1)\}=-3 \sin (3x-1).$

$\dfrac{d}{dx}\left\{\tan \left(1-\dfrac{x}{2}\right)\right\}=-\tfrac{1}{2}\sec^2 \left(1-\dfrac{x}{2}\right).$

And $\displaystyle\int \cos (2x+4)dx = \dfrac{\sin (2x+4)}{2}+c.$

$$\int \sin \left(\frac{x}{2}+1\right)dx = \frac{-\cos \left(\dfrac{x}{2}+1\right)}{\tfrac{1}{2}}+c = -2 \cos \left(\frac{x}{2}+1\right)+c.$$

$$\int \sec^2 (1-x)dx = \frac{\tan (1-x)}{-1}+c = -\tan (1-x)+c.$$

Differentiation of powers of sin x, cos x, and tan x

Ex. 11. *Differentiate* $\sin^3 x$ *i.e.*$(\sin x)^3$ *with respect to* x.

Let $y = \sin^3 x,$

and let $u = \sin x.$ $\therefore \dfrac{du}{dx}=\cos x.$

Then $y = u^3.$ $\therefore \dfrac{dy}{du}=3u^2.$

Hence $\dfrac{dy}{dx}=\dfrac{dy}{du}\times\dfrac{du}{dx}=3u^2\times\cos x = 3 \sin^2 x \cos x.$

i.e. $\dfrac{d}{dx}(\sin^3 x)=3 \sin^2 x \cos x.$

Ex. 12. *Differentiate* $\cos^4 x$ *with respect to* x.

Let $y = \cos^4 x,$

and let $u = \cos x.$ $\therefore \dfrac{du}{dx} = -\sin x.$

$\therefore y = u^4.$ $\therefore \dfrac{dy}{du} = 4u^3.$

Hence $\dfrac{dy}{dx} = \dfrac{dy}{du} \times \dfrac{du}{dx} = 4u^3 \times (-\sin x) = -4 \cos^3 x \sin x.$

i.e. $\dfrac{d}{dx}(\cos^4 x) = -4 \cos^3 x \sin x.$

Ex. 13. Differentiate $\tan^2 \dfrac{x}{2}$ *with respect to x.*

Let $y = \tan^2 \dfrac{x}{2},$

and let $u = \tan \dfrac{x}{2}.$ $\therefore \dfrac{du}{dx} = \tfrac{1}{2} \sec^2 \dfrac{x}{2}.$

$\therefore y = u^2.$ $\therefore \dfrac{dy}{du} = 2u.$

$\therefore \dfrac{dy}{dx} = \dfrac{dy}{du} \times \dfrac{du}{dx} = 2u \times \tfrac{1}{2} \sec^2 \dfrac{x}{2} = \tan \dfrac{x}{2} \sec^2 \dfrac{x}{2},$

i.e. $\dfrac{d}{dx}\left(\tan^2 \dfrac{x}{2}\right) = \tan \dfrac{x}{2} \sec^2 \dfrac{x}{2}.$

(A) EXAMPLES 17c

1. Differentiate the following with respect to x: $\sin(x+3)$, $\cos(x-1)$,

$\tan(x+2)$, $2\sin(3x-1)$, $\tfrac{1}{3}\cos(3x+2)$, $2\sin(1-x)$, $4\cos\left(\dfrac{x}{2}+1\right)$,

$6\tan(1-6x)$, $\sin\left(\pi x - \dfrac{\pi}{2}\right)$, $\cos\left(\dfrac{\pi x}{2} - \dfrac{\pi}{4}\right)$, $\pi \tan(1-\pi x)$.

2. Integrate with respect to t: $\sin(2t+1)$, $\cos(3t-2)$, $\sec^2\left(\dfrac{t}{2}-1\right)$,

$2\sin(1-t)$, $\cos\left(\pi t + \dfrac{\pi}{2}\right)$, $\pi \sin\left(\dfrac{\pi t}{2} - \dfrac{\pi}{4}\right)$.

3. Evaluate

(i) $\displaystyle\int_0^1 2\sin\left(\dfrac{\pi t}{2} - \dfrac{\pi}{4}\right)dt$; (ii) $\pi \displaystyle\int_{-\frac{\pi}{2}}^{\frac{\pi}{2}} \cos\left(t - \dfrac{\pi}{3}\right)dt.$

4. Differentiate with respect to x: $\sin^2 x$, $\cos^2 x$, $\cos^5 x$, $\tan^7 x$, $3\sin^4 x$,

$\frac{1}{2}\cos^4 x$, $\sin^2 2x$, $\cos^2 2x$, $\tan^2 2x$, $2\sin^4\frac{x}{4}$, $3\cos^3\frac{x}{3}$, $\sqrt{\sin x}$, $\sqrt{\cos x}$, $\sqrt{\tan x}$,

$\dfrac{1}{\sin^2 x}$, $\dfrac{1}{\cos^3 2x}$.

5. If $r = \sin^2\theta$, show that $\dfrac{dr}{d\theta} = \sin 2\theta$ and determine the value of $\dfrac{d^2 r}{d\theta^2}$ when $\theta = \dfrac{\pi}{6}$.

6. Find the maximum and minimum values of $\cos^2\theta$.

7. Find the equation of the tangent to the curve $y = 1 - 2\sin^2 x$ at the point where $x = \dfrac{\pi}{8}$.

8. The distance s described by a point in time t is given by the formula $s = 10\cos(2t - 1)$. Find one value of t for which the point is at rest.

9. Evaluate the integrals

$$\text{(i)} \int_0^{\frac{\pi}{2}} 2\sin^2 x\,dx; \qquad \text{(ii)} \int_0^{\frac{\pi}{2}} 2\cos^2 x\,dx,$$

using the results, $\cos 2x = 2\cos^2 x - 1 = 1 - 2\sin^2 x$.

10. The portion of the curve $y = \cos x$, between $x = 0$ and $x = \pi$, is rotated about the axis of x. Find the volume of the solid of revolution so formed.

(B) MISCELLANEOUS EXAMPLES

1. Obtain from the first principles the differential coefficients of

$$\text{(i)} \ \sin 2x; \qquad \text{(ii)} \ \cos\left(\frac{x}{2}\right); \qquad \text{(iii)} \ \tan 2x.$$

2. Find $\dfrac{dy}{dx}$ in the following cases

$$\text{(i)} \ y = (1 + \sin x)^2; \qquad \text{(ii)} \ y = \sin^2 x + \cos^2 x;$$

$$\text{(iii)} \ y = \sqrt{\sin^3 x}; \qquad \text{(iv)} \ y = \frac{\cos^2 x - \sin^2 x}{\cos^2 x}.$$

3. Find $\dfrac{d^2 y}{dx^2}$ when (i) $y = \sin nx$; (ii) $y = \cos nx$, where n is a constant.

Deduce the result $\dfrac{d^2 y}{dx^2} + n^2 y = 0$ when $y = \sin nx + \cos nx$.

4. If $y = C \sin(pt + \epsilon)$, where C, p, ϵ are constants, show that $\dfrac{d^2y}{dt^2} + p^2y = 0$.

5. Evaluate

(i) $\displaystyle\int_0^\pi \sec^2 \frac{x}{4}\, dx$;

(ii) $\displaystyle\int_{-\frac{\pi}{2}}^{\frac{\pi}{2}} (\cos x + 2\cos 2x)\, dx$;

(iii) $\displaystyle\int_0^{\frac{\pi}{4}} (1 + \sin x)^2\, dx$;

(iv) $\displaystyle\int_0^{60^\circ} \sin x^\circ\, dx$.

6. Use the result $2 \sin A \cos B = \sin(A + B) + \sin(A - B)$, to evaluate the integral $\displaystyle\int \sin 3x \cos x\, dx$.

7. Evaluate

(i) $\displaystyle\int_0^{\frac{\pi}{3}} \sin 5x \cos x\, dx$;

(ii) $\displaystyle\int_{-\frac{\pi}{4}}^{\frac{\pi}{4}} \sin \frac{3x}{2} \cos \frac{x}{2}\, dx$.

8. Use the identity $2 \cos A \cos B = \cos(A + B) + \cos(A - B)$ to evaluate the integrals

(i) $\displaystyle\int_0^{\frac{\pi}{4}} \cos 3x \cos x\, dx$;

(ii) $\displaystyle\int_{-\frac{\pi}{2}}^{\frac{\pi}{2}} 4 \cos 2x \cos x\, dx$.

9. Find the maximum and minimum values of (i) $\cos x - \sin x$, (ii) $3 \sin x + 4 \cos x$, (iii) $2x - \tan x$, for values of x between 0 and π.

10. Determine the turning points of the curve $y = x - \sin 2x$ for values of x between 0 and π, and discriminate between them. Sketch the curve.

11. Find the value of $\dfrac{dy}{dx}$ in terms of t for the curve $x = \sin 2t$, $y = 2 \cos t$. For what values of t is the tangent to the curve parallel to the axis of x?

12. Find the equation of the tangent to the curve

$$y = 2(\theta - \sin \theta), \quad x = 2(1 - \cos \theta),$$

at the point $\theta = \frac{1}{2}\pi$.

13. Find the lengths of the subnormal and subtangent to the curve $y = 2 \sin \theta$, $x = 3 \cos \theta$, at the point $\theta = \frac{1}{6}\pi$.

14. If $y = \cos x - 1 + \frac{1}{2}x^2$, prove that $\dfrac{dy}{dx}$ is positive for acute values of x. Is y increasing or decreasing as x increases from the value zero? Deduce that $\cos x > 1 - \frac{1}{2}x^2$ for acute values of x.

15. A particle moves in a straight line so that the distance s m, travelled in t seconds, is given by the expression $s = 4t + 2\cos 2t$. Find the velocity and acceleration after $\dfrac{\pi}{2}$ seconds. After what time is the particle first at rest?

16. The velocity of a moving point is given by $v = t + 2\sin \pi t$, where t is the time. Find the distance moved by the point in the first second of its motion.

17. A particle moves along the x-axis so that the distance travelled in time t is given by $x = 2t + \cos 3t$. Find the distance between the first two positions of rest.

18. Sketch the curve $y = 2\cos \dfrac{x}{2}$ for values of x between 0 and π, and evaluate the area contained between this portion of the curve and the axis of x.

19. Find the area of the segment of the curve $y = \tfrac{1}{2}\sin 2x - \sin x$ below the axis of x, when x is positive and acute.

20. The portion of the curve $y = 1 - \cos x$ between $x = 0$ and $x = \dfrac{\pi}{2}$ is rotated about the axis of x. Find the volume of the solid so formed.

CHAPTER 18

GENERAL METHODS OF DIFFERENTIATION

Differentiation of Sums and Differences

By means of several special cases, we concluded that the differential coefficient of a sum, or difference, of a series of terms is equal to the sum, or difference, of the separate differential coefficients; *i.e.* if u, v, w, etc., are functions of x,

$$\frac{d}{dx}(u+v+w\ldots)=\frac{du}{dx}+\frac{dv}{dx}+\frac{dw}{dx}+\ldots$$

We will now give a formal proof of this result.

i.e. **To prove that if u, v, w ... are functions of x, then**

$$\frac{d}{dx}(u+v+w+\ldots)=\frac{du}{dx}+\frac{dv}{dx}+\frac{dw}{dx}+\ldots$$

Let $y = u+v+w+\ldots$

Let x increase by a small amount δx.

Let the corresponding increases in u, v, $w\ldots$ be $\delta u, \delta v, \delta w\ldots$ respectively and let the resulting increase in y be δy.

Then $\qquad y+\delta y = (u+\delta u)+(v+\delta v)+(w+\delta w)+\ldots$

Subtracting, $\qquad \delta y = \delta u + \delta v + \delta w + \ldots$

$$\therefore \frac{\delta y}{\delta x}=\frac{\delta u}{\delta x}+\frac{\delta v}{\delta x}+\frac{\delta w}{\delta x}+\ldots$$

Let $\delta x \to 0$.

Then $\qquad \dfrac{\delta y}{\delta x}\to\dfrac{dy}{dx}, \dfrac{\delta u}{\delta x}\to\dfrac{du}{dx}, \dfrac{\delta v}{\delta x}\to\dfrac{dv}{dx}$, etc.

Hence $\qquad \dfrac{dy}{dx}=\dfrac{du}{dx}+\dfrac{dv}{dx}+\dfrac{dw}{dx}+\ldots$

i.e. $\qquad \dfrac{d}{dx}(u+v+w+\ldots)=\dfrac{du}{dx}+\dfrac{dv}{dx}+\dfrac{dw}{dx}+\ldots$

Similarly $\dfrac{d}{dx}(u-v+w-t\ldots)=\dfrac{du}{dx}-\dfrac{dv}{dx}+\dfrac{dw}{dx}-\dfrac{dt}{dx}\ldots$

e.g. $\dfrac{d}{dx}\left(3x^3-\sin x+\dfrac{1}{x^2}\right)=\dfrac{d}{dx}(3x^3)-\dfrac{d}{dx}(\sin x)+\dfrac{d}{dx}\left(\dfrac{1}{x^2}\right),$

$$=9x^2-\cos x-\dfrac{2}{x^3}.$$

Differentiation of Products and Quotients

To differentiate the product (uv) where u and v are functions of x.

Let $y=uv.$

Let x increase by a small amount δx.

Let the corresponding increases in u and v be δu and δv, and let the resulting increase in y be δy.

Then $y+\delta y=(u+\delta u)(v+\delta v),$

$$=uv+u\delta v+v\delta u+\delta u\delta v.$$

Subtracting, $\delta y=u\delta v+v\delta u+\delta u\delta v.$

$$\therefore\ \dfrac{\delta y}{\delta x}=u\dfrac{\delta v}{\delta x}+v\dfrac{\delta u}{\delta x}+\dfrac{\delta u}{\delta x}\delta v.$$

Let $\delta x\to 0$.

Then $\dfrac{\delta y}{\delta x}\to\dfrac{dy}{dx}$, $\dfrac{\delta v}{\delta x}\to\dfrac{dv}{dx}$, $\dfrac{\delta u}{\delta x}\to\dfrac{du}{dx}$ and $\delta v\to 0$.

Also as $\delta v\to 0$ and $\dfrac{\delta u}{\delta x}\to\dfrac{du}{dx}$, then $\dfrac{\delta u}{\delta x}\delta v\to 0$.

$$\therefore\ \dfrac{dy}{dx}=u\dfrac{dv}{dx}+v\dfrac{du}{dx}.$$

i.e. $\dfrac{d}{dx}(uv)=v\dfrac{du}{dx}+u\dfrac{dv}{dx}.$

This most important result should be remembered in words as follows: the differential coefficient of a product of two terms is equal to the differential coefficient of the 1st term times the 2nd term PLUS the differential coefficient of the 2nd term times the 1st term.

e.g. $\dfrac{d}{dx}(x\sin x)=\dfrac{d}{dx}(x).\sin x+\dfrac{d}{dx}(\sin x).x,$

$$=\sin x+x\cos x.$$

To differentiate the quotient $\left(\dfrac{u}{v}\right)$ where u and v are functions of x.

Let $$y = \frac{u}{v}.$$

Let x increase by a small amount δx.
Let the corresponding increases in u and v be δu and δv and let the resulting increase in y be δy.

Then $$y + \delta y = \frac{u + \delta u}{v + \delta v}.$$

Subtracting, $$\delta y = \frac{u + \delta u}{v + \delta v} - \frac{u}{v},$$

$$= \frac{v(u + \delta u) - u(v + \delta v)}{v(v + \delta v)}.$$

$$\therefore \; \delta y = \frac{v\delta u - u\delta v}{v(v + \delta v)}.$$

$$\therefore \; \frac{\delta y}{\delta x} = \frac{v\dfrac{\delta u}{\delta x} - u\dfrac{\delta v}{\delta x}}{v^2 + v\delta v}.$$

Let $\delta x \to 0$.

Then $\dfrac{\delta y}{\delta x} \to \dfrac{dy}{dx}, \; \dfrac{\delta u}{\delta x} \to \dfrac{du}{dx}, \; \dfrac{\delta v}{\delta x} \to \dfrac{dv}{dx}$, and as $\delta v \to 0$, $v\delta v$ also $\to 0$.

Hence $$\frac{dy}{dx} = \frac{v\dfrac{du}{dx} - u\dfrac{dv}{dx}}{v^2}.$$

i.e. $$\frac{d}{dx}\left(\frac{u}{v}\right) = \frac{v\dfrac{du}{dx} - u\dfrac{dv}{dx}}{v^2}.$$

In words: the differential coefficient of a quotient of two terms is equal to the differential coefficient of the NUMERATOR times the denominator MINUS the differential coefficient of the denominator times the numerator, all divided by the square of the denominator.

E.g. $$\frac{d}{dx}\left(\frac{\sin x}{x}\right) = \frac{\dfrac{d}{dx}(\sin x) \cdot x - \dfrac{d}{dx}(x) \cdot \sin x}{x^2},$$

$$= \frac{x\cos x - \sin x}{x^2}.$$

Ex. 1. *Differentiate* (a) $2x^2 \cos x$; (b) $\dfrac{2}{1-x}$ *with respect to x.*

(a) Let $y = 2x^2 \cos x$.

Taking $(2x^2)$ as the 1st term and $\cos x$ as the 2nd,

we have $\qquad \dfrac{dy}{dx} = \dfrac{d}{dx}(2x^2) \cdot \cos x + \dfrac{d}{dx}(\cos x) \cdot 2x^2,$

$$= 4x \cos x + (-\sin x) \cdot 2x^2,$$

$$= 4x \cos x - 2x^2 \sin x.$$

(b) Let $y = \dfrac{2}{1-x}$.

$$\therefore \dfrac{dy}{dx} = \dfrac{\dfrac{d}{dx}(2) \cdot (1-x) - \dfrac{d}{dx}(1-x) \cdot 2}{(1-x)^2},$$

$$= \dfrac{0 \cdot (1-x) - (-1) \cdot 2}{(1-x)^2},$$

$$= \dfrac{2}{(1-x)^2}.$$

Ex. 2. *Differentiate* $\dfrac{x^2 \sin 2x}{1-x}$ *with respect to x.*

Let $y = \dfrac{x^2 \sin 2x}{1-x}$.

Then, using the rule for differentiating a quotient,

$$\dfrac{dy}{dx} = \dfrac{\dfrac{d}{dx}(x^2 \sin 2x) \cdot (1-x) - \dfrac{d}{dx}(1-x) \cdot x^2 \sin 2x}{(1-x)^2}$$

To obtain $\dfrac{d}{dx}(x^2 \sin 2x)$ we must use the rule for differentiating a product.

Then $\qquad \dfrac{dy}{dx} = \dfrac{\{2x \sin 2x + 2x^2 \cos 2x\}(1-x) - (-1)x^2 \sin 2x}{(1-x)^2},$

$$= \dfrac{2x}{1-x}\{\sin 2x + x \cos 2x\} + \dfrac{x^2 \sin 2x}{(1-x)^2}.$$

Ex. 3. *Find the turning point on the curve* $y = \dfrac{1}{x(x-1)}$.

We have $$y = \frac{1}{x^2 - x}.$$

$$\therefore \frac{dy}{dx} = \frac{0(x^2 - x) - (2x - 1)1}{(x^2 - x)^2},$$

$$= \frac{1 - 2x}{x^2(x - 1)^2}.$$

But for a turning point, $\frac{dy}{dx} = 0$.

$$\therefore 1 - 2x = 0.$$

i.e. $x = \frac{1}{2}$.

To find whether this is a maximum or a minimum point we find the sign of $\frac{d^2y}{dx^2}$ when $x = \frac{1}{2}$.

We have $$\frac{d^2y}{dx^2} = \frac{-2\{x^2 - x\}^2 - (1 - 2x)\dfrac{d}{dx}(x^2 - x)^2}{(x^2 - x)^4}.$$

But when $x = \frac{1}{2}$, $1 - 2x = 0$.

$$\therefore (1 - 2x)\frac{d}{dx}(x^2 - x)^2 = 0.$$

$\therefore$ When $x = \frac{1}{2}$, $\dfrac{d^2y}{dx^2} = -\dfrac{2}{(x^2 - x)^2}$ which is negative.

$\therefore$ $x = \frac{1}{2}$ corresponds to a maximum point on the curve.

i.e.   the point $(\frac{1}{2}, -4)$ is a maximum point on the curve

$$y = \frac{1}{x(x - 1)}.$$

(A) EXAMPLES 18a

1. Differentiate the following functions with respect to x

(i) $3x^4 - x^2 + 2$;   (ii) $3 - \dfrac{2}{x^2}$;   (iii) $x \cos x$;   (iv) $x^2 \sin x$;

(v) $x \tan x$;   (vi) $x(1 - x^2)$;   (vii) $\dfrac{2x}{1 + x}$;   (viii) $\dfrac{3x^2 + 1}{1 - x}$;

(ix) $\sin x \cos x$;   (x) $\dfrac{\tan x}{x}$;   (xi) $\dfrac{x}{\sin x}$;   (xii) $\dfrac{\cos x}{2x^2}$;

(xiii) $\dfrac{1}{(1 - x)^2}$;   (xiv) $x(1 - \sin x)$;   (xv) $x \sin 2x$;   (xvi) $\dfrac{2 \cos 2x}{x}$.

2. Find $\dfrac{dy}{dx}$ in the following cases

(i) $y = x^2 \cos x - x$;

(ii) $y = x - x^3 \tan 2x$;

(iii) $y = \dfrac{1}{x+1} - \dfrac{1}{x+2}$;

(iv) $y = (x+1)(2\sin x - \cos x)$.

3. Find $\dfrac{d^2 r}{dt^2}$ when

(i) $r = \dfrac{t}{1+t}$;

(ii) $r = 2t \sin t$;

(iii) $r = \dfrac{\cos 2t}{t}$.

4. If $p(v+1) = 5$, find the value of $\dfrac{dp}{dv}$ when $v = 2$.

5. Using the relations $\tan \theta = \dfrac{\sin \theta}{\cos \theta}$, $\cot \theta = \dfrac{\cos \theta}{\sin \theta}$, obtain the differential coefficients of $\tan \theta$ and $\cot \theta$ with respect to θ.

6. Use the results $\sec \theta = \dfrac{1}{\cos \theta}$, $\operatorname{cosec} \theta = \dfrac{1}{\sin \theta}$, to prove the results $\dfrac{d}{d\theta}(\sec \theta) = \sec \theta \tan \theta$, $\dfrac{d}{d\theta}(\operatorname{cosec} \theta) = -\operatorname{cosec} \theta \cot \theta$.

7. Find the values of x for which the gradients of the following curves are zero.

(i) $y = x^2(1-x)$.

(ii) $y = \dfrac{1-x}{x^2}$.

(iii) $y = (1+x)(1-x)^2$.

8. If $y = x \cos x$, show that $\dfrac{dy}{dx} = 0$ when $x = \cot x$.

9. Given that $\theta = \pi + x \sin \theta$, find $\dfrac{dx}{d\theta}$.

10. The distance, s, travelled by a body in time, t, is given by the relation $s = t + \dfrac{1}{1+t}$. Find the velocity of the point when $t = 2$.

(B) EXAMPLES 18b

1. Find $\dfrac{dy}{dx}$ in the following cases

(i) $y = \dfrac{3 \sin 3x}{x}$;

(ii) $y = \dfrac{x}{\sqrt{x-1}}$;

(iii) $y = (x^2+1)\sqrt{1-x^2}$;

(iv) $y = 2 \sin 2x \cos 3x$;

(v) $y = x \sin^2 x$;   (vi) $y = (1+x) \cos^3 x$;

(vii) $y = \dfrac{x}{(1+x)(1-2x)}$;   (viii) $y = \dfrac{1+\sin x}{1+\cos x}$;

(ix) $y = \dfrac{\tan x - x}{\tan x + x}$;   (x) $y = \dfrac{\sin 3x}{(1-x)^3}$;

(xi) $y = \cos\left(\dfrac{1}{x}\right)$;   (xii) $y = x^3 \sin\left(2x + \dfrac{\pi}{2}\right)$;

(xiii) $y = x + \sqrt{x^2 + a^2}$, where a is a constant.

2. Find $\dfrac{d^2 y}{dx^2}$ when $y = \dfrac{(x-1)(x-2)}{x}$, and determine the values of $\dfrac{d^2 y}{dx^2}$ when $\dfrac{dy}{dx}$ is zero.

In examples 3–6 find the maximum and minimum values of y.

3. $y = \dfrac{x^2 - 4x + 3}{4x - 13}$.   **4.** $y = \dfrac{x^2 + 2x}{3x - 2}$.

5. $y = \dfrac{x(x+3)}{x-1}$.   **6.** $y = (x+1)^2(x-1)^3$.

7. Find the turning points on the curve $y = \dfrac{x}{(x-1)(x-4)}$, and make a rough sketch of the curve.

8. Find the equations of the tangent and the normal to the curve $y = \dfrac{4x^2}{x-1}$ at the point $(-1, -2)$.

9. Differentiate with respect to x

(i) $x(x+1)(x^2+1)$;   (ii) $\dfrac{x \sin x}{x-1}$;

(iii) $x \sin x \cos x$;   (iv) $\dfrac{x^2 \tan 2x}{1+x}$.

10. If $y = \dfrac{3 \cos kx}{x}$, where k is a constant, prove that

$$x\frac{d^2 y}{dx^2} + 2\frac{dy}{dx} + k^2 xy = 0.$$

11. Given that $x = \dfrac{2t}{1+t^2}$, $y = \dfrac{1-t^2}{1+t^2}$, express $\dfrac{dy}{dx}$ in terms of t.

12. Prove that the function $x\sqrt{1-x^2}$ increases as x increases from zero to a certain value, and find this value of x.

13. Show that the expression $\dfrac{3x^2+3x+1}{x^3}$ decreases for all values of x with the exception of the value $x = 0$.

14. If the semi-vertical angle of a right circular cone of slant height l is θ, show that the volume of the cone is $\frac{1}{3}\pi l^3 \sin^2 \theta \cos \theta$. Determine the value of θ for which the volume is a maximum, taking the slant height as constant.

15. Show that the expression $\dfrac{1+x\tan x}{x}$ is a minimum when $x = \cos x$.

Find, graphically, the approximate value of x between 0 and $\dfrac{\pi}{2}$.

Differentiation of Implicit Functions

If y is given implicitly in terms of x, it is usually difficult to obtain y directly in terms of x.

E.g. if $y^3 + 2yx + 2x^2 - 3y + x - 6 = 0$, it is clear that y cannot be expressed directly in terms of x.

It may be necessary in such a case to determine the differential coefficient $\dfrac{dy}{dx}$, i.e. the gradient of the curve at any point.

As an example let us take the equation $y^2 + x^2 = 16$.

In this simple case we can find y in terms of $x (i.e.\ y = \pm\sqrt{16 - x^2})$ and hence obtain $\dfrac{dy}{dx}$.

The following method, which will apply to all cases of implicit functions, is much simpler.

We have $\qquad y^2 + x^2 = 16.$

Differentiate each term with respect to x.

Then $\qquad \dfrac{d}{dx}(y^2) + \dfrac{d}{dx}(x^2) = \dfrac{d}{dx}(16).$

Now $\qquad \dfrac{d}{dx}(y^2) = \dfrac{d}{dy}(y^2) \times \dfrac{dy}{dx},$ (using the result for the function of a function)

$$= 2y\dfrac{dy}{dx}.$$

$$\therefore\ 2y\dfrac{dy}{dx} + 2x = 0.$$

i.e. $\qquad \dfrac{dy}{dx} = \dfrac{-2x}{2y} = -\dfrac{x}{y}.$

This method of differentiating is known as *Implicit Differentiation*.

Ex. 4. Find $\dfrac{dy}{dx}$ *if*   (i) $3x^2 - 2y = y^2$;

(ii) $\dfrac{x^2}{4} + \dfrac{y^2}{9} = 1$;

(iii) $3xy \quad x^3 - y^3$.

(i) $3x^2 - 2y = y^2$.

Differentiating with respect to x

$$\frac{d}{dx}(3x^2) - \frac{d}{dx}(2y) = \frac{d}{dx}(y^2).$$

$$\therefore\ 6x - 2 \cdot \frac{dy}{dx} = 2y \cdot \frac{dy}{dx}.$$

$$\therefore\ \frac{dy}{dx} = \frac{3x}{1+y}.$$

(ii) $\dfrac{x^2}{4} + \dfrac{y^2}{9} = 1$.

Differentiating with respect to x

$$\frac{d}{dx}\left(\frac{x^2}{4}\right) + \frac{d}{dx}\left(\frac{y^2}{9}\right) = \frac{d}{dx}(1).$$

$$\therefore\ \frac{2x}{4} + \frac{2y}{9} \cdot \frac{dy}{dx} = 0.$$

$$\therefore\ \frac{dy}{dx} = -\frac{9x}{4y}.$$

(iii) $3xy - x^3 = y^3$.

Differentiating with respect to x

$$3 \cdot \frac{d}{dx}(xy) - \frac{d}{dx}(x^3) = \frac{d}{dx}(y^3).$$

Differentiating the product xy, we have

$$\frac{d}{dx}(xy) = \frac{d}{dx}(x) \cdot y + x \cdot \frac{d}{dx}(y),$$

$$= y + x\frac{dy}{dx}.$$

$$\therefore\ 3\left(y + x\frac{dy}{dx}\right) - 3x^2 = 3y^2\frac{dy}{dx}.$$

$$\therefore \ y - x^2 = \frac{dy}{dx}(y^2 - x).$$

$$\therefore \ \frac{dy}{dx} = \frac{y - x^2}{y^2 - x}.$$

Ex. 5. *Find the equations of the tangent and the normal to the curve* $4x^2 - 7y^2 = 36$ *at the point* $(4, 2)$.

We have $\qquad\qquad 4x^2 - 7y^2 = 36.$

Differentiating with respect to x,

$$8x - 14y \cdot \frac{dy}{dx} = 0.$$

$$\therefore \ \frac{dy}{dx} = \frac{8x}{14y} = \frac{4x}{7y}.$$

$\therefore$ When $x = 4$, $y = 2$, $\dfrac{dy}{dx} = \frac{16}{14} = \frac{8}{7}.$

$\left.\begin{array}{l}\therefore \ \text{Gradient of the tangent at the point } (4, 2) = \frac{8}{7} \\[4pt] \therefore \ \text{Gradient of the normal at the point } (4, 2) = -\frac{7}{8}\end{array}\right\}$

$\therefore$ Equation of the tangent at $(4, 2)$ is $y - 2 \ = \frac{8}{7}(x - 4).$

$$i.e. \quad 7y - 14 = 8x - 32,$$

or $\qquad\qquad\qquad\qquad 7y = 8x - 18.$

Equation of the normal at $(4, 2)$ is

$$y - 2 = -\tfrac{7}{8}(x - 4).$$

$$i.e. \quad 8y - 16 = -7x + 28,$$

or $\qquad\qquad\qquad\qquad 8y + 7x = 44.$

Ex. 6. *The volume of metal in a hollow sphere is constant. If the inner radius is increasing at the rate of* $1 \ cm \ s^{-1}$, *find the rate of increase of the outer radius when the radii are* $4 \ cm$ *and* $8 \ cm$ *respectively.*

Let $V \ cm^3 = $ volume of the metal.

Let the inner and outer radii be r and R respectively.

Then $\qquad\qquad\qquad V = \tfrac{4}{3}\pi R^3 - \tfrac{4}{3}\pi r^3.$

Differentiating implicitly with respect to t, the time,

we have $\qquad\qquad \dfrac{dV}{dt} = \tfrac{4}{3}\pi\left\{3R^2\dfrac{dR}{dt} - 3r^2\dfrac{dr}{dt}\right\}.$

But as V is constant, $\qquad \dfrac{dV}{dt} = 0.$

and also $\dfrac{dr}{dt}$ = rate of increase of $r = 1$ cm s^{-1}.

$$\therefore \ 0 = \tfrac{4}{3}\pi \left\{ 3R^2\dfrac{dR}{dt} - 3r^2 \cdot 1 \right\}.$$

$$\therefore \ \dfrac{dR}{dt} = \dfrac{r^2}{R^2}.$$

$\therefore$ When $\left. \begin{array}{l} r = 4 \text{ cm} \\ R = 8 \text{ cm} \end{array} \right\}$, $\dfrac{dR}{dt} = \tfrac{1}{4}$ cm s^{-1}.

$\therefore$ The rate of increase of the outer radius $= \tfrac{1}{4}$ cm s^{-1}.

Maximum and Minimum Values of an Implicit Function

Ex. 7. Find the maximum and minimum values of y, *where* $x^2 - y^2 + 10x - 5y + 19 = 0$.

$$x^2 - y^2 + 10x - 5y + 19 = 0. \tag{1}$$

Differentiating implicitly with respect to x,

$$2x - 2y\dfrac{dy}{dx} + 10 - 5\dfrac{dy}{dx} = 0. \tag{2}$$

Hence $\dfrac{dy}{dx} = 0$, when $2x - 2y(0) + 10 - 5(0) = 0$.

$$i.e. \quad 2x + 10 = 0,$$

$$\therefore \ x = -5.$$

Differentiating (2) with respect to x,

$$2 - 2\left\{ y\dfrac{d^2y}{dx^2} + \dfrac{dy}{dx}\cdot\dfrac{dy}{dx} \right\} - 5\dfrac{d^2y}{dx^2} = 0. \quad \begin{array}{l}[\textit{Note} \text{ the use of the product} \\ \text{rule in differentiating the} \\ \text{term } y \cdot \dfrac{dy}{dx}\Big]. \end{array}$$

$$\therefore \text{ When } \dfrac{dy}{dx} = 0, \quad \dfrac{d^2y}{dx^2} = \dfrac{2}{5 + 2y}.$$

So, in order to determine the value of $\dfrac{d^2y}{dx^2}$ when $x = -5$, we must first determine the value of y.

Substituting $x = -5$ in (1), we have $25 - y^2 - 50 - 5y + 19 = 0$,

$$i.e. \quad y^2 + 5y + 6 = 0.$$

$$\therefore \ y = -2 \text{ and } -3.$$

When $y = -2$, $\dfrac{d^2y}{dx^2} = \dfrac{2}{5-4} = 2$ (*i.e.* positive),

and when $y = -3$, $\dfrac{d^2y}{dx^2} = \dfrac{2}{5-6} = -2$ (*i.e.* negative).

Hence when $y = -2$, $\dfrac{dy}{dx} = 0$ and $\dfrac{d^2y}{dx^2}$ is positive.

Hence   y has a minimum value -2,

and similarly   y has a maximum value -3.

(B) EXAMPLES 18c

1. Find $\dfrac{dy}{dx}$ in the following cases

     (i) $x^2 - 2y^2 = 4$;   (ii) $y^3 - x^3 = 1$;

     (iii) $x^2 + y^2 + 2x - 3 = 0$;   (iv) $y^2 = 4x$;

     (v) $y^3 = 3(x-2)$;   (vi) $xy = 1$;

     (vii) $2x^2 - xy + y^2 + 3y - 4 = 0$;   (viii) $\dfrac{1}{x} + \dfrac{1}{y} = 2$;

     (ix) $x^3y^2 - x = 0$;   (x) $\sqrt{x} + \sqrt{y} = 2$.

2. Find $\dfrac{dr}{d\theta}$ if (i) $r^2 \cos \theta = c$ {a constant};

     (ii) $r\{\sin 2\theta + \cos 2\theta\} = r + r \sin \theta + 1$.

3. Find the gradients of the following curves

     (i) $y^2 = 8x$ at the point $(2, 4)$;

     (ii) $\dfrac{x^2}{9} + \dfrac{y^2}{16} = 1$ at the point $(0, 4)$;

     (iii) $x^2 + 2y^2 = 3$ at the point $(1, 1)$;

     (iv) $\dfrac{x^2}{25} - \dfrac{y^2}{16} = 1$ at the point $(5\sqrt{2}, 4)$.

4. If $x^2 + 2xy + 3y^2 = 1$, find $\dfrac{dy}{dx}$ and prove that $\dfrac{d^2y}{dx^2} = \dfrac{-2}{(x+3y)^3}$.

5. Find the value of $\dfrac{d^2s}{dt^2}$ when $s = -2$, if $2s^2t^3 + 2s + 3 = 0$.

6. Find the gradients of the curves $y^2 = x$ and $x^2 = 8y$ at their points of intersection. Hence find the angles at which the curves cut.

7. Find the points of intersection of the curves $xy = 1$, $2x^2 - y^2 = 1$, and determine the angles at which they cut each other.

8. Find the equations of the tangent and normal to the curve $4x^2 - 9y^2 = 36$ at the point $(3\sqrt{2}, 2)$.

9. If $\sin y = x$, find $\dfrac{dy}{dx}$

    (i) in terms of y;   (ii) in terms of x.

    [When $\sin y = x$ we write $y = \sin^{-1} x$.]

10. When $y = \cos^{-1} x$ [*i.e.* $\cos y = x$], find $\dfrac{dy}{dx}$ in terms of x.

11. Given that $\sin y = u$ where $u = 2x^2 - 1$. Find $\dfrac{dy}{dx}$ in terms of x.

12. Express $\dfrac{dv}{dt}$ in terms of $\dfrac{dv}{ds}$ and $\dfrac{ds}{dt}$. Hence express the acceleration of a body in the form $v\dfrac{dv}{ds}$, where s is the distance travelled and v the velocity. If the velocity of a body, v, is given by the expression $v^2 = \omega(a^2 - x^2)$, where x is the distance travelled and ω and a are constants, prove that the retardation is proportional to the distance travelled.

13. Find the maximum and minimum values of y if

(i) $x^2 - 2y^2 + 6x - 3y + 18 = 0$;   (ii) $x^2 + y^2 - 6x - 8y + 16 = 0$.

14. If $r^2 = 9 \cos 2\theta$, find the values of θ for which r is a maximum or a minimum.

15. The volume of metal in a hollow circular cylinder of height h cm is constant. If the outer radius increases at the rate of 2 cm min^{-1}, find the rate of increase of the inner radius when the radii are 3 cm and 5 cm. [h is constant.]

16. The volume of a right circular cone is constant and equal to 30 cm^3. If the height of the cone is decreasing at the rate of 3 cm s^{-1}, find the rate at which the radius of the base is increasing when the height is 9 cm.

17. The ends A, B of a rod **AB** of length 2 m move on two perpendicular lines, **OA**, **OB**. A is moving away from O with a velocity of 6 cm s^{-1}. Find the velocity of B when OB $= 1$ m.

18. A bird A is flying horizontally, 150 m above the ground, at a speed of 10 m s^{-1}. Find the rate at which its distance from a point C on the ground is increasing when CB $= 200$ m, where B is the point on the ground vertically below A.

19. The distance, v of the image from a concave mirror, radius r, of an object distant u from the mirror is given by the formula $\dfrac{1}{v}+\dfrac{1}{u}=\dfrac{2}{r}$.

If the object approaches the mirror at a speed of 5 cm s^{-1}, find the rate at which the image moves away from the mirror when $r = 12$ cm and $v = 8$ cm.

REVISION PAPERS IV

PAPER A (1)

1. If x is an acute angle whose sine is $\frac{3}{5}$, find, without using tables, the values of $\cos x$, $\sin 2x$, $\cos 2x$.

2. Find the smallest angle and largest altitude of the triangle whose sides are of lengths 6 cm, 7 cm, 10 cm.

3. (a) Differentiate with respect to x

    (i) $x - 3 \sin 2x$;   (ii) $3x^3 \cos x$.

  (b) Evaluate

    (i) $\displaystyle\int_0^{\frac{\pi}{6}} \sin 3x\, dx$;   (ii) $\displaystyle\int (\cos x + \cos 2x)\, dx$.

4. Find the equations of the tangent and normal to the curve $y = 2 \sin \dfrac{x}{2}$ at the point where $x = \dfrac{\pi}{3}$.

5. The area contained between the curve $y = 2 \sec x$, the x-axis and the ordinates $x = -\dfrac{\pi}{4}$, $x = \dfrac{\pi}{4}$, is rotated about the axis of x. Find the volume swept out.

PAPER A (2)

1. (i) Find the possible values of $\tan \frac{1}{2}\theta$, if $\tan \theta = \frac{5}{12}$.

  (ii) Solve the equation $5 \sin 2\theta = 2 \sin \theta$ giving all solutions between $0°$ and $180°$.

2. Solve the triangles **ABC** in which $\hat{\mathbf{B}} = 20°$, $\mathbf{AC} = \frac{1}{2}\mathbf{AB}$ and $\mathbf{BC} = 6$ cm.

3. If **A** and **B** are acute angles such that $\sin \mathbf{A} = \frac{3}{5}$, $\cos \mathbf{B} = \frac{12}{13}$, find the value of $\tan(\mathbf{A} - \mathbf{B})$.

4. (a) Differentiate with respect to x

    (i) $2 \tan 3x$;   (ii) $\dfrac{2x - 1}{x + 1}$.

  (b) Find the minimum value of the expression $(x - 2 \cos x)$, where x is measured in radians.

5. Evaluate the following integrals

(i) $\displaystyle\int (2\sin x - x)dx$; 　　(ii) $\displaystyle\int_0^{\frac{\pi}{4}} 2\cos 4x dx$;

(iii) $\displaystyle\int_{-\frac{\pi}{2}}^{\frac{\pi}{2}} \frac{\sec x + 1}{\sec x} dx$.

PAPER A (3)

1. Write down expressions for $\sin(A+B)$ and $\sin(A-B)$ and deduce the value of $\sin(A+B)+\sin(A-B)$. Use this result to evaluate $\sin 195° + \sin 75°$ without using tables.

2. In a triangle ABC, AB = 8, AC = 6 and angle ABC = 35°. Show that two triangles can be constructed with these data and find the difference between the two possible values of angle ACB.

3. Differentiate with respect to x

(i) $\dfrac{3}{x} + \tan x$; 　　(ii) $\dfrac{1+x^2}{3x-1}$;

(iii) $\sin 2x \cos 2x$; 　　(iv) $\tan^3 x$.

4. If $y = 2\cos 2x$, find the approximate change in y due to an increase in x from $\dfrac{\pi}{6}$ to $\left(\dfrac{21\pi}{120}\right)$.

5. (i) If $\dfrac{dy}{dx} = 1 + \sin x$ and $y = 0$ when $x = 0$, find y in terms of x.

(ii) Find the area included between the curve $y = 3\sin\dfrac{x}{3}$, the x-axis and the ordinates $x = 0$, $x = 3\pi$.

PAPER A (4)

1. If $\tan x = \dfrac{a}{b}$, prove that $\cos 2x = \dfrac{b^2 - a^2}{b^2 + a^2}$.

2. Solve the equation $2\cos 2\theta = 1 - 4\cos\theta$, giving only the acute value of θ.

3. A plane is inclined to the horizontal at an angle whose sine is $\frac{1}{3}$. At an angle of 45° with the line of greatest slope is drawn another line on the plane. Find, to the nearest degree, the angle that this line makes with the horizontal.

4. (*a*) Differentiate with respect to x

    (i) $(3x^2 - x + 1)(\sin x)$;   (ii) $3 \sin^3 x$.

    (*b*) Find the value of $\tan x$ for which $\dfrac{dy}{dx}$ is zero where $y = 1 + 2 \sin x + 3 \cos x$.

5. Evaluate

    (i) $\displaystyle\int_0^{\frac{\pi}{2}} (\sin x + \sin 2x)dx$;   (ii) $\displaystyle\int_{-\frac{\pi}{2}}^{\frac{\pi}{2}} (1 + \cos x)^2 dx$.

PAPER A (5)

1. If $\tan \alpha = \frac{3}{4}$ and α is an angle in the third quadrant, find the values of $\sin \alpha$, $\cos \alpha$, $\tan 2\alpha$, $\cos 2\alpha$.

2. Find the area of the triangle **ABC** in which $a = 10$ cm, $\mathbf{B} = 41°$, $\mathbf{C} = 67°$.

3. (i) Factorise $\sin 3x - \sin x$.

    (ii) Without using tables, prove that $\dfrac{\cos 75° + \cos 15°}{\sin 75° - \sin 15°} = \sqrt{3}$.

4. The distance s travelled by a point in time t is given by the relation $s = 10 \sin \pi t$. Find (*a*) the maximum value of s; (*b*) the velocity of the point when $t = \frac{1}{3}$.

5. Evaluate

    (i) $\displaystyle\int_0^{\frac{\pi}{4}} 2 \sec^2 x dx$;   (ii) $\displaystyle\int_{-\frac{\pi}{2}}^{\frac{\pi}{2}} (\sin^2 x + \cos^2 x)dx$.

PAPER A (6)

1. If θ and ϕ are acute angles such that $\tan \theta = \frac{1}{7}$ and $\tan \phi = \frac{1}{3}$, find the value of $\tan(\theta + 2\phi)$. What is the value of $\theta + 2\phi$?

2. Find the lengths of the diagonals of the parallelogram **ABCD** in which $\mathbf{AB} = 7$ cm, $\mathbf{BC} = 5$ cm and $\cos \mathbf{A\hat{B}C} = -\frac{1}{3}$.

3. (i) If $p(v + 2) = 1 - v$, find the value of $\dfrac{dp}{dv}$ when $v = 1$.

    (ii) Differentiate $(x^2 \tan x)$ with respect to x.

4. Given that $x = 2(\theta - \sin \theta)$, $y = 2(1 - \cos \theta)$, find $\dfrac{dx}{d\theta}$, $\dfrac{dy}{d\theta}$, and deduce that $\dfrac{dy}{dx} = \cot \dfrac{\theta}{2}$.

5. Find the area included between the curve $y = \cos 2x$, the x-axis and the ordinates $x = -\dfrac{\pi}{4}$, $x = \dfrac{\pi}{4}$.

PAPER B (1)

1. By means of an isoceles right-angled triangle show that $\sin 45° = \frac{1}{2}\sqrt{2}$.

A, **D**, **B** are points in a straight line such that $AD = 2$. $DB = 3$. At **D** a perpendicular **DC** is drawn to **AB**, and $DC = 1$. If angle $CAB = x°$ and angle $CBD = y°$, show that $x + y = 45°$. Also show that $5 \cos (x - y) = 7 \sin (x + y)$. (J.M.B.)

2. In the convex quadrilateral **ABCD**, $AB = 5$ cm, $BC = 8$ cm. $CD = 3$ cm, and $DA = 3$ cm. If the diagonal $BD = 7$ cm, find the angle **A**, and prove that the quadrilateral is cyclic. Show also that $\cos D\hat{C}A = \frac{13}{14}$. (J.M.B.)

3. An aeroplane leaves Bristol at noon and flies due north at a speed of 800 km h^{-1}; half an hour after it starts another aeroplane leaves London and flies in a direction $10°$ south of west at 900 km h^{-1}. Taking Bristol to lie 150 km west of London, find the distance between the aeroplanes at 2.15 on that afternoon. (J.M.B.)

4. Assuming the rule for differentiating a product, prove that, if u and v are functions of x, $\dfrac{d}{dx}(u^m v^n) = u^{m-1} v^{n-1}\left\{ mv\dfrac{du}{dx} + nu\dfrac{dv}{dx}\right\}$; m and n being positive integers.

Hence find the differential coefficients of

$$\text{(i) } x^7(2x - 1)^4; \qquad \text{(ii) } \sin^4 x \cos^4 x. \qquad \text{(J.M.B.)}$$

5. (i) If $y = \dfrac{\sin 2x}{4} + \dfrac{x}{2}$, show that $\dfrac{dy}{dx} = \cos^2 x$.

(ii) The arc of the curve $y = 1 + \cos x$ between $x = 0$ and $x = \dfrac{\pi}{2}$ is rotated about the axis of x. prove that the volume thus obtained is approximately $13 \cdot 7$. (take $\pi = 3 \cdot 14$.) (J.M.B.)

PAPER B (2)

1. (i) Prove that $(\sec \theta - \tan \theta)^2 = \dfrac{1 - \sin \theta}{1 + \sin \theta}$.

(ii) Solve the equation $\sin \theta = \cos 2\theta$, giving values of θ between $0°$ and $180°$.

2. (i) With the usual notation, prove that $\cos C = \dfrac{a^2+b^2-c^2}{2ab}$.

(ii) **ABC** is a triangle having **AB** $= 11$, **BC** $= 10$, **CA** $= 7$. Calculate the angle **ACB** to the nearest minute. If **M** is the mid-point of **BC** show that **AM** $= \sqrt{60}$. (J.M.B.)

3. For the curve $y = x \cos x$ (x being in radians) prove that the gradient vanishes when $\cot x - x$. By plotting the curves $y = \cot x$ and $y = x$ show that $x = 0{\cdot}86$ is an approximate solution of this equation.

4. (i) Differentiate with respect to x

$$(a)\ \frac{\sqrt{x^2+1}}{(1-x)^3}; \qquad\qquad (b)\ \sqrt{\sin 2x}.$$

(ii) Find the gradient of the curve $y = \dfrac{3x+2}{(x+1)^2}$ at the point $x = -\tfrac{1}{3}$.

5. A point moves in a straight line so that its distance x m from the origin after t s is given by the equation

$$x = 2\sin t + \cos t.$$

If after time t its velocity is v and acceleration a, show that $v^2 + x^2 = 5$ and $x + a = 0$.

What is the greatest distance of the point from **O**?

Show that after 3 s its velocity is approximately $-2{\cdot}12$ m s^{-1}, and find its acceleration at this time. (1 radian $= 57{\cdot}3°$.) (J.M.B.)

PAPER B (3)

1. Prove that the value of the expression $a\cos x + b\sin x$ cannot be greater than $\sqrt{a^2+b^2}$.

Find all the angles between $0°$ and $360°$ which satisfy the equation

$$9\cos 2x + 40\sin 2x = 30. \qquad\qquad (\text{J.M.B.})$$

2. In the triangle **ABC**, $a = 38{\cdot}37$ cm, $b = 21{\cdot}06$ cm, $C = 41° 32'$. Find the remaining side and angles.

3. (i) If $\dfrac{a\cos\theta}{x\sin\theta + y\cos\theta} = k$, find the value of $\tan\theta$ in terms of a, k, x, y.

(ii) A rectangular piece of cardboard **ABCD**, in which **AB** $= 3$**BC**, is fixed against a vertical wall. The heights, in cm, of **A**, **B**, **C** above the ground are respectively 41, 62 and 104. Find the inclination of **AB** to the horizontal.

4. Find the derivatives of $\dfrac{2x+3}{(x-1)^3}$ and $(x^2+1)\sin 2x$ with respect to x.

The internal and external radii of a hollow sphere at a certain instant are 4 cm and 10 cm respectively, and the internal radius is increasing at a rate of 1 mm s^{-1}. At what rate must the external radius be increasing at that instant if the volume of the hollow sphere is unchanged?

(J.M.B.)

5. Show that the result of differentiating $\dfrac{x}{2} - \dfrac{\sin 4x}{8}$ is $\sin^2 2x$.

The area in the first quadrant bounded by the curve $y - 2 - \sin 2x$ and the ordinates $x = 0$ and $x = \dfrac{\pi}{4}$ is rotated about the x-axis. Find the volume thus obtained.

(J.M.B.)

PAPER B (4)

1. If $\tan \dfrac{\theta}{2} = t$, show that $\sin \theta = \dfrac{2t}{1+t^2}$, $\cos \theta = \dfrac{1-t^2}{1+t^2}$.

Hence, or otherwise, solve the equation $4 \sin x = 3 + 2 \cos x$, giving values of x between $0°$ and $360°$.

2. In a triangle **ABC**, $b = 4$ m, $c = 5$ m, **A** $= 55°$. The line **AD** bisects the angle **A** and meets **BC** at **D**. Calculate the lengths of **BC** and **AD**.

(J.M.B.)

3. In any triangle **ABC**, prove that

$$a^2 = b^2 + c^2 - 2bc \cos \mathbf{A}.$$

A tower is north of a point **A** and west of a point **B**. The angles of elevation of the top of the tower, **C**, are α at **A** and β at **B**. Prove that $\cos \mathbf{ACB} = \sin \alpha \sin \beta$.

(L.U.)

4. (i) Differentiate with respect to x

(a) $(1-x^3)\sqrt{x+1}$;   (b) $\sec x \tan x$.

(ii) Find the maximum and minimum values of $\dfrac{9}{x} + \dfrac{4}{1-x}$.

5. Sketch the curve $y = \sin 2x$ for values of x between 0 and π.

Find the area of the segment cut off on the curve by the straight line $y = \frac{1}{2}$.

PAPER B (5)

1. The internal bisectors of the angles of a triangle **ABC** meet at **P**. Calculate, correct to the nearest metre, the length of **PC** if **A** $= 105° \, 10'$, **B** $= 58° \, 20'$, $a = 1000$ m.

(J.M.B.)

2. A person stands between two chimneys **AB** and **CD**, 100 m apart, in the line **AC** joining their bases, and observes that the elevation of the top of **AB** is 60°. After walking 75 m in a direction at right angles to **AC**, he finds that the elevation of the top of **CD** is 30°. If the height of **AB** is known to be 50 m, find the height of **CD**. (J.M.B.)

3. Find the values of x between 0 and $\dfrac{\pi}{2}$ which satisfy the equation $\cos x = \sin 2x$.

Hence find the maximum and minimum values of $\sin x + \frac{1}{2}\cos 2x$.

4. (i) Differentiate with respect to x

$$(a)\ \frac{1-x^3}{\sqrt{x}}; \qquad\qquad (b)\ (x^3-1)\cos 6x.$$

(ii) if $x = \dfrac{3m}{1+m^3}$, $y = \dfrac{3m^2}{1+m^3}$, find $\dfrac{dy}{dx}$ in terms of m.

5. A point starts from the origin and moves along the x-axis in such a way that its velocity after time t is $\pi\,(\sin \pi t + \cos \pi t)$. Find the distance travelled in time t and also find the acceleration of the point when $t = 1$.

PAPER B (6)

1. (i) Given that $a = \sin\theta + \sin 2\theta$, $b = \cos\theta + \cos 2\theta$, prove that $(a+b)^2 = 4\cos\frac{1}{2}\theta(1+\sin 3\theta)$.

(ii) Solve the equation $\sin 4x + \sin 2x = \cos 4x + \cos 2x$ for values of x in the interval $0° \leqslant x \leqslant 180°$.

2. The triangle **ABC** has **A** = 12° and **C** = 28°. Prove that the sum of **AB** and **BC** exceeds **AC** by rather more than 5 per cent. of **AC**. (J.M.B.)

3. From a stationary balloon at an altitude of 4000 m two places **P** and **Q** at sea-level are observed, **P** bears due west and **Q** bears N. 14° W. The angles of depression of **P** and **Q** are 35° 10′ and 27° 30′ respectively. Find the distance **PQ**.

4. (i) Differentiate with respect to x

$$(a)\ x^3\sin(2x+1); \qquad\qquad (b)\ \frac{2x+1}{(x+1)(x+2)}.$$

(ii) In a triangle **ABC** the angles **B** and **C** are equal. Prove that the maximum value of $\cos\mathbf{A} + \cos\mathbf{B}$ is $\frac{9}{8}$.

5. Evaluate (i) $\displaystyle\int\left(x+\frac{1}{x}\right)^2 dx;$ \qquad (ii) $\displaystyle\int(1+\cos x)^2 dx.$

The gradient of a curve at any point (x, y) varies directly as $(x-2)$. It passes through the point $(10, -9)$ and its gradient there is -3. Find the equation of the curve, the co-ordinates of the points where it crosses the x-axis, and the area between the curve and the x-axis. (L.U.)

Chapter 19

THE STRAIGHT LINE AND THE CIRCLE

The Straight Line

The following results related to the straight line have already been established.

(i) *an equation of the first degree represents a straight line and conversely;*

(ii) *the equation $y = mx + c$ represents a straight line with gradient m;*

(iii) *the equation of the straight line passing through the point (h, k) and having gradient m is $y - k = m(x - h)$;*

(iv) *the equation of the straight line passing through the points (x_1, y_1), (x_2, y_2) is*

$$\frac{y - y_1}{y_2 - y_1} = \frac{x - x_1}{x_2 - x_1};$$

(v) *if m, m' are the gradients of perpendicular lines, then $mm' = -1$.*

Point dividing a line in a given ratio

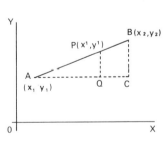

Fig. 128

Take the case of internal division and let the point $P(x', y')$ divide the line joining $A(x_1, y_1)$ and $B(x_2, y_2)$ in the ratio $m : n$ (Fig. 128).

As PQ is parallel to BC,

$$\frac{AP}{PB} = \frac{AQ}{QC} = \frac{m}{n}.$$

But $AQ = x' - x$, $\quad QC = x_2 - x'$.

$$\frac{x' - x_1}{x_2 - x'} = \frac{m}{n},$$

$$nx' - nx_1 = mx_2 - mx',$$

$$x'(m + n) = mx_2 + nx_1,$$

$$\mathbf{x' = \frac{mx_2 + nx_1}{m + n};} \quad \text{similarly} \quad \mathbf{y' = \frac{my_2 + ny_1}{m + n}.}$$

N.B.   (i) If $m = n$, **P** is the mid-point of **AB** and has coordinates $[\frac{1}{2}(x_1 + x_2), \frac{1}{2}(y_1 + y_2)]$.

(ii) If **P** divides **AB** *externally* in the ratio $m : n$, the result obtained holds so long as either m or n is taken as negative.

Ex. 1. Given the points **A**$(-2, 3)$, **B**$(1, -1)$. *Find* (*i*) *the co-ordinates of the point* **P** *which divides* **AB** *internally in the ratio* $1 : 2$; (*ii*) *the co-ordinates of the point* **Q** *which divides* **AB** *externally in the ratio* $5 : 3$.

(i) Taking $m = 1$, $n = 2$, $x = \dfrac{1(1) + 2(-2)}{1 + 2}$; $y = \dfrac{1(-1) + 2(3)}{1 + 2}$,

$$= -1. \qquad\qquad = \tfrac{5}{3}.$$

So **P** is the point $(-1, \tfrac{5}{3})$.

(ii) Taking $m = 5$, $n = -3$,

$$x = \frac{5(1) + (-3)(-2)}{5 + (-3)}, \qquad y = \frac{5(-1) + (-3)(3)}{5 + (-3)},$$

$$= 5\tfrac{1}{2}. \qquad\qquad = -7.$$

So **Q** is the point $(5\tfrac{1}{2}, -7)$.

Ex. 2. Find the ratio in which the point **P**$(4, 5)$ *divides the line joining the points* **A**$(-2, 1)$, **B**$(1, 3)$.

Let the ratio be $m : n$.

Taking the x-co-ordinate of **P**, $4 = \dfrac{m(1) + n(-2)}{m + n}$,

$$4m + 4n = m - 2n,$$

$$3m = -6n.$$

$$m : n = -2 : 1.$$

So **P** divides **AB** externally in the ratio $2 : 1$.

Length of the perpendicular from a point to a line

General case. To find the perpendicular distance of the point (h, k) *from the line* $ax + by + c = 0$.

As the y co-ordinate of **P** and **R** is k, the x co-ordinate of **R** is

$$\frac{-bk - c}{a}. \quad \text{(Fig. 129)}$$

$$\text{Length } \mathbf{RP} = h - \left(\frac{-bk - c}{a}\right),$$

$$= \frac{ah + bk + c}{a}.$$

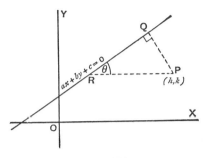

FIG. 129

But $\tan \theta = \dfrac{-a}{b}$; $\therefore \sin \theta = \pm \dfrac{a}{\sqrt{a^2+b^2}}$.

$$\therefore \text{PQ} = \text{PR} \sin \theta = \pm \frac{ah+bk+c}{a} \cdot \frac{a}{\sqrt{a^2+b^2}}$$

$$= \pm \frac{ah+bk+c}{\sqrt{a^2+b^2}}.$$

So the numerical value of the length of the perpendicular from the point (h, k) to the line ax + by + c = 0 is given by the result,

$$\textbf{perpendicular} = \pm \frac{\textbf{ah+bk+c}}{\sqrt{\textbf{a}^2+\textbf{b}^2}}.$$

Sign of the perpendicular

For points which lie on opposite sides of a line the perpendiculars are of opposite signs. It is usual to take the perpendicular from the origin as positive. If only the numerical length of a perpendicular is required, a negative sign is ignored.

Ex. 3. Find the length of the perpendicular from the point (1, 0) *to the line* $3y - 4x + 1 = 0$.

Substituting in the general result, $a = -4$, $b = 3$, $c = 1$ and $h = 1$, $k = 0$, we get,

$$\text{perpendicular} = \pm \frac{-4+1}{\sqrt{16+9}} = \pm \left(\frac{-3}{5}\right) = \tfrac{3}{5} \text{ numerically.}$$

Ex. 4. Find the coordinates of the image Q *of the point* P(1, 3) *in the line* $2x - y - 4 = 0$.

Length of perpendicular from P to
the given line

$$= \pm \frac{2(1)+(-1)(3)-4}{\sqrt{4+1}},$$

$$= \sqrt{5} \text{ numerically.}$$

$$\therefore PQ = 2\sqrt{5}.$$

Gradient of PQ $= -\frac{1}{2}$,
so $\quad \tan \theta = \frac{1}{2}$ (Fig. 130).

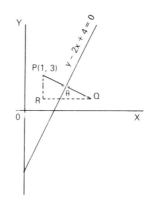

Hence
$$\sin \theta = \frac{1}{\sqrt{5}}, \quad \cos \theta = \frac{2}{\sqrt{5}}.$$

$$\therefore PR = PQ \sin \theta = 2;$$

$$RQ = PQ \cos \theta = 4.$$

So Q is the point $(5, 1)$.   FIG. 130

Angle between two straight lines

Let the lines $y = mx + c$, $y = m'x + c$ make angles of α, β with the positive direction of the x-axis. Then $\tan \alpha = m$; $\tan \beta = m'$. If θ is an angle between the lines,

$$\theta = \alpha - \beta.$$

$$\therefore \tan \theta = \tan(\alpha - \beta),$$

$$= \frac{\tan \alpha - \tan \beta}{1 + \tan \alpha \tan \beta},$$

$$= \frac{m - m'}{1 + mm'}.$$

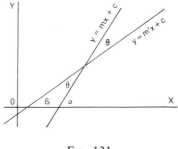

FIG. 131

So the tangent of one of the angles between lines with gradients m and m' is $\dfrac{m - m'}{1 + mm'}$. The tangent of the other angle is $\dfrac{m' - m}{1 + mm'}$.

Ex. 5. Find the acute angle between the lines $2y - x = 3$, $3y + 4x = 5$.
The gradients of the lines are $\frac{1}{2}$ and $-\frac{4}{3}$.

Taking
$$m = \tfrac{1}{2}, m' = -\tfrac{4}{3},$$

$$\tan \theta = \frac{\tfrac{1}{2} - (-\tfrac{4}{3})}{1 + (\tfrac{1}{2})(-\tfrac{4}{3})} = \frac{\tfrac{1}{2} + \tfrac{4}{3}}{1 - \tfrac{2}{3}},$$

$$= \frac{\tfrac{11}{6}}{\tfrac{1}{3}} = \tfrac{11}{2} \quad \text{or} \quad 5 \cdot 5.$$

As the tangent is positive, θ is the required acute angle. From tables, acute angle $= 79° 42'$.

Ex. 6. Find the equations of each of the lines drawn through the point $(-2, -1)$ which are inclined at 45° to the line $y - 2x = 3$.

Take the gradient of the given line, 2, as the value of m, and let m' be the gradient of one of the required lines.

Then,
$$1 = \tan 45° = \frac{2 - m'}{1 + 2m'},$$
$$1 + 2m' = 2 - m',$$
$$m' = \tfrac{1}{3}.$$

The gradient of the second line is most easily obtained by noting that the required lines are perpendicular. So other gradient is -3. Equations of the two lines are

$$y + 1 = \tfrac{1}{3}(x + 2) \quad \text{and} \quad y + 1 = -3(x + 2),$$

i.e.
$$3y = x - 1 \quad \text{and} \quad y + 3x = -7.$$

(A) EXAMPLES 19a

1. Find the co-ordinates of the point which divides the line joining the points $(-3, 1)$, $(4, 4)$ internally in the ratio $3 : 5$.

2. What is the perpendicular distance of the origin from the line $3y - 4x + 10 = 0$?

3. If θ is the acute angle between the lines $x - y + 1 = 0$, $2x + y - 1 = 0$, find the value of $\tan \theta$.

4. Prove that the lines $2y - x - 5 = 0$, $2y + x - 5 = 0$ and $y + 2x = 5$ are equidistant from the origin.

5. Find the value of the obtuse angle between the lines $4y - 3x + 1 = 0$, $2y + x - 3 = 0$.

6. A triangle has vertices $A(-1, 2)$, $B(1, 5)$ and $C(3, 2)$. Find (i) the co-ordinates of A' the midpoint of BC; (ii) the co-ordinates of the centroid G of the triangle which is the point dividing AA' internally in the ratio $2 : 1$.

7. In parallelogram $ABCD$, vertices A, B have co-ordinates $(1, 4)$, $(3, -1)$ respectively and the point of intersection of the diagonals AC, BD has coordinates $(2, 0)$. Find the co-ordinates of the vertices C, D.

8. Show that the point $(-1, 2)$ is equidistant from the lines with equations $3x + 4y = 20$, $5x - 12y = 10$.

9. Determine whether the points $(3, 2)$, $(-1, 0)$ are on the same or opposite sides of the line $4y - x = 3$.

10. Find the acute angles between each of the following pairs of lines:—

(i) $y = x$ and $y = 2x$; (ii) $y = 2x + 1$ and $y = 3x - 2$;
(iii) $2y = x + 1$ and $3y = x - 1$; (iv) $y + 2x = 0$ and $y + 3x = 0$.

11. Find the co-ordinates of the points which divide the line joining the points $(-4, 1)$, $(2, 3)$ internally and externally in the ratio $3:5$.

12. A triangle has vertices $A(0, 2)$, $B(3, 5)$ and $C(5, -1)$. Find (i) the co-ordinates of the point of trisection of the side AB nearer A; (ii) the length of the perpendicular from B to AC; (iii) the angle BAC.

13. A line AB is produced to C so that $2BC = AB$. Given that the co-ordinates of A, B are $(-2, 0)$, $(3, 2)$ respectively, determine the co-ordinates of C.

14. Find the angle subtended at the origin by the line joining the points with co-ordinates $(1, 4)$, $(4, 1)$.

15. A circle with centre $(1, 5)$ touches the line $y - 2x + 2 = 0$. What is the length of the radius of the circle?

16. A line of gradient m is inclined to the line $2y - x + 1 = 0$ at an angle whose tangent is $\frac{1}{3}$. Find the two possible values of m.

17. Find the equation of the straight line drawn through the origin which divides the line joining the points $(-1, 2)$, $(3, 4)$ internally in the ratio $3:2$.

18. Using a diagram, show that the origin lies on the same side of the parallel lines $y - 2x - 1 = 0$, $y - 2x - 3 = 0$. Find the perpendicular distances of the origin from these lines and hence determine the distance between them.

19. Show that the line $3y - 3x = 5$ is equally inclined to each of the lines $y = 2x + 1$, $2y - x = 4$.

20. P divides AB internally in the ratio $1:2$ and Q divides AB externally in the ratio $5:3$. Given that A, B have co-ordinates $(0, 2)$, $(3, 1)$ respectively, find the tangent of the angle POQ, where O is the origin.

21. Determine the distance between the parallel lines $3y - 4x = 6$, $3y - 4x = -4$.

22. Find the equations of the two lines through the point $(3, -1)$ which make an angle of $45°$ with the line $2x = y + 2$.

Loci

The *locus* of a moving point P is the path traced out by P. The equation of the locus of P is normally taken to be the (x, y) equation of the line or curve on which P always lies.

In some cases, a locus is defined by what are called *parametric equations*; examples of these will be given in the next chapter.

One important geometrical locus is the circle and an introduction to the co-ordinate geometry of this curve follows.

The Circle

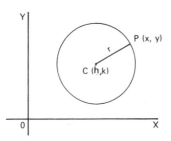

FIG. 132

Definition. A circle is the locus of a point which moves at a constant distance from a fixed point.

Consider the fixed point C (h, k) and let the moving point $P(x, y)$ be at a constant distance r from C.

Then $PC^2 = (x - h)^2 + (y - k)^2 = r^2$. So the relationship between x, y and given constants is

$$(x - h)^2 + (y - k)^2 = r^2.$$

This is the equation of the circle centre (h, k), radius r.

Ex. 7. Find the equation of a circle centre $(-2, 1)$, radius 3.

Using the result $\qquad (x - h)^2 + (y - k)^2 = r^2,$

equation is $\qquad (x + 2)^2 + (y - 1)^2 = 3^2,$

i.e. $\quad x^2 + y^2 + 4x - 2y - 4 = 0.$

General equation of a circle

The equation of the circle centre (h, k), radius r, is

$$(x - h)^2 + (y - k)^2 = r^2,$$

or $\qquad x^2 + y^2 - 2hx - 2ky + h^2 + k^2 - r^2 = 0.$

Conversely, an equation of the form

$$x^2 + y^2 + 2gx + 2fy + c = 0,$$

where g, f and c are constants, will usually represent a circle although there are exceptional cases when this is not so; these special cases will be disregarded.

The equation $x^2 + y^2 + 2gx + 2fy + c = 0$, where g, f, c are constants is called *the general equation* of the circle. It is important to note that in this equation,

(i) the coefficients of x^2 and y^2 are the same; they need not necessarily be unity;

(ii) there is no term in xy.

e.g. the equations

$$x^2 + y^2 - 4x + 5y - 1 = 0,$$
$$2x^2 + 2y^2 + 6x - 7y + 5 = 0,$$

both represent circles.

Centre and radius of a circle whose equation is given in the general form

Ex. 8. Find the centres and radii of the circles with equations (i) $x^2 + y^2 - 4x + 5y - 1 = 0$; (ii) $2x^2 + 2y^2 + 6x - 7y + 5 = 0$.

The equations must be expressed in the form

$$(x-h)^2 + (y-k)^2 = r^2.$$

This is done by completing the square for the x and y terms separately.

(i) Equation is $x^2 - 4x + y^2 + 5y = 1$

But $x^2 - 4x = (x-2)^2 - 2^2$ and $y^2 + 5y = (y + \frac{5}{2})^2 - (\frac{5}{2})^2$.

So equation is $(x-2)^2 - 4 + (y + \frac{5}{2})^2 - \frac{25}{4} = 1$,

or $(x-2)^2 + (y + \frac{5}{2})^2 = \frac{45}{4}$.

Comparing this equation with the standard form, we see that the centre is the point $(2, -\frac{5}{2})$ and the radius is $\frac{1}{2}\sqrt{45}$ or $\frac{3}{2}\sqrt{5}$.

(ii) First divide by 2 to make the coefficients of x^2 and y^2 unity,

giving $x^2 + y^2 + 3x - \frac{7}{2}y + \frac{5}{2} = 0$.

Completing the squares as in (i) above,

$$(x + \frac{3}{2})^2 - \frac{9}{4} + (y - \frac{7}{4})^2 - \frac{49}{16} = -\frac{5}{2},$$
$$(x + \frac{3}{2})^2 + (y - \frac{7}{4})^2 = \frac{45}{16}.$$

Hence the centre is the point $(-\frac{3}{2}, \frac{7}{4})$ and the radius is $\frac{1}{4}\sqrt{45}$ or $\frac{3}{4}\sqrt{5}$.

Determination of the equation of a given circle

In order to obtain the equation of a particular circle it is necessary to find the co-ordinates of the centre and the length of the radius. It is important to remember some of the simple geometrical properties of a circle such as the centre is the midpoint of a diameter; the centre lies on the perpendicular bisector of a chord; a tangent and a radius are at right angles. In some cases, a diagram will be helpful.

Ex. 9. Find the equation of the circle, centre (2, 3), *which touches the line* $2x - y = 6$.

As the perpendicular from the centre to the given line is a radius of the circle,

$$\text{radius} = \pm\frac{2(2) + (-1)(3) - 6}{\sqrt{4+1}} = \sqrt{5}, \text{ numerically.}$$

Equation of circle is $(x-2)^2 + (y-3)^2 = 5$,

i.e. $x^2 + y^2 - 4x - 6y + 8 = 0$.

Ex. 10. Find the equation of the circle which touches the y-axis at the point A(0, 3) *and passes through the point* B(3, 1).

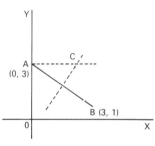

Fig. 133

The centre C is the point of intersection of the line through A perpendicular to the y-axis and the perpendicular bisector of AB.

Equation of AC is $y = 3$.

Equation of the perpendicular bisector of AB

is $\qquad y - 2 = \frac{3}{2}(x - \frac{3}{2})$,

i.e. $4y - 6x + 1 = 0$.

Solving this equation with the equation $y = 3$, we get $x = \frac{13}{6}$. So C is the point $(\frac{13}{6}, 3)$.

Radius of circle = AC = $\frac{13}{6}$.

Equation of circle is $(x - \frac{13}{6})^2 + (y - 3)^2 = (\frac{13}{6})^2$,

i.e. $3(x^2 + y^2) - 13x - 18y + 27 = 0$.

(A) EXAMPLES 19b

In ex. 1–10, sketch and find the equation of the circle in each case.

1. Centre $(1, 1)$; radius 2.   **2.** Centre $(0, 2)$; radius 1.

3. Centre $(0, 0)$; radius 3.   **4.** Centre $(-1, 3)$; radius $\sqrt{3}$.

5. Centre $(0, -2)$; radius $2\sqrt{5}$.   **6.** Centre $(-3, 4)$; radius 5.

7. Centre $(4, 0)$ and passing through the origin.

8. Centre $(-3, -4)$ and passing through the origin.

9. Centre $(-1, 2)$ and passing through the point $(4, 1)$.

10. Centre $(-3, 1)$ and passing through the point $(1, -2)$.

11. AB is a diameter of a circle. Find the equation of the circle in each of the following cases:—

     (i) A$(2, 0)$, B$(4, 2)$;   (ii) A$(-1, -1)$, B$(5, -3)$.
     (iii) $(-2, 4)$, B$(0, -2)$;   (iv) A$(0, -3)$, B$(3, 0)$.

12. A circle, centre $(2, 4)$, touches the x-axis. Find (i) the radius; (ii) the equation of the circle.

13. A circle with centre $(-1, 2)$ touches the y-axis. Find its equation.

14. Find the centre and the radius of each of the following circles:—

     (i) $(x - 1)^2 + (y + 2)^2 = 16$;   (ii) $(x + 2)^2 + (y - 3)^2 = 1$;

(iii) $4x^2 + 4(y-2)^2 = 9$;   (iv) $(x + \frac{3}{2})^2 + y^2 = 5$;
(v) $x^2 + y^2 - 2x - 4y = 0$;   (vi) $x^2 + y^2 - 8x + 7 = 0$;
(vii) $x^2 + y^2 - 6x + 2y - 6 = 0$;   (viii) $x^2 + y^2 - x + y = 0$;
(ix) $x^2 + y^2 + 3x + 2y + 1 = 0$;   (x) $2x^2 + 2y^2 - 8x + 6y + 5 = 0$;
(xi) $2(x^2 + y^2) - 5x + 3y - 1 = 0$;   (xii) $3(x^2 + y^2) + 8x - 2y - 6 = 0$.

15. Show that the circle $x^2 + y^2 - 4x - 4y + 4 = 0$ touches both axes.

16. Find the equation of the circle which is concentric with the circle $x^2 + y^2 + 2x - 4y = 0$ and has a radius of 5 units.

17. Find the equation of the circle which has for a diameter the portion of the line $x + 4y = 8$ lying in the positive quadrant.

18. Obtain the equation of each of the following circles:—
(i) centre $(2, 0)$ and touching the y-axis; (ii) centre $(0, 0)$ and touching the line $3x - 4y + 5 = 0$; (iii) centre $(2, -1)$ and touching the line $x + 2y + 3 = 0$; (iv) concentric with the circle $x^2 + y^2 + 4x + 2 = 0$ and touching the line $y = 3$.

19. Find the equation of the diameter of the circle $x^2 + y^2 - 2x + 4y - 1 = 0$ which passes through the point $(2, -1)$.

20. Show that the circle with centre $(4, -1)$ and radius $\sqrt{2}$ units touches the line $x + y = 1$.

21. Prove that the line $3x + 4y = 25$ is a tangent to the circle $x^2 + y^2 = 25$.

22. A circle touches the x-axis at the point $(4, 0)$ and passes through the point $(0, 4)$. Find its equation.

23. Find the coordinates of the points of intersection of the circles $x^2 + y^2 + 2x - 4y = 0$, $x^2 + y^2 + 4x - 2y = 0$ and hence find the equation of the circle which has for diameter the common chord of the two given circles.

24. Find the equations of the two circles which touch the line $3x - 4y + 2 = 0$ at the point $(2, 2)$ and have a radius of 5 units.

25. Show that the circles $x^2 + y^2 - 2x - 4y - 4 = 0$, $x^2 + y^2 - 8x - 12y + 48 = 0$, touch each other externally.

26. Find the equation of the circle which passes through the origin and the points $(3, 0)$, $(0, 4)$.

Tangents to a circle

The following problems associated with tangents to a circle will be considered:
   (i) length of a tangent from a point to a circle;
   (ii) equation of the tangent to a circle at a given point on the circle;
   (iii) equations of the tangents to a circle from a point outside the circle.

Length of a tangent from a point to a circle

Ex. 11. *Find the length of a tangent drawn from the point* P$(-1, 2)$ *to the circle* $x^2 + y^2 - 4x + 2y + 3 = 0$.

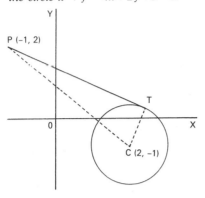

The equation of the circle is

$$(x-2)^2 - 4 + (y+1)^2 - 1 = -3,$$

i.e. $(x-2)^2 + (y+1)^2 = 2.$

So the centre C is the point $(2, -1)$ and the radius is $\sqrt{2}$. In right-angled triangle PTC,

$$PT^2 = PC^2 - CT^2.$$

But $PC^2 = 3^2 + 3^2 = 18,$

and $\qquad CT^2 = 2.$

Fig. 134

$$\therefore PT^2 = 18 - 2 = 16.$$

Length of tangent from P = 4 units.

Equation of the tangent at a given point on a circle

Two examples are given, one using an elementary method dependent upon the perpendicular property of tangent and radius and the other using the more general calculus method.

Ex. 12. *Find the equation of the tangent to the circle* $x^2 + y^2 + 3x - 4y + 3 = 0$ *at the point* P$(0, 3)$.

The centre C of the circle is the point $(-\frac{3}{2}, 2)$.

$$\text{Gradient of radius } CP = \frac{3-2}{0-(-\frac{3}{2})} = \tfrac{2}{3},$$

So gradient of tangent at P $\qquad = -\tfrac{3}{2}.$
Equation of tangent at P is $\qquad y - 3 = -\tfrac{3}{2}(x - 0),$

i.e. $2y + 3x = 6.$

Ex. 13. *Obtain the equation of the tangent to the circle* $x^2 + y^2 - 6x + 3y - 5 = 0$ *at the point* $(-1, -2)$.

To find $\dfrac{dy}{dx}$, differentiate the equation of the circle implicitly with respect to x, giving

$$2x + 2y\frac{dy}{dx} - 6 + 3\frac{dy}{dx} = 0,$$

$$\frac{dy}{dx} = \frac{6 - 2x}{3 + 2y}.$$

At the point $(-1, -2)$, $\qquad$ gradient $= \dfrac{6+2}{3-4} = -8$.

Equation of tangent is $\qquad y + 2 = -8(x + 1)$,

$\qquad$ *i.e.* $\quad y + 8x + 10 = 0$.

Equations of the tangents to a circle from a point outside the circle

Ex. 14. *Find the equations of the tangents to the circle* $x^2 + y^2 + 2x - 4y - 4 = 0$ *from the point* $P(5, \frac{5}{4})$.

The equation of the circle can be written $(x + 1)^2 + (y - 2)^2 = 9$, so the centre is $C(-1, 2)$ and the radius is 3 units.

To find the gradients of the tangents PT_1, PT_2, we get the value of $\tan \theta$, where 2θ is the angle between the tangents.

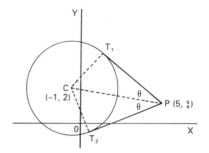

Fig. 135

In triangle PCT_1,

$$PT_1^2 = PC^2 - CT_1^2,$$

and

$$PC^2 = 6^2 + (\tfrac{3}{4})^2 = \tfrac{585}{16}, \ CT_1^2 = 9.$$

$$\therefore \ PT_1^2 = \tfrac{585}{16} - \tfrac{144}{16};$$

i.e. $\ PT_1 = \tfrac{21}{4}$.

$$\therefore \ \tan \theta = \frac{CT_1}{PT_1} = \frac{3}{\tfrac{21}{4}} = \tfrac{4}{7}.$$

Now let the gradient of one of the tangents be m.

Using the result $\tan \theta = \dfrac{m_1 - m_2}{1 + m_1 m_2}$, with $m_1 = m$, and $m_2 =$ gradient of CP which equals $-\frac{1}{8}$,

we have $\qquad\qquad \dfrac{4}{7} = \dfrac{m + \tfrac{1}{8}}{1 - \tfrac{1}{8}m}$,

$$4 - \tfrac{1}{2}m = 7m + \tfrac{7}{8},$$

$$m = \tfrac{5}{12}.$$

From the diagram, Fig. 135, this is clearly the gradient of PT_2. To find m', the gradient of PT_1, use the $\tan \theta$ result with $m_1 = -\frac{1}{8}$, $m_2 = m'$, giving

$$m' = -\tfrac{3}{4}.$$

Equation of PT_1 is $\qquad\qquad y - \tfrac{5}{4} = -\tfrac{3}{4}(x - 5)$,

$\qquad$ *i.e.* $\quad 4y + 3x = 20$.

Equation of PT_2 is $\qquad y - \frac{5}{4} = \frac{5}{12}(x - 5),$

$\qquad i.e. \quad 12y - 5x + 10 = 0.$

(A) EXAMPLES 19c

1. Find the length of the tangents which can be drawn from the point P to the given circle in each of the following cases:

(i) $P(3, 4)$, circle $x^2 + y^2 = 4$; \qquad (ii) $P(0, 0)$, circle $(x - 5)^2 + y^2 = 9$;

(iii) $P(-1, 1)$, circle $(x - 2)^2 + (y + 1)^2 = 2$;

(iv) $P(2, 2)$, circle $(x - 3)^2 + (y - 1)^2 = 1$;

(v) $P(-1, 0)$, circle $x^2 + y^2 - 4x + 2y + 2 = 0$;

(vi) $P(-2, 3)$, circle $x^2 + y^2 + 2x = 0$;

(vii) $P(0, 3)$, circle $x^2 + y^2 - 5x + 2y + 1 = 0$;

(viii) $P(0, 0)$, circle $2(x^2 + y^2) - 9x + 4 = 0$.

2. Find the equations of the tangents to the following circles at the points specified:—

(i) $x^2 + y^2 = 3$, $(1, 1)$; \qquad\qquad (ii) $x^2 + y^2 = 25$, $(-3, 4)$;

(iii) $(x - 1)^2 + (y + 2)^2 = 5$, $(2, 0)$; \qquad (iv) $(x + 3) + y^2 = 25$, $(1, -3)$;

(v) $x^2 + y^2 - 5x + 4y = 0$, $(0, 0)$;

(vi) $x^2 + y^2 + 2x - 4y - 11 = 0$, $(-1, -2)$;

(vii) $x^2 + y^2 - 6x + 4y + 3 = 0$, $(0, -1)$;

(viii) $2(x^2 + y^2) - x + 4y - 15 = 0$, $(3, 0)$.

3. Find the equations of the normals to the circles in ques. 2 at the specified points.

4. Prove that the tangents to the circles $x^2 + y^2 = 6$, $x^2 + y^2 - 6x = 0$ from the point $(1, 4)$ are equal in length.

5. AB is a diameter of the circle $(x - 2)^2 + (y - 3)^2 = 25$ and **A** is the point $(5, -1)$. Find the equations of the tangents to the circle at A and B.

6. Find the equations of the tangents to the circle $x^2 + y^2 + 2x - 4y - 3 = 0$ at the points where it cuts the x-axis.

7. Find the equations of the tangents to the circle $x^2 + y^2 = 8$ which are parallel to the diameter $x + y = 0$.

8. Obtain the equation of the tangent to the circle $x^2 + y^2 - 3x + 4y - 16 = 0$ at the point $(-1, 2)$ and also the equation of the parallel tangent.

9. Tangents are drawn from the point $(0, 1)$ to the circle $(x - 3)^2 + (y - 3)^2 = 4$. Show that one of the tangents is parallel to the x-axis and state its length. If 2θ is the angle between the tangents, show that $\tan \theta = \frac{2}{3}$.

10. Find the co-ordinates of the point on the circle $x^2 + y^2 - 6x - 6y +$

$16 = 0$ which is nearest the origin and find the equation of the tangent at this point.

11. A circle has centre $(2, 2)$. One tangent to the circle from the point $(0, 1)$ has a gradient of 2. Find (i) $\tan \theta$, where 2θ is the angle between the tangents from $(0, 1)$ to the circle; (ii) the gradient of the second tangent.

12. Find the gradients of the tangents which can be drawn from the origin to the circle $(x - 5)^2 + y^2 = 9$.

(B) MISCELLANEOUS EXAMPLES

1. Find the distances of the points $(2, 1)$, $(-3, 2)$ from the line $2y - 3x = 1$ and determine whether the points are on the same or opposite sides of the line.

2. Find the ratio in which the line joining the points $(1, 1)$, $(4, 3)$ is divided by the line $2x + 3y - 7 = 0$.

3. Obtain the equations of the two lines through the point $(-1, -1)$ which each make an angle of $45°$ with the line $x - 2y = 1$.

4. Calculate the co-ordinates of the point **C** in which the line joining the points **A**$(2, 0)$, **B**$(4, 2)$ meets the line $2x - y = 5$. Show that **C** is the midpoint of **AB** and verify this result by showing that the lengths of the perpendiculars from **A** and **B** to the line $2x - y = 5$ are equal in length and that **A** and **B** are on opposite sides of the line.

5. Prove that the line $x + 8y - 15 = 0$ bisects one of the angles between the lines $5x - 12y - 10 = 0$, $3x + 4y - 20 = 0$.

6. Find the image point of the point $(3, 3)$ in the line $y - 2x = 2$.

7. The line $4x + 3y = 24$ cuts the x-axis at **A** and the y-axis at **B**. Find the equation of the perpendicular bisector of **AB**. Given that **P** is the point in the first quadrant which lies on the perpendicular bisector of **AB** and is such that the area of triangle **APB** is 50 units^2, calculate the co-ordinates of **P**.   (J.M.B.)

8. The acute angle between the lines $y = x$, $y = 3x$ is θ and the acute angle between the lines $y = x$, $y + 7x = 0$ is φ. Find $\tan \theta$ and $\tan \varphi$. Hence prove that $\varphi = 2\theta$.   (J.M.B.)

9. Write down the equation of the line passing through the point **A**$(1, 2)$ and having a gradient m. If this line meets the lines $2y = 3x - 5$, $x + y = 12$ at **P** and **Q** respectively, find the x-co-ordinates of **P** and **Q**. Given that **AQ** $= 2$**AP**, find the value of m and the co-ordinates of **P** and **Q**.   (C.)

10. Calculate the co-ordinates of the foot of the perpendicular from the point $(-4, 2)$, to the line $3x + 2y = 5$ and hence find the equation of the smallest circle passing through the point $(-4, 2)$ and having its centre on the line $3x + 2y = 5$.   (C.)

11. Prove that the point $(2, -1)$ is equidistant from the lines $7x + 4y + 3 = 0$, $8x - y - 4 = 0$ and find the equation of the circle centre $(2, -1)$ which touches these two lines. (c.)

12. Find the equation of the circle centre $(3, 1)$ which touches the line $5x - 12y + 10 = 0$. Also find the equation of the tangent to this circle at the point $(4, 1)$.

13. The diameter **AB** of the circle $x^2 + y^2 - 6x + 2y + 5 = 0$ is produced to **C** such that **BC** = 2**AB**. Given that **A** is the point $(1, -2)$, find the coordinates of **C**.

14. Find the gradients of the tangents to the circle $x^2 + y^2 - 4x - 3y - 4 = 0$ at the points where it cuts the y-axis. Hence find the acute angle between these tangents.

15. Find the equation of the circle which passes through the points $(2, 5)$, $(4, 3)$ and has its centre on the line $x + y = 3$.

16. Find the centre and radius of the circle $5(x^2 + y^2) - 4x - 22y + 20 = 0$ and show that the tangent at the point $(\frac{6}{5}, \frac{8}{5})$ has a gradient of $\frac{4}{3}$ and that it passes through the origin.

17. Prove that the circles $x^2 + y^2 - 4x + 2y - 4 = 0$, $x^2 + y^2 - 10x - 6y + 30 = 0$ touch each other externally. Find the co-ordinates of the point of contact by using the fact that it divides the line of centres internally in the ratio of the radii.

18. The line $x = 3$ cuts the circle $x^2 + y^2 - 8x - 10y - 9 = 0$ at the points **P** and **Q**. Obtain (i) the equations of the tangents to the circle at **P** and **Q**; (ii) the acute angle between these tangents.

19. The equations of two circles are $x^2 + y^2 - 16x - 10y + 8 = 0$ and $x^2 + y^2 + 6x - 4y - 36 = 0$. Find (i) the radii; (ii) the distance between the centres. Hence show that the radii to the circles at each common point are at right angles.

20. Find the equation of the circle passing through the points $(2, 0)$, $(8, 0)$ and touching the y-axis on the positive side of the origin. Also find the coordinates of the point on this circle which is furthest from the point $(2, 0)$.

21. Obtain the equations of the tangents from the origin to the circle with equation $x^2 + y^2 - 4x - 4y + 4 = 0$.

22. Show that circles whose equations are of the form $(x - a)^2 + (y - a)^2 = a^2$, where $a \neq 0$, touch both axes. Use this result to obtain the equations of the circles which touch both axes and pass through the point $(2, 1)$. Also find the equation of the common chord of these circles. (J.M.B.)

23. A circle passes through the points $(1, 0)$, $(4, 0)$ and $(0, 2)$, find its equation. Show that the circle touches both the y-axis and the line $3x + 4y = 3$.

24. Show that the line $y = 4$ is one of the tangents to the circle $x^2 + y^2 - 8x - 4y + 16 = 0$ from the point $(-1, 4)$ and find the equation of the other tangent.

25. Two circles have centres $A(1, 3)$ and $B(6, 8)$; they intersect at the points $C(2, 6)$ and D. Find the equations of the circles and also that of the common chord CD. Verify that the point $(6, 2)$ lies on the common chord and that the tangents to the circles from this point are equal in length.

26. The line $3x + 4y = 8$ is a common tangent to two circles whose centres are at $(0, 1)$ and $(3, 2)$. Find the equations of the circles and show that another common tangent is parallel to the x-axis. (L)

27. Prove that the tangents to the circle $x^2 + y^2 + 2x - 4y + 3 = 0$ from the point $(1, 2)$ are at right angles.

28. Find the equation of the circle, radius 5 units and lying in the first quadrant, which touches the circle $x^2 + y^2 - 4x - 60 = 0$ externally and also touches the x-axis.

CHAPTER 20

LOCI AND PARAMETRIC EQUATIONS

Loci

As already defined in the previous chapter *the locus* of a moving point is the path traced out by it. In finding the (x, y) equation of a locus it is necessary to obtain the relationship connecting the x and y co-ordinates of the moving point. We will consider two types of loci,

type (i), where a geometrical relationship determines the locus as, for example, in the case of a circle,

type (ii), where the co-ordinates of the moving point are given in terms of a variable called a *parameter*. The co-ordinates are called *parametric co-ordinates*,

e.g. The moving point could have co-ordinates $(t^2, 2t)$, where the parameter t can take all real values.

Examples 1 and 2 below illustrate the method of dealing with type (i) loci whilst examples 3 and 4 show the method of handling type (ii).

Ex. 1. **A** *and* **B** *are the points* $(-1, 0)$ *and* $(1, 0)$. *A point* **P** *moves so that* $3\mathbf{PA} = 5\mathbf{PB}$, *show that the locus of* **P** *is a circle.*

Let **P** be the point (x, y).

Then $\mathbf{PA} = \sqrt{[(x+1)^2 + y^2]}$ and $\mathbf{PB} = \sqrt{[(x-1)^2 + y^2]}$.

$$\therefore\ 3\sqrt{[(x+1)^2 + y^2]} = 5\sqrt{[(x-1)^2 + y^2]}$$

Squaring, $\quad 9[x^2 + 2x + 1 + y^2] = 25[x^2 - 2x + 1 + y^2],$

rearranging, $\quad\cdot\quad 16x^2 + 16y^2 - 68x + 16 = 0,$

$$4(x^2 + y^2) - 17x + 4 = 0.$$

This is the equation of a circle and hence the locus of **P** is a circle.

Ex. 2. *Find the equation of the locus of a point* **P** *which moves such that its distance from the origin is equal to its perpendicular distance from the line* $x - y - 2 = 0$.

Let **P** be the point (x, y).

Distance of **P** from the origin $= \sqrt{(x^2 + y^2)}$, and perpendicular distance of **P** from the given line $= \dfrac{x - y - 2}{\sqrt{2}}$.

$$\therefore\ \sqrt{(x^2 + y^2)} = \frac{x - y - 2}{\sqrt{2}},$$

squaring, $$x^2 + y^2 = \frac{(x-y-2)^2}{2},$$

$$2x^2 + 2y^2 = x^2 - 2xy + y^2 - 4x + 4y + 4,$$

i.e. $x^2 + 2xy + y^2 + 4x - 4y - 4 = 0$, the equation

of the locus of P.

Ex. 3. *The co-ordinates of a moving point P are $(1-2t, 2t-t^2)$, where t is a parameter. Find the equation of the locus of P.*

Let P be the point (x, y),

then $$x = 1 - 2t, \qquad y = 2t - t^2.$$

To find the relationship between x and y, it is necessary to eliminate t from these equations.

From the linear equation, $t = \dfrac{1-x}{2}$; substituting in the other equation,

$$y = 2\left(\frac{1-x}{2}\right) - \left(\frac{1-x}{2}\right)^2,$$

$$4y = 4 - 4x - (1 - 2x + x^2),$$

i.e. $4y = 3 - 2x - x^2$, the equation of the locus of P.

Ex. 4. *Find the equation of the line which passes through the point $(3, 1)$ and has gradient m. If this line meets the x-axis at P and the y-axis at Q, find the co-ordinates of M, the midpoint of PQ in terms of m. Hence find the equation of the locus of M as m varies.*

Equation of the line is $y - 1 = m(x - 3)$,

or, $$y = mx + 1 - 3m.$$

For P, $y = 0$, so $$0 = mx + 1 - 3m, \qquad x = \frac{3m-1}{m}.$$

For Q, $x = 0$, so $$y = 1 - 3m.$$

∴ P, Q are the points $\left(\dfrac{3m-1}{m}, 0\right)$, $(0, 1-3m)$.

∴ M is the point $\left(\dfrac{3m-1}{2m}, \dfrac{1-3m}{2}\right)$.

As we require the equation of the locus of M, let its co-ordinates be (x, y).

$$\therefore \quad x = \frac{3m-1}{2m}, \qquad y = \frac{1-3m}{2}.$$

From the second equation, $m = \dfrac{1-2y}{3}$.

Substituting for m in the first equation after removing the fraction,

we have, $\qquad 2\left(\dfrac{1-2y}{3}\right)x = 3\left(\dfrac{1-2y}{3}\right) - 1,$

i.e. $\qquad x + 3y = 2xy,$ the equation of the locus of P.

(A) EXAMPLES 20a

In each of ex. 1–10, find the equation of the locus of the moving point P(x, y).

1. P moves at a constant distance of 2 units from the origin.

2. P moves at a fixed distance of 3 units from the x-axis.

3. P moves in the first quadrant equidistant from the x- and y-axes.

4. P moves equidistant from the points $(-1, 0)$, $(0, 1)$.

5. P moves such that its distance from the point $(1, 1)$ is constant and equal to 1 unit.

6. P moves equidistant from the points $(-2, 1)$, $(3, 4)$.

7. P moves such that its distances from the x- and y-axes are in the ratio $2:1$.

8. P moves such that its distance from the origin is equal to its distance from the line $x = -2$.

9. P moves such that the sum of the squares of its distances from the origin and the point $(2, 0)$ is equal to 4.

10. P moves such that $AP^2 + BP^2 + CP^2 = 25$, where A, B, C are the points $(-1, 0)$, $(0, 1)$, $(1, 0)$.

In each of ex. 11–20, find the (x, y) equation of the locus of the point P whose co-ordinates are given in terms of the parameter t.

11. P is the point $(2t, 4t)$.   **12.** P is the point $(1-t, 2t)$.

13. P is the point $(3-4t, 1+7t)$.   **14.** P is the point $(2t^2, 1-t)$.

15. P is the point $(1, 2t+3)$.   **16.** P is the point $\left(4t, \dfrac{4}{t}\right)$.

17. P is the point (t, t^2+2t).   **18.** P is the point $\left(1+t, \dfrac{2}{t}\right)$.

19. P is the point $(8t^3, 4t^2)$.   **20.** P is the point $\left(\dfrac{1-t}{t}, 2t-1\right)$.

21. Find the equation of the locus of a point which moves such that its distance from the origin is twice its distance from the point $(4, 0)$.

22. A, B have co-ordinates $(2t, t^2)$ and $(6t, 4t^2)$, where t is a parameter. Find the (x, y) equation of the locus of the midpoint M of AB as t varies.

23. Prove that, as t varies, the point $(3t+2, t+1)$ lies on the line $3y - x = 1$.

24. Show that the locus of the midpoint of the line joining the points $(4+t, 3t^2)$, $(2-t, 2t^2)$ is a line parallel to the y-axis.

Parametric equations of a curve

It is often desirable to express the co-ordinates of a variable point P in terms of a parameter, say t, so that corresponding to one position of P there is one particular value of t and conversely. In this event, the curve which is the locus of P can be defined by two equations of the form

$$x = f(t), \quad y = g(t).$$

These are *the parametric equations of the curve.* Such equations are useful in deriving general properties of particular curves. In many instances it is possible to derive the (x, y) equation of the curve by eliminating t but this is not always the case. The elementary examples which follow illustrate the method of solving problems by using only the parametric equations of a given curve.

Ex. 5. Find the co-ordinates of the points of intersection of the line $x - 4y - 12 = 0$ *and the curve with equations* $x = 4t^2$, $y = 2t$.

Substituting for x and y in terms of t in the equation of the line,

$$4t^2 - 8t - 12 = 0,$$

$$4(t-3)(t+1) = 0,$$

so $t = 3, -1.$

∴ the points on the curve which also lie on the line are the points where the parameter t has the values 3 and -1.
I.e. the points of intersection have co-ordinates $(36, 6)$ and $(4, -2)$.

Ex. 6. Find the equations of the tangent and normal to the curve with equations $x = 2t^3$, $y = 3t^2$ *at the point* $P(2m^3, 3m^2)$.

To find the gradient, $\dfrac{dy}{dx}$, of the tangent, the result

$$\frac{dy}{dx} = \frac{dy}{dt} \div \frac{dx}{dt} \text{ is used.}$$

So here, $$\frac{dy}{dx} = 6t \div 6t^2 = \frac{1}{t}.$$

At the point P, $t = m$ and $\dfrac{dy}{dx} = \dfrac{1}{m}.$

Equation of tangent at P is $y - 3m^2 = \dfrac{1}{m}(x - 2m^3),$

i.e. $my - x = m^3.$

Equation of normal at P is $y - 3m^2 = -m(x - 2m^3),$

i.e. $y + mx = 3m^2 + 2m^3.$

Ex. 7. The tangent at any point P *on the curve with equations* $x = ct,$ $y = \dfrac{c}{t},$ *where* c *is a constant, meets the* x- *and* y-*axes at* A *and* B *respectively. Show that* P *is the midpoint of* AB *and find the equation of the locus of* M, *the midpoint of* AP, *as* P *moves on the curve.*

Take P as the point parameter p, $\left(cp, \dfrac{c}{p}\right).$

$$\frac{dy}{dx} = \frac{dy}{dt} \div \frac{dx}{dt} = -\frac{c}{t^2} \cdot \frac{1}{c} = -\frac{1}{t^2}.$$

So the gradient of the tangent at P $= -\dfrac{1}{p^2}.$

Equation of tangent at P is $y - \dfrac{c}{p} = -\dfrac{1}{p^2}(x - cp),$

i.e. $p^2 y + x = 2cp.$

This line meets the x-axis at the point $A(2cp, 0)$ and the y-axis at the point $B\left(0, \dfrac{2c}{p}\right).$

The midpoint of AB has co-ordinates $\left(cp, \dfrac{c}{p}\right)$; so it is the point P.

The co-ordinates of M, the midpoint of AP, are $\left(\dfrac{3cp}{2}, \dfrac{c}{2p}\right).$

So taking M as the point (x, y), $x = \dfrac{3cp}{2}$, $y = \dfrac{c}{2p}.$

Eliminating p, $xy = \frac{3}{4}c^2$ the equation of the locus of M.

(A) EXAMPLES 20b

In ex. 1–6, find the co-ordinates of the point P in each case.

1. P is the point where $t = 2$ on the curve $x = 4t^2 - 1$, $y = 2t.$

2. P is the point where $t = -1$ on the curve $x = t + \dfrac{1}{t}$, $y = t - \dfrac{1}{t}$.

3. P is the point where $\theta = \frac{1}{2}\pi$ on the curve $x = 4 \cos \theta$, $y = 2 \sin \theta$.

4. P is the point with parameter 3 on the curve $x = t^3$, $y = t^2$.

5. P is the point with parameter -2 on the curve $x = t^2 + 2t$, $y = 2t^2 - t$.

6. P is the point with parameter $\frac{1}{4}\pi$ on the curve $x = 2 \cos \theta + \sin \theta$, $y = 2 \sin \theta - \cos \theta$.

7. Find the value of the parameter t at the point where the locus $x = 4t - 1$, $y = 2 - t$ meets the x-axis. Hence determine the co-ordinates of this point.

8. Find the co-ordinates of the point where the curve $x = 2t - 1$, $y = 3t^2$ meets the y-axis.

9. Find the co-ordinates of the points of interection of the curve $x = t - t^2$, $y = t + t^2$ and the line $y = 2$.

10. The curve $x = 2 \cos \theta$, $y = \sin \theta$ meets the x-axis at **A** and **B**. Find the length **AB**.

11. Find the co-ordinates of the point of intersection of the locus $x = 3 - 4t$, $y = 1 + t$ and the line $x + 3y = 1$.

12. By showing that the gradient $\dfrac{dy}{dx}$ is constant, establish that the locus $x = 2t - 3$, $y = 4t + 1$ is a straight line.

In ex. 13–18, find, in each case, the gradient of the curve at the point stated.

13. Curve $x = 3t$, $y = t^2 - 4$; point where $t = 3$.

14. Curve $x = 4t$, $y = \dfrac{4}{t}$; point where $t = -1$.

15. Curve $x = 2t^3$, $y = 3t^2$; point where $t = 2$.

16. Curve $x = 4 \cos \theta$, $y = 2 \sin \theta$; point where $\theta = \frac{1}{4}\pi$.

17. Curve $x = t - \dfrac{1}{t}$, $y = t + \dfrac{1}{t}$; point where $t = 1$.

18. Curve $x = a(\theta + \sin \theta)$, $y = a(1 - \cos \theta)$; point where $\theta = \frac{1}{2}\pi$.

19. Find the equation of the tangent to the curve $x = 4t$, $y = 2t^2$ at the point $(4m, 2m^2)$.

20. Find the equation of the normal to the curve $x = 2t$, $y = \dfrac{2}{t}$ at the point $\left(2p, \dfrac{2}{p}\right)$.

21. Show that the equation of the tangent to the curve $x = 4t^3$, $y = 3t^4$ at the point $P(4p^3, 3p^4)$ is $px - y = p^4$. Without further working, write down the equation of the tangent at the point $Q(4q^3, 3q^4)$.

(B) EXAMPLES 20c

In ex. 1–5, find, in each case, the equation of the locus of the moving point P.

1. P moves such that its distance from the line $x = -a$ is equal to its distance from the point $(a, 0)$.

2. P moves such that its distances from the points $(-3, 0)$, $(3, 0)$ are in the ratio $2:5$.

3. P moves such that its distance from the point $(2, 0)$ is twice its distance from the y-axis.

4. P moves such that the lines joining it to the points $(-2, 1)$, $(0, 3)$ are perpendicular.

5. P moves such that the lines joining it to the points $(-1, 0)$, $(2, 0)$ are inclined at an angle whose tangent is 2, taking the y-co-ordinate of P as positive.

In each of ex. 6–11, find the (x, y) equation of the locus of the point P whose parametric co-ordinates are as given.

6. $P(t^2 + t, t^2 - t)$.   **7.** $P(1 - t^2, t + t^2)$.   **8.** $P\left(t + \dfrac{1}{t}, t - \dfrac{1}{t}\right)$.

9. $P\left(\dfrac{t}{1+t}, \dfrac{1}{1+t}\right)$.   **10.** $P(3 \cos \theta, 2 \sin \theta)$.   **11.** $P(2 \sin \theta, \cos 2\theta)$.

12. Find the equation of the tangent at the point $P(ap^2, 2ap)$ on the curve $x = at^2$, $y = 2at$ and write down the equation of the tangent at the point $Q(aq^2, 2aq)$.

13. Find the gradient of the tangent to the curve $x = 2 + \cos 2t$, $y = 1 - \sin 2t$ at the point with parameter $\frac{1}{6}\pi$.

14. Obtain the equation of the locus of the midpoint of the line joining the point $A(7, 8)$ to the variable point $P(2t + 3, t - 9)$ and find the value of t when AP is perpendicular to this locus.

15. Find the equation of the locus of a point which moves such that its distance from the point $(2, 1)$ is equal to its distance from the x-axis and verify that the point $(2+t, \frac{1}{2}(1+t^2))$ lies on the locus for all values of t.

16. T is a variable point with co-ordinates $(t, 0)$ and R is the point $(3, 1)$. The perpendicular to TR at T meets the y-axis at Q and P is the midpoint of TQ. Find the equation of the locus of P as t varies.

17. A, B are the points $(0, 2)$, $(0, 4)$ respectively and O in the origin. A variable point P moves such that $OP^2 + 2AP^2 = BP^2 + OA^2 + OB^2$. Show that the locus of P is a circle, centre O, and find its radius.

18. Show that the point $P\left(2+\dfrac{1}{t}, 4t\right)$ lies on the curve $y = \dfrac{4}{x-2}$ for all values of t. Show also that the equation of the tangent to the curve at the point $P(3, 4)$ is $y + 4x = 16$. This tangent meets the x-axis at B and the midpoint of PB is Q. Find, in terms of t, the co-ordinates of Q and hence find the equation of the locus of Q as t varies. (J.M.B.)

19. The tangent to the curve $xy = c^2$, where c is a constant, at the point $P\left(cp, \dfrac{c}{p}\right)$ meets the axes at T and N. Prove that $TN = 2OP$, where O is the origin.

20. Show that the point with parametric co-ordinates $\left(\dfrac{t}{1+t^3}, \dfrac{t^2}{1+t^3}\right)$ always lies on the curve $x^3 + y^3 = xy$.

21. The tangent at the point $P(ap^2, 2ap)$ on the curve $x = at^2$, $y = 2at$ meets the axis of y at T. Find the co-ordinates of M, the midpoint of PT, in terms of p, and hence find the equation of the locus of M as p varies.

22. Show that the equation of the tangent to the curve $x = 3t^2$, $y = 2t^3$ at the point $P(3p^2, 2p^3)$ is $px - y = p^3$.
If Q is the point $(3q^2, 2q^3)$, find the co-ordinates of the point of intersection of the tangents to the curve at P and Q. If these tangents are at right angles, what is the connection between p and q?

23. Prove that the gradient of the curve $x = a(\theta + \sin \theta)$, $y = a(1 - \cos \theta)$ at the point parameter α is $\tan \frac{1}{2}\alpha$.

24. Parameters p and q are such that $p + q = 1$. Find the equation of the locus of a moving point whose co-ordinates are $(p^2 + q^2, pq)$.

25. Write down the equation of a line of gradient m which passes through the point $(1, 1)$. If this line cuts the x- and y-axes at A and B respectively and P is the point which divides AB internally in the ratio $1:2$, show that, as m varies, P moves on the curve $3xy - x - 2y = 0$.

26. Find the equation of the normal to the curve $x = 2t^2$, $y = 4t$ at the point $P(2p^2, 4p)$. This normal meets the x-axis at L and Q is the point on PL produced such that $PQ = \frac{3}{2}PL$. Find the equation of the locus of Q as p varies.

27. The points $P(ap^2, 2ap)$, $Q(aq^2, 2aq)$ move on the curve $x = at^2$, $y = 2at$ and $p + q = 2$. Show that the gradient of the chord PQ is constant and find the equation of the locus of the midpoint of PQ.

28. Find the equation of the tangent to the curve $xy = 1$ at the point $\left(t, \dfrac{1}{t}\right)$. This tangent cuts the x-axis at A and the y-axis at B; C is the point on AB such that $AC : CB = 3 : 2$. Show that the locus of C as t varies is the curve with equation $xy = \frac{24}{25}$.

CHAPTER 21

ELEMENTARY PERMUTATIONS, COMBINATIONS
AND PROBABILITY

A *combination* is a selection in which a number of quantities are chosen without any attention being paid to the order in which the choice is made. A *permutation* is an arrangement in which a number of quantities are chosen with attention being paid to the order of choice.

Ex. 1. *Find the number of combinations of two which can be made from a group of four children.*

Let the children in the group be designated *a*, *b*, *c*, *d*.
Then the following combinations of two can be chosen

ab, ac, ad, bc, bd, cd.

So the number of combinations of two from the group of four = 6.

Ex. 2. *From a group of four children, in how many ways can a captain and a vice-captain be chosen?*

Two children have to be chosen from four as in ex. 1. However in this case the order of choice is involved as it is necessary to distinguish between captain and vice-captain so the question is one of permutations.
For any selection or combination of two children, say *ab*, there are two arrangements or permutations, *ab* and *ba*—i.e. captain *a*, vice-captain *b* and vice-versa.
As there are 6 combinations of two there will be 6×2 permutations of two. Hence the officials can be chosen in 12 ways.
The previous examples are intended as a help in distinguishing between combinations and permutations. In these examples, the number of permutations was found by using the number of combinations but it is more usual to work the other way round as is illustrated in the following example.

Ex. 3. *Find the number of (i) permutations, (ii) combinations of two quantities which can be made from six different quantities.*

(i) In finding the number of permutations or arrangements of two, think of filling two vacant places by using the six available quantities.
The first place can be filled by choosing any one of the six—i.e. in 6 ways. Once the first place is filled there are only five quantities left and so the second place can be filled in 5 ways.

As each way of filling the first place can be associated with all five ways of filling the second place, it follows that both places can be filled in $6 \times 5 = 30$ ways.

∴ The number of permutations of two from a group of six = 30.

(iii) In finding the number of combinations or selections of two, note that each combination of two leads to 2 permutations of two.

Hence number of permutations of two = 2 (number of combinations of two).

$$\therefore \text{ number of combinations of two} = \tfrac{30}{2} = 15.$$

(r, s) principle. **If one thing can be done in r ways and a second thing can be done in s ways, then both things can be done in r × s ways.**

This important principle was used in the first part of ex. 3. When from the results that the first place could be filled in 6 ways and the second place in 5 ways, it was deduced that both places could be filled in 6×5 ways.

Extension of the (r, s) principle. This principle can be extended to cover any number of operations, for if one thing can be done in p ways, a second in q ways, a third in r ways and so on, the number of ways of performing all the operations is $p \times q \times r \times \ldots$.

Ex. 4. In how many ways can a group of two, one man and one woman, be selected from 8 men and 6 women?

The man can be chosen in 8 ways and the woman in 6 ways.

∴ number of ways of choosing a couple = $8 \times 6 = 48$.

Ex. 5. In how many ways can the results of six football matches be forecast?

The result of each match can be forecast in 3 ways—home win, draw or away win.

So the results of the six matches can be forecast in

$$3 \times 3 \times 3 \times 3 \times 3 \times 3 = 3^6 \text{ or } 729 \text{ ways.}$$

(A) EXAMPLES 21a

1. One process can be performed in 4 ways and a second process can be performed in 3 ways. In how many ways can both processes be performed?

2. If there are 5 routes between towns **A** and **B** and 4 routes between towns **B** and **C**, how many different ways are there from **A** to **C** via **B**?

3. In how many ways can one boy and one girl be chosen from a class of 12 boys and 14 girls?

4. How many ways are there of forecasting the results of 3 football matches?

5. Using the letters a, b and c, how many different groups of two letters can be formed? How many different arrangements of two letters are there?

6. There are 4 roads connecting towns **A** and **B**. In how many ways can a man travel from **A** to **B** and return by a different route?

7. In how many ways can six people make use of two vacant seats in a railway compartment?

8. In a three-figure number, the units digit can be chosen in 5 ways, the tens digit in 9 ways and the hundreds digit in 4 ways. How many different numbers can be formed?

9. At a golf club meeting there are six candidates for the posts of captain and vice-captain. In how many ways can the posts be filled?

10. The integers 1, 3, 5, 7, 9 are used to form two-digit numbers. How many different numbers can be formed if (i) no integer can be repeated in any one number; (ii) repetition is permitted?

11. Given three envelopes with different addresses and three different letters, in how many ways can the letters be put into the envelopes, one in each?

12. A committee is to be formed consisting of one representative from each of three streets. If the numbers of residents in the three streets are 11, 14 and 10, in how many ways can the committee be formed?

13. In how many ways can two prizes be awarded to a form of 20 pupils if (i) any pupil can win only one prize; (ii) any pupil can win both prizes?

14. In how many ways can a consonant and a vowel be chosen from the letters of the word *forecast*? How many different two-letter words can be formed each containing one consonant and one vowel?

15. There is one vacant room in each of four hotels. In how many different ways can three travellers be accommodated?

16. There are three vacant places on a bookshelf. In how many ways can three different books be placed on the shelf?

17. Two straight lines intersect at **O**; vertices **A, B, C** lie on one line and vertices **P, Q, R** on the other. How many different triangles can be formed with one vertex at **O** and one vertex on each of the given lines?

Permutations of unlike quantities

Ex. 6. How many arrangements of three can be made using seven different quantities?

To obtain an arrangement of three think of filling three vacant places with quantities chosen from the given group of seven.

The first place can be filled in 7 ways as there are seven quantities to choose from. Once this place is filled only six quantities remain and so there are 6 ways of filling the second place. Similarly there will be 5 ways of filling the third place.

∴ number of ways of filling all three places is $7 \times 6 \times 5$ or 210.

I.e. the number of arrangements, or permutations, of three which can be made using seven different quantities is 210.

Ex. 7. In how many ways can five different quantities be arranged amongst themselves?

Here all five quantities are used in each arrangement and so there are five vacant places to be filled using the given group of five.

The first place can be filled in 5 ways, the second in 4 ways, the third in 3 ways, the fourth in 2 ways and the fifth in 1 way.

Hence the number of ways of arranging five different quantities amongst themselves is

$$5 \times 4 \times 3 \times 2 \times 1 \quad \text{or} \quad 120.$$

Notation. The continued product $5 \times 4 \times 3 \times 2 \times 1$ is called *factorial* five and is written as 5!

Generally, $n! = n(n-1)(n-2) \ldots 3.2.1.$

The number of arrangements or permutations of seven unlike quantities taken three at a time is written $_7P_3$.

From ex. 6., $_7P_3 = 7.6.5 = 210.$

Generally,

$_nP_r$ = the number of permutations of n unlike quantities

taken r at a time,

$= n(n-1)(n-2) \ldots r$ **products,**

$$= \frac{n!}{(n-r)!}.$$

Ex. 8. Evaluate (i) $_6P_2$; (ii) $_4P_4$.

$_6P_2 = 6.5$ (2 products) $= 30.$

$_4P_4 = 4.3.2.1$ (4 products) $= 4!$ or   24.

Ex. 9. In how many ways can the letters of the word ACORN be arranged to make different five-letter words?

Here we require the number of arrangements of five different quantities taken altogether.

Number of different words $= {_5P_5} = 5!$ or 120.

Ex. 10. How many three digit numbers can be made using the integers 1, 2, 3, 4, 5, 6 *when* (i) *no repetition is allowed*; (ii) *repetition is allowed?*

(i) In this case, we require the number of arrangements of three which can be obtained using six different quantities—*i.e.* $_6P_3$ which equals $6.5.4$ or 120.

(ii) When repetitions are allowed, proceed from first principles and think of filling three vacant places using the integers from 1 to 6.

The first place can be filled in 6 ways and so can the second and the third as repetition is permitted.

So with repetition, $6 \times 6 \times 6$ or 216 three digit numbers can be obtained.

(A) EXAMPLES 21b

1. In how many ways can the letters *a*, *b*, *c* be arranged?

2. Find the number of permutations of five unlike things taken two at a time.

3. How many different three-digit numbers can be formed from the integers 3, 4, 5, no repetition being allowed.

4. Evaluate (i) 5!; (ii) $\dfrac{6!}{4!}$; (iii) $_6P_3$; (iv) $_4P_4$.

5. There are ten standing passengers in a bus and three seats become vacant. In how many ways can the seats be occupied?

6. How many different three-letter words can be formed by using the letters of the word *card*?

7. In how many ways can eight people arrange themselves in a row?

8. Find the number of permutations of eight unlike things taken four at a time?

9. How many (i) three-digit numbers; (ii) four-digit numbers; (iii) five-digit numbers can be formed using the integers 1, 2, 3, 4, 5 without repetitions.

10. In how many ways can six letters be put into six differently addressed envelopes, one in each?

11. There are five differently coloured flags. How many different signals consisting of three flags can be made?

12. Two examination papers have to be set on a particular date (one in the morning and the other in the afternoon.) In how many ways can this be done if there are ten papers to choose from?

13. Evaluate 8! $-6.7!$

14. How many four-letter words can be obtained by using (i) the letters *a*, *b*, *c*, *d*; (ii) the letters *a*, *b*, *c*, *d*, *e*? Hence find the number of four letter words which can be obtained from the letters *a*, *b*, *c*, *d*, *e* and which contain the letter *e*.

15. There are eighteen members of a committee. In how many ways can representatives be chosen to attend three conferences one to each, assuming that no member can attend more than one conference.

16. In how many different orders can eight records be played on a record-player? How many arrangements are possible if there is time only to play four?

17. How many four digit numbers can be formed by using the integers 1, 2, 3, 4, 5, 6 when repetition is allowed?

18. In how many ways can eight people seat themselves in a railway compartment with eight seats? If four of the people wish to sit facing forward and the other four wish to sit facing backward, in how many ways can this be done?

19. Four children and six adults are in a waiting room. There are two seats available for children and four seats for adults. In how many ways can the seats be occupied?

20. There are six differently coloured jars with lids to match. In how many ways can the jars be arranged on a shelf? In how many ways can the lids be put on the jars? Deduce the number of possible arrangements of jars and lids on the shelf.

21. Five married couples attend a party together and take part in a dance competition. In how many different ways can the men and women pair up? If one particular couple insist on dancing together, how many pairings are possible?

Permutations of a number of quantities not all different

Ex. 11. *How many different arrangements can be made by using all the letters a, a, b, c?*

Think of the two a's as different letters a_1, a_2.

Then there are four different letters which can be arranged in 4! ways.

There will be two different arrangements in which the positions of the b and the c will be unchanged, e.g. a_2cba_1, and a_1cba_2. These arrangements become identical when the suffixes are removed from the a's.

So the number of possible arrangements of the given letters is equal to the number of arrangements when all the letters are different (4!) divided by the number of ways of arranging a_1 and a_2 in two places (2!).

$$\text{Number of different arrangements} = \frac{4!}{2!} = 12.$$

Ex. 12. *How many different numbers can be formed by using all the digits* $1, 2, 2, 2, 3, 3$?

Proceed as in the previous example and imagine the three 2's to be distinguishable, say $2_1, 2_2, 2_3$ and similarly think of the two 3's as $3_1, 3_2$.

Then there are six different digits which can be arranged to give 6! different numbers.

Writing down any one of these numbers, say $2_12_213_12_33_2$, it will be seen that without changing the positions of the digits, the three 2's can be interchanged in 3! ways and the two 3's can be interchanged in 2! ways. So there will be 3! 2! different numbers with the digits positioned as shown. When the suffixes are dropped these 3! 2! numbers become identical.

So the number of different numbers which can be formed

$$= \frac{6!}{3!\,2!} = 60.$$

General result

The following general result follows by using the method illustrated in the previous examples:—

The number of permutations of n quantities taken altogether when p are alike of one kind, q alike of a second kind, r alike of a third kind and so on is

$$\frac{n!}{p!\,q!\,r!\ldots}.$$

(A) EXAMPLES 21c

1. Evaluate (i) $\dfrac{5!}{3!}$; (ii) $\dfrac{6!}{4!\,2!}$; (iii) $\dfrac{8!}{(4!)^2}$; (iv) $\dfrac{9!}{(3!)^3}$.

In ex. 2–7, find, in each case, the number of different arrangements possible using all the quantities stated.

2. Four quantities of which two are identical and the others different.

3. Five quantities of which three are alike and the others different.

4. Seven quantities of which two are alike of one kind, two alike of a second kind and the others different.

5. Six quantities of which two are alike of one kind, two are alike of a second kind and two alike of a third kind.

6. The letters a, a, a, b, b, c, c, d.

7. The letters x, x, x, x, y, y, y, y.

8. How many different numbers can be formed by using all the digits 1, 2, 2, 3, 3, 4, 5?

9. How many different words can be obtained by using all the letters of the word *tomato*?

10. Ten balls are placed in a straight groove. How many different arrangements are possible if six of the balls are black and the remainder white? Assume the balls are identical in size.

11. Of six signal flags, two are white, two blue and two red. How many different signals can be made using all the flags?

12. In how many ways can the letters of the word *substitution* be arranged amongst themselves?

13. How many different nine-digit numbers can be formed using the integers 1, 1, 1, 2, 2, 3, 3, 3, 4 in each number?

14. In how many ways can six dots and eight dashes be arranged in a row?

15. How many additional four-digit numbers can be obtained by using the digits in the number 3224?

16. In how many ways can the letters of the word *accommodation* be arranged amongst themselves?

Combinations of unlike quantities

As already defined, a combination or selection is a set of quantities chosen from a given group without regard to the order of the quantities. The following examples illustrate how the number of combinations in a particular case is found by first finding the number of permutations in that case.

Ex. 13. *How many selections of three letters can be chosen from the letters* a, b, c, d, e, f?

Let the number of possible selections be x.

Then each selection consists of three different letters which can be arranged amongst themselves in 3! ways.

Hence the x selections or combinations lead to $x \times 3!$ arrangements.

But the number of arrangements of three letters chosen from six different letters is $_6P_3$ or $6 . 5 . 4$.

$$\therefore x . 3! = 6 . 5 . 4,$$
$$x = 20.$$

I.e. The number of selections of three letters chosen from the given group of six is 20.

General result

Using the method of the previous example, the following general result can be obtained:—
the number of combinations of n unlike quantities taken r at a time

$$= \frac{_nP_r}{r!} = \frac{n(n-1)(n-2)\ldots r \text{ products}}{r!}.$$

Notation

The number of combinations of n unlike quantities taken r at a time is written $_nC_r$.

So $\quad _nC_r = \dfrac{_nP_r}{r!} = \dfrac{n(n-1)(n-2)\ldots r \text{ products}}{r!} = \dfrac{n!}{(n-r)!\, r!}$.

Ex. 14 A committee is to be chosen from a group of ten men and eight women. In how many different ways can the committee be selected if it must contain four men and four women?

The men can be chosen in $_{10}C_4$ ways and the women in $_8C_4$ ways.
∴ the number of ways of choosing the committee $= {}_{10}C_4 \times {}_8C_4$,

$$= \frac{10.9.8.7}{4!} \times \frac{8.7.6.5}{4!},$$
$$= 210 \times 70,$$
$$= 14700.$$

(A) EXAMPLES 21d

In ex. 1–10, find the number of possible selections in each case.

1. Three letters chosen from the letters a, b, c, d, e.

2. Two digits chosen from the digits 1, 3, 5, 7, 9.

3. Two aces chosen from the four aces in a pack of cards.

4. Three toothbrushes chosen from a rack containing seven differently coloured brushes.

5. Four books chosen from a pile of ten different books.

6. Three records chosen from an album of eight records.

7. Five people chosen from a committee of eight.

8. Four football matches chosen from a list of twelve matches.

9. A committee of six chosen from a group of fifteen people.

10. A hand of five cards chosen from a suit of thirteen cards.

11. Evaluate (i) $_3C_1$; (ii) $_5C_2$; (iii) $_7C_3$; (iv) $_8C_8$.

12. How many different triangles can be drawn with five given points as vertices, no three of these points being collinear?

13. In how many ways can a group of four men be selected to play a game of tennis if there are twelve men wishing to play?

14. In how many ways can a committee of five men and four ladies be selected from a group of eight men and six ladies?

15. A group of eight pupils is to be divided into groups of five and three. In how many different ways can this be done?

16. A form of twenty-five pupils contains thirteen girls. In how many ways can a group of twelve pupils be selected if it must contain equal numbers of boys and girls?

17. From a group consisting of eight men, six women and six children, a selection of ten people has to be made. In how many ways can this be done if (i) there are no restrictions; (ii) the selected group must contain four children; (iii) the selected group must contain four children and four women?

18. In how many ways can a group of fifteen people be divided into groups of seven and eight? In how many ways can a group of eight be divided into groups of five and three? Deduce the number of ways of dividing the original group into groups of seven, five and three.

Harder examples of permutations and combinations

Ex. 15. How many odd numbers can be formed by using all the digits 1, 2, 3, 4, 5?

The units digit must be 1, 3 or 5 and so this digit can be chosen in 3 ways.
The remaining digits can be arranged in 4! ways.
Hence the required number is 3 . 4! or 72.

Ex. 16. In how many ways can a committee of six be chosen from a group of six men and eight women if it must contain at least two men and two women?

The total of six can be made up of (i) four men, two women, (ii) three men, three women or (iii) two men, four women.

(i) Number of ways of choosing four men and two women $= {}_6C_4 \times {}_8C_2$,
$$= 15 \times 28 = 420.$$
(ii) Number of ways of choosing three men and three women $= {}_6C_3 \times {}_8C_3$,
$$= 20 \times 56 = 1120.$$
(iii) Number of ways of choosing two men and four women $= {}_6C_2 \times {}_8C_4$,
$$= 15 \times 70 = 1050.$$
$\therefore$ The number of ways of choosing the committee $= 420 + 1120 + 1050$
$$= 2590.$$

Ex. 17. There is a row of seven chairs. In how many ways can a man, a woman and a child be seated if (i) the woman and the child sit together; (ii) all three people sit on adjoining chairs?

(i) There are six adjoining pairs of chairs so the pair of chairs on which the woman and the child sit can be selected in 6 ways. On any pair of chairs the woman and the child can arrange themselves in 2! ways.

$\therefore$ the woman and the child can be seated in 6 . 2! or 12 ways.

There are five seats left for the man and so he can seat himself in 5 ways.

Hence the total number of ways of seating all three $= 12 \times 5 = 60$.

(ii) There are five adjoining sets of three chairs so the trio of chairs on which the three people sit can be selected in 5 ways. On any trio of chairs, the three can arrange themselves in 3! ways.

Hence the total number of ways of seating all three $= 5 . 3! = 30$.

Ex. 18. In how many ways can the letters of the word dragoon *be arranged if the two o's must not come together?*

Number of possible arrangements without restriction $= \dfrac{7!}{2!} = 2520$. Now treating the two o's as a single letter, the number of arrangements when the o's are together

$$= 6! = 720$$

∴ the number of arrangments with the two o's not together

$$= 2520 - 720 = 1800.$$

(B) EXAMPLES 21e

1. Using the letters a, b, c, d, e, f without repetition, how many selections of four letters can be obtained including (i) either a or b; (ii) both a and b?

2. In how many ways can seven people be arranged in a row if (i) the youngest must be at one end; (ii) the youngest and the oldest must be at the ends?

3. Given three consonants and three vowels, all different, how many six-letter words can be formed with the vowels occupying the even positions?

4. A group of five children has to be chosen from a form of eight boys and twelve girls. In how many ways can the selection be made if the group must contain at least two boys and two girls?

5. In how many ways can five people be seated in a row of five chairs if two particular people must (i) sit together; (ii) not sit together?

6. Using the letters a, a, b, c, d, e, how many groups of three containing (i) no a's; (ii) one a; (iii) two a's can be obtained? Deduce the total number of different groups of three letters which can be obtained.

7. How many even four-digit numbers can be formed by using the integers $2, 3, 4, 5$ without repetition? How many of these numbers will be less than 3000?

8. Four boys and three girls are arranged in a line so that each girl has a boy on either side of her. How many different arrangements are possible?

9. In how many ways can a sub-committee consisting of a chairman, a vice-chairman and a secretary together with four other members be selected from a group of twelve councillors?

10. How many different arrangements can be made using all the letters a, a, a, b, c, d? In how many of these will the three a's not be together?

11. Find the number of ways of dividing seventeen different things into groups of four, six and seven.

12. How many numbers greater than 3000 can be formed by using some or all the integers $1, 2, 3, 4, 5$ without repetition?

13. In how many ways can a group of four be chosen from eight children if the youngest and oldest are not both included?

14. Two straight lines **OA, OB** intersect at **O**. On **OA** are marked five points and an **OB** are marked four points. These points can be vertices of triangles. How many triangles can be drawn (i) with **O** as one vertex; (ii) without **O** as one vertex?

15. How many arrangements of the letters of the word *evening* can be made in which (i) the two n's are together; (ii) the two n's are not together?

16. There are eight cards on each of which one of the integers $2, 3, 4, 5, 6, 7, 8, 9$ is printed. In how many ways can a group of three cards be chosen (i) to include just one card marked with an odd number; (ii) to include at least one card marked with an odd number?

17. How many different five-digit numbers are there? How many of these are divisible by 5?

18. The integers $1, 2, 3, 4, 5, 6$ are written down at random to form a six-digit number. Find (i) how many such numbers are possible; (ii) how many of the numbers will be odd; (iii) how many of the numbers are divisible by four (*i.e.* the number formed by the tens and units digits must be divisible by four).

19. In how many ways can the letters of the word *omelette* be arranged? In how many of the arrangements will no two e's come together?

20. How many numbers divisible by five and greater than (i) 500000; (ii) 400000 can be formed by using the digits $0, 1, 2, 3, 4, 5$ without repetition?

21. Out of sixteen players available to form a cricket team of eleven, two can keep wicket, five can bowl and the remainder are batsmen. In how many ways can the team be selected if it must contain one wicket-keeper and at least three bowlers?

22. How many four-digit numbers are there which do not have four different digits?

Probability or chance

Ex. 19. *Out of a class of twelve boys and eight girls, two pupils are chosen at random. What is the chance that they are both boys?*

The number of ways of choosing two pupils out of the twenty available is $_{20}C_2$ or 190.

The number of ways of choosing two boys out of the twelve available is $_{12}C_2$ or 66.

So out of a total of 190, there are 66 ways in which two boys can be selected.

The chance, or probability, of two boys being chosen is the ratio of the number of ways in which this event can happen to the total number of ways in which the selection can be made,

$$i.e. \quad \text{chance of two boys} = \tfrac{66}{190} \quad \text{or} \quad \tfrac{33}{95}.$$

Definition

The mathematical probability or chance of a particular event happening is the ratio of the number of ways in which this event can happen to the total number of events.

E.g. when a coin is spun it can fall in 2 ways, this is the total number of events. It can fall heads uppermost in 1 way and so the probability of it falling heads uppermost is $\tfrac{1}{2}$.

Ex. 20. *The digits* 1, 2, 3, 4, 5 *are written down in a random order to form a five-digit number. What is the chance that the number is odd?*

For an odd number the units digit must be 1, 3 or 5. So there are 3 ways in which the number will be odd. The units digit can be any one of the given digits and so there are 5 ways of putting down the units digit.

$$\therefore \text{ the chance of obtaining an odd number} = \tfrac{3}{5}.$$

Ex. 21. *Three men and two women seat themselves at random on five chairs. Find the chance of a particular man and a particular woman sitting on adjoining chairs.*

The number of ways of sitting with the particular couple an adjoining chairs is 2 . 4!. So the required event can happen in 2 . 4! ways. The total number of ways in which the people can be seated is 5!

$$\therefore \text{ the chance of a particular couple sitting together} = \frac{2 \cdot 4!}{5!} = \tfrac{2}{5}.$$

(A) EXAMPLES 21f

Find the chance of the event happening in each of the following examples.

1. Spinning a coin and getting a tail uppermost.

2. Drawing a red ball out of a bag containing one red, one blue and one black ball.

3. Choosing one particular person from a group of ten.

4. Drawing an ace out of a pack of fifty two cards which contains four aces.

5. An event can happen in three ways and fail to happen in ten ways.

6. Obtaining an even number when the digits 3, 4, 5, 6, 7 are put in a row.

7. Of drawing the favourite in a race with five tickets out of a total of seventy five.

8. A particular person being at one end of a line when twelve people are arranged at random.

9. Obtaining two heads when two coins are spun.

10. Getting a number divisible by five when the digits 0, 1, 2, 3, 4, 5, 6 are put in a row.

11. Choosing a girl in a random choice of one child from a group of eight boys and five girls.

12. Drawing two white balls from a bag containing two white and six coloured balls.

13. The a and b coming together when the letters a, b, c, d, e are written down at random.

14. A boy and his sister being chosen in a random selection of two children, one boy and one girl, from a group of ten boys and eight girls.

15. Obtaining a total of (i) twelve; (ii) eleven when two dice are thrown.

16. Choosing two men in a random choice of two from a list of surnames of ten men and fifteen women.

Multiple probabilities

If the chance of one event happening is $\dfrac{a_1}{b_1}$, a second event happening is $\dfrac{a_2}{b_2}$, a third event happening is $\dfrac{a_3}{b_3}$ and so on, the chance of all the events happening is $\dfrac{a_1}{b_1} \times \dfrac{a_2}{b_2} \times \dfrac{a_3}{b_3} \times \dots$.

Ex. 22. In a selected group of one hundred people, sixty have brown hair and forty-five have blue eyes. Find the chance that a particular member of the group has brown hair and blue eyes.

The chance that a particular person has brown hair $= \frac{60}{100}$ or $\frac{3}{5}$.
The chance that a particular person has blue eyes $= \frac{45}{100}$ or $\frac{9}{20}$.

∴ the chance that a particular person has both brown hair and blue eyes $= \frac{3}{5} \times \frac{9}{20} = \frac{27}{100}$.

Ex. 23. A bag contains seven white and three black balls. (i) If three balls are drawn out of the bag, what is the probability that they are all white? (ii) If the balls are drawn out individually and then replaced in the bag, find the chance that the drawing gives two white and one black in that order.

(i) The number of ways of drawing three out of the ten balls

$$= {}_{10}C_3 = 120.$$

The number of ways of drawing three out of the seven white balls

$$= {}_7C_3 = 35.$$

So the chance of drawing three white balls $= \frac{35}{120} = \frac{7}{24}$.

(ii) When a ball is drawn from the bag, the chance of drawing a white ball is clearly $\frac{7}{10}$ and the chance of drawing a black ball is $\frac{3}{10}$.

So the chance of drawing a white ball followed by a second white ball and then by a black ball $= \frac{7}{10} \times \frac{7}{10} \times \frac{3}{10} = \frac{147}{1000}$.

(B) EXAMPLES 21g

1. A coin is spun three times, what is the probability that the head will fall uppermost on all three occasions?

2. Three balls are drawn from a bag containing six white balls, four blue balls and two red balls. Find the chance that (i) all three balls are white; (ii) the balls consist of two red and one blue?

3. Two cards are drawn at random from a pack of fifty-two containing four aces, what is the chance that both cards are aces?

4. The letters of the word *acorn* are arranged at random, what is the probability that the two vowels come together?

5. A die with six faces is thrown twice. Find the probability of obtaining a total of ten.

6. Four children have to be chosen at random out of a group of twenty children, what is the chance that two particular children are included?

7. Event **A** happens in five cases out of twelve, event **B** happens in three cases out of eight and event **C** happens in two cases out of six. What is the chance of events **A** and **B** happening and event **C** not happening?

8. A bag contains ten white balls and six black balls. A ball is drawn and returned to the bag on three occasions. What is the chance that all three balls are black? If on being drawn, a ball is not returned to the bag, what is the chance of drawing two white and one black ball in that order?

9. In a selected group of forty people, twenty-five have brown eyes, fifteen have grey hair and sixteen wear glasses. Find the chance that a particular member of the group has brown eyes and grey hair and does not wear glasses.

10. What is the chance of the two o's coming together when the letters of the word *saloon* are arranged amongst themselves?

11. A coin is spun three times. What is the chance that it will fall heads, heads and tails in that order? What is the chance of getting two heads and one tail in any order?

12. A five-digit number is obtained by putting the digits $1, 2, 2, 3, 4$ down at random. What is the chance of the number being greater than 30000?

13. A pack of fifty-two cards contains twenty honour cards. Show that a selection of three cards from the pack can be made in 22100 different ways and that 4960 of these will contain no honour card. Deduce the probability that a selection of three cards contains at least one honour.

14. Four plants are chosen at random out of eighteen, of which ten are red-flowering and the others white-flowering. What is the chance that at least three of the plants are red?

15. When the letters a, a, b, c, d, e are arranged in a row what is the chance of the a's being at the ends?

16. If the chance of a particular server at tennis winning a point is $\frac{2}{3}$, find the chance that, of four successive points, the server (i) wins all four; (ii) wins three.

CHAPTER 22

INDICES AND LOGARITHMS. EXPONENTIAL AND LOGARITHMIC FUNCTIONS

Laws of indices. The basic laws of indices are

(i) $x^a \times x^b = x^{a+b}$;
(ii) $x^a \div x^b = x^{a-b}$;
(iii) $(x^a)^b = x^{ab}$.

These laws, which can readily be established when a and b are positive integers, are taken as being true for all indices. By using them, meanings can be obtained for indices other than positive integers as follows:—

$$x^0 = 1; \quad x^{p/q} = [\sqrt[q]{x}]^p \quad \text{or} \quad \sqrt[q]{x^p}; \quad x^{-n} = \frac{1}{x^n}.$$

Ex. 1. (*i*) *Express* $\sqrt{(x^a \times x^b)}$ *as a power of* x; (*ii*) *evaluate* $\sqrt[3]{9} \times \sqrt[6]{9}$ *without using tables.*

(i) $\sqrt{(x^a \times x^b)} = \sqrt{x^{a+b}} = (x^{a+b})^{\frac{1}{2}} = x^{\frac{1}{2}(a+b)}$.
(ii) $\sqrt[3]{9} \times \sqrt[6]{9} = 9^{\frac{1}{3}} \times 9^{\frac{1}{6}} = 9^{\frac{1}{2}} = 3$.

Ex. 2. *Solve the simultaneous equations* $3^x . 9^{2y} = 27$, $2^x . 4^{-y} = \frac{1}{8}$.

$$3^x . 9^{2y} = 3^x . 3^{4y} = 3^{x+4y},$$

$$2^x . 4^{-y} = 2^x . 2^{-2y} = 2^{x-2y}.$$

$$\therefore 3^{x+4y} = 27 = 3^3, \quad \text{so} \quad x + 4y = 3;$$

$$2^{x-2y} = \frac{1}{8} = 2^{-3}, \quad \text{so} \quad x - 2y = -3.$$

The solution of these simultaneous equations gives $x = -1$, $y = 1$.

Logarithms

Definition. The logarithm of a number n *to a given base* a, *assumed positive, is the index* x *of the power to which the base must be raised in order to equal the number,*

$$\text{i.e.} \quad a^x = n.$$

N.B. As a is a positive number, for all real values of x the number n must be positive. *Hence the logarithm of a negative number does not exist in real terms.*

Notation. The logarithm of n to base a is written $\log_a n$.

Common logarithms. Logarithms to base 10 are called *common logarithms* and the notation lgn is sometimes used instead of $\log_{10} n$.

Ex. 3. *Evaluate* (i) $\log_4 16$; (ii) $\log_9 3$; (iii) $\log_2 \frac{1}{8}$; (iv) $\log_a 1$; (v) $\log_{10} 100$.

(i) As $\quad 16 = 4^2$; $\qquad \log_4 16 = 2$;
(ii) as $\quad 3 = 9^{\frac{1}{2}}$, $\qquad \log_9 3 = \frac{1}{2}$:
(iii) as $\quad \frac{1}{8} = 2^{-3}$; $\qquad \log_2 \frac{1}{8} = -3$;
(iv) as $\quad 1 = a^0$; $\qquad \log_a 1 = 0$;
(iv) as $\quad 100 = 10^2$; $\quad \log_{10} 100 = 2$.

Laws of logarithms. Corresponding to the three laws of indices there are three laws of logarithms:—

(i) $\log_a(mn) = \log_a m + \log_a n$;

(ii) $\log_a \left(\dfrac{m}{n}\right) = \log_a m - \log_a n$;

(iii) $\log_a(m^n) = n \log_a m$.

The method of proving these laws will be illustrated by the proof of law (i).

Write $\qquad\qquad \log_a m = p \quad$ and $\quad \log_a n = q$.

Then $\qquad\qquad m = a^p \quad$ and $\quad n = a^q$.

$$\therefore \ mn = a^p \times a^q = a^{p+q},$$

So $\qquad\qquad \log_a(mn) = p + q = \log_a m + \log_a n$.

Ex. 4. *Express* $\log_a \sqrt{(x^p y^q)}$ *in terms of* $\log_a x$ *and* $\log_a y$.

$$\log_a \sqrt{(x^p y^q)} = \log_a (x^p y^q)^{\frac{1}{2}} = \tfrac{1}{2} \log_a (x^p y^q),$$
$$= \tfrac{1}{2}[\log_a x^p + \log_a y^q] = \tfrac{1}{2}[p \log_a x + q \log_a y].$$

Ex. 5. *Find* x *if* $2 \log_3 x = 2 - \log_3 4$.

We get the equation into the form $\log_3 a = \log_3 b$, from which $a = b$.

$$2 \log_3 x = \log_3 x^2; \qquad 2 - \log_3 4 = \log_3 9 - \log_3 4 = \log_3 \tfrac{9}{4}.$$
$$\therefore \ \log_3 x^2 = \log_3 \tfrac{9}{4},$$
$$x^2 = \tfrac{9}{4}; \quad x = \tfrac{3}{2}$$

N.B. the solution $x = -\tfrac{3}{2}$ does not apply as the logarithm of a negative number is not real.

(A) EXAMPLES 22a

1. Evaluate (i) 20^0; (ii) $36^{-\frac{1}{2}}$; (iii) $64^{\frac{2}{3}}$; (iv) $81^{-\frac{3}{4}}$; (v) $(\tfrac{1}{3})^{-2}$.

2. Simplify (i) $a^{\frac{1}{3}} \times a^{\frac{2}{3}}$; (ii) $(a^{\frac{2}{3}})^3 \div (a^{\frac{1}{3}})^2$; (iii) $\sqrt[3]{a} \times \sqrt[6]{a}$; (iv) $\sqrt{(a^{\frac{4}{3}} b^{\frac{2}{3}})}$.

3. Express the following as powers of 2:– (i) $2^4 \times 2^7$; (ii) $(2^a)^2 \div 2^b$; (iii) $2 \cdot 4^6$; (iv) $2^x \cdot 8^y$; (v) $4^{2a} \div 8^b$; (vi) $\sqrt[3]{4} \times \sqrt[4]{8}$.

4. Solve for x (i) $2^x = \frac{1}{16}$; (ii) $3^x = 1$; (iii) $10^{x+1} = 0 \cdot 001$; (iv) $3^{5x} = 27^{x+1}$; (v) $4^x \cdot 16^{x-1} = 64$.

5. Express the following in terms of 2^x:– (i) 4^x; (ii) 2^{2x}; (iii) 2^{x+1}; (iv) $2^{2x+\frac{1}{2}}$.

6. Evaluate (i) $\sqrt[3]{2} \times \sqrt[3]{4}$; (ii) $\sqrt{27} \div \sqrt{3}$; (iii) $\sqrt{8} \div \sqrt[6]{8}$.

7. Expand (i) $(a^{\frac{2}{3}} + b^{\frac{1}{3}})^2$; (ii) $(a^{\frac{1}{3}} + b^{\frac{1}{3}})(a^{\frac{1}{3}} - b^{\frac{1}{3}})$.

8. Simplify (i) $\sqrt[3]{(a^{\frac{6}{5}}b^{\frac{3}{2}})}$; (ii) $\sqrt{a} \times \sqrt[4]{a}$; (iii) $\sqrt{(a^{4p}b^{2p-2})}$.

9. What are the logarithms to base 2 of (i) 8; (ii) 32; (iii) 64; (iv) $\frac{1}{2}$; (v) $\frac{1}{16}$; (vi) $\sqrt{2}$; (vii) $\sqrt[3]{4}$; (viii) $\sqrt{8}$?

10. Find $\log_{10}n$ when n has the values (i) 10000; (ii) $0 \cdot 1$; (iii) $\sqrt{10}$; (iv) $10^{1 \cdot 5}$.

11. Find the values of (i) $\log_2 16$; (ii) $\log_3 81$; (iii) $\log_{\frac{1}{2}}2$; (iv) $\log_{64}8$; (v) $\log_5 1$; (vi) $\log_4 \frac{1}{64}$; (vii) $\log_2 2\sqrt{2}$.

12. Find n if (i) $\log_2 n = 4$; (ii) $\log_5 n = 2$; (iii) $\log_3 n = -3$; (iv) $\log_4 n = -\frac{1}{2}$; (v) $\log_{10} n = 0$.

13. If $\log_x 16 = 2$, what is the value of x?

14. Given $\log_{10}5 = 0 \cdot 69897$, evaluate (i) $\log_{10}25$; (ii) $\log_{10}2$; (iii) $\log_{10}20$.

15. If $\log_{10}5 = 0 \cdot 69897$ and $\log_{10}7 = 0 \cdot 84510$, find the values of (i) $\log_{10}35$; (ii) $\log_{10}14$; (iii) $\log_{10}0 \cdot 28$.

16. Find the values of $\log_2 16$ and $\log_{16}2$.

17. Express each of the following as a single logarithm:–

(i) $\lg 2 + \lg 6$ where $\lg n = \log_{10}n$; (ii) $\log_3 12 - \log_3 2$; (iii) $4 \log_5 2 - 3 \log_5 3$;

(iv) $\frac{1}{2}\log_3 16 + \frac{1}{3}\log_3 8$; (v) $2 \log_5 15 - 3 \log_5 4 + \frac{3}{2}\log_5 16$; (vi) $\frac{1}{3}\log_2 27 - \frac{3}{2}\log_2 49$;

(vii) $\log_2 5 + 1$; (viii) $3 - \log_4 8$; (ix) $2 + \lg 2$. $(\lg n = \log_{10}n)$.

18. Solve (i) $\log_2 x = \log_2 20 - \log_2 5$; (ii) $\log_4 x = 2 \log_4 5 - 1$;
(iii) $\log_3 x = 2 - \log_3 6$; (iv) $2 \log_{10}x = 1 + \log_{10}5 - \log_{10}2$.

19. Express in terms of $\log a$, $\log b$ and $\log c$ (i) $\log a^2 b^3 c$; (ii) $\log 1/abc$; (iii) $\log \sqrt{(a^2 bc^3)}$.

20. Using the notation $\lg n = \log_{10}n$ and given that $\lg 2 = 0 \cdot 301030$, $\lg 3 = 0 \cdot 477121$ and $\lg 7 = 0 \cdot 845089$, evaluate (i) $\lg 12$; (ii) $\lg 84$; (iii) $\lg 0 \cdot 128$; (iv) $\lg \sqrt{\frac{56}{27}}$.

21. What are the logarithms to base 2 of 2^a and $2^{\log_2 3}$? Deduce the value of $2^{\log_2 3}$.

22. Evaluate (i) $10^{\log_{10}2}$; (ii) $5^{\log_5 4}$; (iii) $a^{\log_a b}$.

Change of base rule

To express $\log_a n$ in terms of $\log_b n$ and $\log_b a$.

Let $x = \log_a n$, so $\qquad\qquad a^x = n$.

Taking logarithms to base b, $\log_b a^x = \log_b n$,

$$x \log_b a = \log_b n.$$

$$\therefore \; x = \log_a n = \frac{\log_b n}{\log_b a}.$$

Ex. 6. *If* $\log_2 6 = n$, *find the value of* $\log_8 6$.

We have $\qquad\qquad \log_8 6 = \frac{\log_2 6}{\log_2 8}.$

But $\log_2 8 = 3$, so $\qquad \log_8 6 = \frac{n}{3}.$

Special case.

In the result $\qquad \log_a n = \frac{\log_b n}{\log_b a}$, put $n = b$.

Then $\qquad\qquad \log_a b = \frac{\log_b b}{\log_b a} = \frac{1}{\log_b a},$

i.e. $\qquad \log_a b = \frac{1}{\log_b a}.$

Ex. 7. *Prove that* $\log_2 \frac{1}{320} + \dfrac{1}{\lg 2} + 5 = 0$.

As $\qquad\qquad \lg 2 = \log_{10} 2 = \dfrac{1}{\log_2 10},$

given expression $\qquad = \log_2 \frac{1}{320} + \log_2 10 + 5,$

$$= \log_2 \tfrac{1}{32} + 5,$$

$$= -5 + 5 = 0.$$

(B) EXAMPLES 22b

1. Without using tables, evaluate $\log_4 8$ and $\log_8 4$.

2. If $\dfrac{\log a}{p} = \dfrac{\log b}{q} = \dfrac{\log c}{r} = \log x$, express $\dfrac{b^3}{a^2 c}$ as a power of x.

3. Simplify (i) $(a^{\frac{1}{3}} + b^{\frac{1}{3}})(a^{\frac{2}{3}} - a^{\frac{1}{3}} b^{\frac{1}{3}} + b^{\frac{2}{3}})$; (ii) $(2^{2n+1} - 2^{2n-1}) \div 4^n$.

4. Find, correct to three significant figures, (i) $\log_3 4$; (ii) $\log_2 10$; (iii) $\log_{2.5} 7.5$.

5. Simplify without the use of tables (i) $\dfrac{\log_a 8}{\log_a 2}$; (ii) $\dfrac{\log_a \sqrt{3}}{\log_a 27}$; (iii) $2\log_{10}3 + \dfrac{1}{\log_3 10}$.

6. Given that $u^t = 49$, find $\log_7 u$ in terms of t.

7. Express $\log_4(xy)$ in terms of $\log_2 x$ and $\log_2 y$.

8. Given $\log_3 6 = n$, find in terms of n (i) $\log_3 18$; (ii) $\log_3 2$; (iii) $\log_2 3$.

9. If $\log_8 a = x$, show that $3x \log_a 2 = 1$.

10. Given $(\log_3 x)^2 = 4$, find the two possible values of x.

11. If $u = \log_9 x$, find, in terms of u, (i) x; (ii) $\log_9(3x)$; (iii) $\log_x 81$.

12. Find x, if $\log_{10}4 + 2\log_{10}x = 2$.

13. Solve, for x and y, the equations $2^x . 4^{-y} = 2$, $3^{-x} . 9^{2y} = 3$.

14. Find a and b if $\log_2(ab^2) = 4$ and $\log_2 a . \log_2 b = 2$.

15. Solve the simultaneous equations, $\log_{10}x + \log_{10}y = 1$, $\log_{10}x - \log_{10}y = \log_{10}2\cdot5$.

16. Find the possible values of x if $9\log_x 5 = \log_5 x$.

17. Express $\log_2 3$ and $\log_4 3$ as logarithms to base 8. Hence find n if $\log_8 n = \log_2 3 + 2\log_4 3$.

18. Solve the simultaneous equations, $x^{\frac{1}{2}} + y^{\frac{2}{3}} = 6$, $x^{-\frac{1}{2}} + y^{-\frac{2}{3}} = \frac{3}{4}$.

19. Find x if $\log_4 2 . \log_4(\frac{1}{2}x) = \log_4 x$.

20. The variables x and y are known to be connected by an equation of the form $y = ax^n$, where a and n are constants. Show that the graph of $\log_{10}y$ against $\log_{10}x$ will be a straight line. Simultaneous values of x and y are given below:

x	1·6	2·65	3·8	4·7	5·7
y	6	12	16·5	23·5	26·5

Show that there is an approximate relationship of the form $y = ax^n$ and estimate the values of the constants a and n.

21. Variables x and y are connected by the relationship $y = ax^n$, where a and n are constants. If $y = 45$ when $x = 8$ and $y = 20$ when $x = 27$, calculate the values of a and n.

22. Nm^3 of water were measured as flowing per second over a weir when the difference of water levels was xm, The following results were obtained:

x	1·2	1·4	1·6	1·8	2·0
N	6·3	9·2	12·8	17·3	22·4

Prove that N and x are connected by an equation of the form $N = ax^n$ and find approximate values of the constants a and n.

23. The following experimental values of x and y were obtained:

x	2	3	4	5	6
y	45	100	370	710	2293.

Plot $\log_{10} y$ against x and establish that x and y are connected by an equation of the form $y = an^x$. Use your graph to estimate values of the constants a and n.

24. The relationship between x and y is of the form $y = ax^n$ and corresponding values corrected to the nearest whole number are given below.

x	4	8	12	16
y	12	34	62	96.

By plotting $\log_{10} y$ against $\log_{10} x$, estimate the values of the constants a and n.

Exponential functions

An exponential function is of the form a^x, where a is a positive constant and x is a variable.

The function 2^x is an exponential function with 2 as base. Exponential functions with the same base can be combined by the index laws.

Ex. 8. If $2^x = y$, express the following functions in terms of y, (i) 2^{2x}; (ii) 4^x; (iii) $8^{x+\frac{1}{3}}$; (iv) $4^{\frac{1}{2}-x}$.

(i) $2^{2x} = 2^x \times 2^x = y^2$.
(ii) $4^x = (2^2)^x = 2^{2x} = y^2$.
(iii) $8^{x+\frac{1}{3}} = 8^x \cdot 8^{\frac{1}{3}} = 2^{3x} \cdot 2 = 2y^3$.
(iv) $4^{\frac{1}{2}-x} = 4^{\frac{1}{2}} \div 4^x = 2 \div 2^{2x} = 2/y^2$.

Graphs of exponential functions

First take an exponential function with base greater than one, say 2^x.

Let $y = 2^x$.

x	-4	-3	-2	-1	0	1	2	3	4
y	$\frac{1}{16}$	$\frac{1}{8}$	$\frac{1}{4}$	$\frac{1}{2}$	1	2	4	8	16

The shape of the graph of $y = 2^x$ for x in the range ± 4 is shown in Fig. 136.

For values of x greater than 4, the graph climbs increasingly steeply as y tends to ∞. For values of x less than -4, the graph approaches closer and closer to the x-axis as y tends to zero.

Now consider the exponential function $(\frac{1}{2})^x$, with a base less than one.

Let $y = (\frac{1}{2})^x = 2^{-x}$.

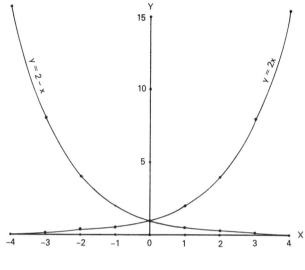

FIG. 136

The previous table of values can be used with the positive and negative values of x interchanged.

The graph of $y = (\frac{1}{2})^x$ is the reflection of the graph of $y = 2^x$ in the y-axis (Fig. 136).

From these two examples the important fact emerges that *an exponential function, a^x, is positive for all real values of the variable x.*

The exponential function e^x

The exponential function with base e, where e is an irrational constant approximately equal to $2 \cdot 718$, is of the greatest importance in mathematics. The function e^x is referred to as *the exponential function*, exp x. At the end of the chapter some basic calculus results involving the constant e, the Naperian constant, will be given without proof.

Values of e^x and e^{-x} for a range of values of x are tabulated in standard books of mathematical tables.

The graphs of $y = e^x$ and $y = e^{-x}$ closely resemble the graphs of $y = 2^x$ and $y = 2^{-x}$ (Fig. 136). In particular it should be remembered that e^x is positive for all real values of x.

Exponential equations

The methods of solving the following exponential equations should be noted.

Ex. 9. Solve the equation $7^x \cdot 8^{2x-1} = 6 \cdot 3$, giving x correct to 3 s.f.

Equations involving one exponential function or a product or quotient of exponential functions are solved by taking logarithms, base 10, of both sides of the equations. Using this process here we get

$$x \lg 7 + (2x - 1) \lg 8 = \lg 6 \cdot 3,$$
$$i.e. \quad 0 \cdot 8451x + 0 \cdot 9031(2x - 1) = 0 \cdot 7993,$$
$$2 \cdot 6513x = 1 \cdot 7024$$
$$x = \frac{1 \cdot 7024}{2 \cdot 6513} - 0 \cdot 642, \text{ to } 3 \text{ s.f.}$$

Ex. 10. *Solve the equation* $3^{2x} - 3^{x+2} + 8 = 0$.

It is not possible to take logarithms in this case because of the signs between the terms.

The equation is solved by writing $y = 3^x$ and noting that $3^{2x} = y^2$ and $3^{x+2} = 3^x \cdot 3^2 = 9y$.

So
$$y^2 - 9y + 8 = 0,$$
$$(y - 8)(y - 1) = 0,$$
$$y = 1, 8.$$
$$\therefore \ 3^x = 1, \text{ giving } x = 0,$$

or $3^x = 8$, an exponential equation solved by taking logarithms to base 10.

We have $x \lg 3 = \lg 8; \quad x = 1 \cdot 89 \quad \text{to} \quad 3 \text{ s.f.}$

$\therefore$ Solutions of the given equation are $x = 0$ and $1 \cdot 89$.

Logarithmic functions

The basic logarithmic function is $\log_a x$, where the base a is positive and usually considered to be greater than 1.

If $y = \log_a x$ then $x = a^y$, so logarithmic and exponential functions are *inverse functions*. Logarithmic functions to the same base can be combined by the ordinary laws of logarithms. The change of base rule also applies,

So
$$\log_a x = \frac{\log_b x}{\log_b a}.$$

Ex. 11. *If* $y = \log(1 + 6x + 8x^2) - \log(1 + 5x + 4x^2) + \log(1 + x) - \log(1 + 2x)$, *the logarithms being to the same base, prove that y is independent of x.*

Using the laws of logarithms,

$$y = \log \frac{(1 + 6x + 8x^2)(1 + x)}{(1 + 5x + 4x^2)(1 + 2x)},$$
$$= \log \frac{(1 + 4x)(1 + 2x)(1 + x)}{(1 + 4x)(1 + x)(1 + 2x)},$$
$$= \log 1 = 0.$$

Hence y is independent of x.

Graphs of logarithmic functions

To draw the graph of $y = \log_a x$, for any particular value of a, a table of values is formed in the usual manner.

For values of $a > 1$, the shape of the graph of $y = \log_a x$ is similar to that of the graph of $y = \log_2 x$ shown in Fig. 137.

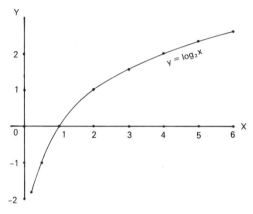

FIG. 137

Two important facts about the function $\log_a x$ should be noted:

(i) the function only exists for values of $x > 0$;
(ii) the function is zero when $x = 1$.

Naperian or natural logarithms

Naperian or *natural logarithms* are logarithms to base e. They are used for all theoretical purposes especially in the calculus. Common logarithms, base 10, are of course used for computational purposes.

Notation. $\log_e x$ is frequently written with the base omitted or more simply as ln x. The notation log x will be used here.

Naperian or natural logarithms are tabulated in standard books of tables. The method of using such tables is shown in the following example.

Ex. 12. Use tables to evaluate (i) log 15; (ii) log 0·54.

(i) As Naperian logarithm tables only give the logarithms of numbers from 1 to 10, we write 15 as $1·5 \times 10$.

$$\log 15 = \log 1·5 + \log 10,$$
$$= 0·4055 + 2·3026 = 2·7081.$$

(ii) Writing $0\cdot54$ as $5\cdot4 \div 10$,

$$\log 0\cdot54 = \log 5\cdot4 - \log 10,$$
$$= 1\cdot6864 - 2\cdot3026 = -0\cdot6162.$$

N.B. It should be noted that the "bar notation" is not used with Naperian logarithms.

(A) EXAMPLES 22c

[where necessary take e as $2\cdot718$]

1. Using tables, evaluate (i) $e^{1\cdot5}$; (ii) $e^{0\cdot5}$; (iii) $e^{-0\cdot4}$; (iv) e^{3}; (v) $\sqrt[4]{e}$, correct to 4 s.f.

2. Given that $\cosh x = \frac{1}{2}(e^x + e^{-x})$ and $\sinh x = \frac{1}{2}(e^x - e^{-x})$, find the values of (i) $\cosh 0$; (ii) $\sinh 0$; (iii) $\cosh 1$; (iv) $\sinh 0\cdot5$, where necessary to 3 s.f.

3. Using tables, evaluate to 4 s.f., (i) $\log 2$; (ii) $\log 4\cdot5$; (iii) $\log 25$; (iv) $\log 0\cdot4$.

4. Show that $\log_a \dfrac{1}{x} = -\log_a x$.

5. Evaluate without the use of tables (i) $\log e^2$; (ii) $\log \sqrt{e}$; (iii) $\log \dfrac{1}{e}$; (iv) $\frac{1}{2}\log_a a^3$.

6. Show that the results $e^a = b$ and $\log b = a$ are equivalent and hence find x if (i) $e^x = 2$; (ii) $e^{-x} = 1$; (iii) $e^{5x} = 2$; (iv) $e^{-10x} = 0\cdot5$.

7. What is the natural logarithm of $e^{\log x}$? What is the value of the function?

8. Evaluate (i) $e^{\log 3}$; (ii) $e^{\log x}$; (iii) $e^{2\log 2}$; (iv) $e^{\frac{1}{2}\log x}$; (v) $e^{-\log x}$.

9. Express as single logarithmic functions:– (i) $\log(x+1) - 3\log(1-x) + 2\log x$; (ii) $\log \sqrt{(x^2-1)} + \frac{1}{2}\log \dfrac{x+1}{x-1}$.

10. Find x in terms of e if (i) $\log x = 3$; (ii) $\log x = -2$; (iii) $\log x = 1 + \log 2$; (iv) $\log(\log x) = 0$.

11. Solve (i) $e^x(1-2x) = 0$; (ii) $e^{-x}(1+x) = 0$; (iii) $xe^x = e^x$.

12. Solve the following equations giving x correct to 3 s.f.:— (i) $2^x = 3$; (ii) $4^x = 21$; (iii) $5^{2x} = 8$; (iv) $3^{x+1} = 18$.

(B) EXAMPLES 22d

Solve the equations in ex. 1–3, giving solutions correct to 3 s.f.

1. $2 = 5e^{0\cdot1t}$ **2.** $3 = 2e^{0\cdot5t}$ **3.** $6 = 7te^{0\cdot5}$.

4. The number of bacteria in a culture at time t was given by $n = n_0 e^{5t}$. Find (i) the number present at time $t = 0$; (ii) the value of t when the colony was double its initial size.

5. Given $\log_b(xy^3) = m$ and $\log_b(x^3y^2) = n$, find the value of $\log_b \sqrt{(xy)}$ in terms of m and n.

6. Solve the following equations correct to 3 s.f.:   (i) $3^{x+1} - 4^{x-1}$; (ii) $7^x \cdot 8^{2x-1} = 6 \cdot 3$; (iii) $2^x \cdot 3^{2x} = 5$.

7. By writing $y = 2^x$, solve the equation $2^{2x} - 5 \cdot 2^x + 6 = 0$.

8. Using the same axes, sketch the graphs of $y = e^x$ and $y = e^{-x}$ for values of x in the range ± 3. Use these graphs to sketch the graph of $y = \cosh x = \frac{1}{2}(e^x + e^{-x})$ over the same range.

9. A 5g sample of radioactive substance decays according to the formula $y = 5e^{-kt}$, where y is the quantity remaining at time t min. If half the initial sample decays in 10 min., find the amount remaining after 20 min.

10. Use the substitution $u = 3^x$ to solve the equation $3^{2x+1} + 3^2 = 3^{x+3} + 3^x$.

11. Sketch the graphs of $y = e^{-x^2}$ (a normal probability curve) for values of x from -3 to 3.

12. Solve for x and y (i) $x - y = 1$, $2^x \cdot 3^y = 432$; (ii) $\log x + \log y = \log 10$, $\log x - \log y = \log 2 \cdot 5$.

13. Draw the graph of $y = \log x$ for values of x from $\frac{1}{2}$ to 6. Using the same axes draw the graph of $xy = 1$ and hence find an approximate solution of the equation $x \log x = 1$.

14. Solve the equation $e^x - e^{-x} = 6$, taking e as $2 \cdot 718$.

15. From the definitions, $\cosh x = \frac{1}{2}(e^x + e^{-x})$ and $\sinh x = \frac{1}{2}(e^x - e^{-x})$, show that $\cosh(-x) = \cosh x$ and $\sinh(-x) = -\sinh x$. Also verify that $\sinh 2x = 2 \sinh x \cosh x$.

16. Solve, correct to 3 s.f., the equations $4^x = 5^y$, $2 \cdot 4^x = 7^y$.

Derivatives of logarithmic and exponential functions

The following basic results are given without proof: —

(i) $\dfrac{d}{dx}(\log x) = \dfrac{1}{x}$;

(ii) $\dfrac{d}{dx}(e^x) = e^x$.

The following worked examples show how these results can be extended by using the rules of differentiation.

Ex. 13. *Differentiate with respect to* x (*i*) *log* (1+x); (*ii*) *log* (x²+1); (*iii*) *log* (sin x); (*iv*) x² *log* x.

(i)   Let $y = \log (1+x)$.

Using the function of a function rule with $u = 1 + x$,

We have $$\frac{dy}{dx} = \frac{d}{du} (\log u) \cdot \frac{d}{dx} (1+x),$$

$$= \frac{1}{u} \cdot 1 = \frac{1}{1+x}.$$

(ii)   Let $y = \log (x^2 + 1)$.

Using the function of a function rule with $u = x^2 + 1$,

We have $$\frac{dy}{dx} = \frac{d}{du} (\log u) \cdot \frac{d}{dx} (x^2 + 1),$$

$$= \frac{1}{u} \cdot 2x = \frac{2x}{x^2 + 1}.$$

(iii)   Let $y = \log (\sin x)$.

Using the function of a function rule with $u = \sin x$,

We have $$\frac{dy}{dx} = \frac{d}{du} (\log u) \cdot \frac{d}{dx} (\sin x),$$

$$= \frac{1}{u} \cdot \cos x = \frac{\cos x}{\sin x} \quad \text{or} \quad \cot x.$$

(iv)   Let $y = x^2 \log x$.

By the product rule,

$$\frac{dy}{dx} = 2x \log x + x^2 \cdot \frac{1}{x},$$

$$= x(2 \log x + 1).$$

Ex. 14. *Differentiate with respect to* x (*i*) e²ˣ; (*ii*) e⁻ˣ; (*iii*) e³ˣ²; (*iv*) x²e⁻²ˣ.

(i) Using the function of a function rule with $u = 2x$,

$$\frac{d}{dx} (e^{2x}) = \frac{d}{du} (e^u) \cdot \frac{d}{dx} (2x),$$

$$= e^u \cdot 2 = 2e^{2x}.$$

(ii) Using the function of a function rule with $u = -x$,

$$\frac{d}{dx} (e^{-x}) = \frac{d}{du} (e^u) \cdot \frac{d}{dx} (-x),$$

$$= e^u \cdot (-1) = -e^{-x}.$$

(iii) Using the function of a function rule with $u = 3x^2$,

$$\frac{d}{dx}(e^{3x^2}) = \frac{d}{du}(e^u) \cdot \frac{d}{dx}(3x^2),$$

$$= e^u \cdot 6x = 6xe^{3x^2}.$$

(iv) By the product rule,

$$\frac{d}{dx}(x^2 e^{-2x}) = 2xe^{-2x} + x^2(-2e^{-2x}),$$

$$= 2e^{-2x}(x - x^2).$$

Ex. 15. Find the minimum value of x log x.

$$\text{Let} \quad y = x \log x.$$

$$\frac{dy}{dx} = 1 \log x + x \cdot \frac{1}{x}$$

$$= \log x + 1.$$

For maximum or minimum, $\frac{dy}{dx} = 0.$

$$\therefore \ \log x + 1 = 0,$$

$$\log x = -1; \quad x = e^{-1} \quad \text{or} \quad \frac{1}{e}.$$

Differentiating again, $\quad \dfrac{d^2y}{dx^2} = \dfrac{1}{x}.$

So when $x = e^{-1}$, $\quad \dfrac{d^2y}{dx^2} = e,$ which is positive.

$\therefore$ the function is a minimum when $x = e^{-1}$.

$$\text{Minimum value} = \frac{1}{e} \log\left(\frac{1}{e}\right) = -\frac{1}{e}.$$

Integral results

$$\text{As} \quad \frac{d}{dx}(\log x) = \frac{1}{x},$$

$$\int \frac{dx}{x} = \log x + c.$$

More generally, using the function of a function rule with $u = ax + b$,

$$\frac{d}{dx}[\log(ax+b)] = \frac{a}{ax+b}.$$

$$\therefore \ \int \frac{dx}{ax+b} = \frac{1}{a} \log(ax+b) + c.$$

$$\text{As} \quad \frac{d}{dx}(e^x) = e^x,$$

$$\int e^x \, dx = e^x + c.$$

More generally, using the function of a function rule with $u = ax + b$,

$$\frac{d}{dx}(e^{ax+b}) = ae^{ax+b}$$

$$\therefore \int e^{ax+b} dx = \frac{1}{a} e^{ax+b} + c.$$

Ex. 16. *Evaluate* (i) $\displaystyle\int_1^3 \frac{dx}{2x}$; (ii) $\displaystyle\int_2^3 \frac{3dx}{2x-1}$.

(i) $\displaystyle\int_1^3 \frac{dx}{2x} = \frac{1}{2}[\log x]_1^3 = \frac{1}{2}\log 3 - \frac{1}{2}\log 1,$

$$= \frac{1}{2}\log 3.$$

(ii) $\displaystyle\int_2^3 \frac{3dx}{2x-1} = 3[\frac{1}{2}\log(2x-1)]_2^3 = \frac{3}{2}\log 5 - \frac{3}{2}\log 3,$

$$= \frac{3}{2}\log \frac{5}{3}.$$

Ex. 17. *Integrate* e^{2x+1} *and evaluate* $\displaystyle\int_0^2 \frac{dx}{e^x}$.

$$\int e^{2x+1} dx = \frac{1}{2} e^{2x+1} + c.$$

$$\int_0^2 \frac{dx}{e^x} = \int_0^2 e^{-x} dx$$

$$= \left[\frac{e^{-x}}{-1} \right]_0^2 = (-e^{-2}) - (-e^0),$$

$$= -\frac{1}{e^2} + 1.$$

(A) EXAMPLES 22e

Differentiate the following logarithmic functions with respect to x.

1. $\log 2x$.	**2.** $\log(x+2)$.	**3.** $\log(3x+1)$.
4. $\log(2x-5)$.	**5.** $\log(1-x)$.	**6.** $\log(4-3x)$.
7. $\log(x^2+1)$.	**8.** $\log(x^2-x+1)$.	**9.** $\log(3-x^2)$.
10. $\log\left(\frac{1}{x}\right)$.	**11.** $\log(x^3+2)$.	**12.** $\log\left(\frac{1}{x+1}\right)$.

Differentiate the following exponential functions with respect to x.

13. $3e^x$.

14. e^{3x}.

15. e^{-3x}.

16. $2e^{2x}$.

17. $4e^{-x}$.

18. e^{x+1}.

19. e^{4x-1}.

20. $3e^{1-2x}$.

21. $-e^{-4x}$.

22. $\dfrac{1}{e^{2x}}$.

23. $\dfrac{1}{e^{x-1}}$.

24. $\dfrac{3}{2e^{3x}}$.

25. Integrate with respect to x (i) $\dfrac{1}{2x}$; (ii) $\dfrac{4}{3x}$; (iii) $\dfrac{1}{x-1}$; (iv) $\dfrac{1}{x+4}$; (v) $\dfrac{1}{x-7}$; (vi) $\dfrac{1}{2x+1}$; (vii) $\dfrac{2}{3+x}$; (viii) $\dfrac{1}{4-x}$; (ix) $\dfrac{1}{2(x+1)}$; (x) $\dfrac{3}{1-3x}$; (xi) $\dfrac{1}{4x+3}$; (xii) $\dfrac{2}{3(2x-5)}$.

26. Integrate with respect to x (i) $2e^x$; (ii) $\frac{1}{2}e^x$; (iii) e^{2x}; (iv) $2e^{-x}$; (v) $4e^{4x}$; (vi) e^{-2x}; (vii) $3e^{6x}$; (viii) e^{x-2}; (ix) e^{1-x}; (x) e^{2x+3}; (xi) $\dfrac{1}{e^{2x}}$; (xii) $\dfrac{1}{e^{1-x}}$.

Evaluate the following definite integrals leaving answers as logarithms or in terms of e.

27. $\displaystyle\int_1^4 \dfrac{dx}{x}$.

28. $\displaystyle\int_{-3}^0 \dfrac{dx}{x+4}$.

29. $\displaystyle\int_2^3 \dfrac{dx}{2x-1}$.

30. $\displaystyle\int_0^1 e^{2x}dx$.

31. $\displaystyle\int_{-2}^{-1} e^{-x}dx$.

32. $\displaystyle\int_0^2 e^{-2x}dx$.

33. $\displaystyle\int_{-2}^{-1} \dfrac{dx}{1-x}$.

34. $\displaystyle\int_{-1}^1 \dfrac{dx}{3+x}$.

35. $\displaystyle\int_0^1 \dfrac{2dx}{3x+4}$.

36. $\displaystyle\int_{-\frac{1}{2}}^1 e^{2x+1}dx$.

37. $\frac{1}{2}\displaystyle\int_0^1 (e^x+e^{-x})dx$.

38. $\frac{1}{2}\displaystyle\int_0^1 (e^{2x}-e^{-2x})dx$.

(B) EXAMPLES 22f

Differentiate the following functions with respect to x.

1. $x^3 \log x$.

2. xe^{3x}.

3. $\log(\cos x)$.

4. $e^x \sin x$.

5. $3x^2 e^{-2x}$.

6. $e^{\sin x}$.

7. $\dfrac{x}{e^x}$.

8. $\log \dfrac{(x^2+1)}{x}$.

9. $\dfrac{1}{e^x+1}$.

10. $\log\left(\dfrac{x}{x+1}\right)$.

11. xe^{-x^2}.

12. $\log\sqrt{(x^2+1)}$.

13. Find the gradient of the curve $y = \log x$ at the point where it cuts the x-axis.

14. Show that the function $\sinh x$, which is equal to $\frac{1}{2}(e^x - e^{-x})$, increases for all values of x.

15. Find the area bounded by the arc of the curve $xy = 1$ between $x = 1$ and $x = 3$, the x-axis and the ordinates $x = 1$, $x = 3$.

16. Find the turning point on the curve $y = x^2 \log x$ and state whether it is a maximum or a minimum point.

17. If $s = te^{2t}$, find the value of $\dfrac{d^2s}{dt^2}$, when $t = 0$.

18. Show that $y = (3 + 2x)e^{-2x}$ satisfies the differential equation $\dfrac{d^2y}{dx^2} + 4\dfrac{dy}{dx} + 4y = 0$.

19. Find the minimum value of the function xe^x.

20. Find the area bounded by the curve $y(1 + x) = 1$, the x- and y-axes and the ordinate $x = 2$.

21. Show by division, that $\dfrac{2x - 1}{x + 2} = 2 - \dfrac{5}{x + 2}$ and use this result to evaluate $\displaystyle\int_{-1}^{1} \dfrac{2x - 1}{x + 2}\,dx$.

22. Using the method of the previous example, evaluate (i) $\displaystyle\int_{0}^{1} \dfrac{x\,dx}{x + 1}$; (ii) $\displaystyle\int_{0}^{1} \dfrac{x^2\,dx}{x + 1}$.

23. The area bounded by the curve $y = e^x$, the axes and the ordinate $x = 1$ is rotated about the x-axis. Find the volume of the solid of revolution.

24. Evaluate (i) $\displaystyle\int_{0}^{1} (e^x + 1)^2\,dx$; (ii) $\displaystyle\int_{-1}^{1} (e^x + e^{-x})^2\,dx$.

CHAPTER 23

FURTHER ALGEBRAIC METHODS

Remainder theorem

Ex. 1. *Find the remainder when the polynomial* $f(x) = x^3 + x^2 - x + 2$ *is divided by* $(x - 1)$.

The process of division can be undertaken as shown, giving a quotient of $x^2 + 2x + 1$ and a remainder of 3.

If, however, only the remainder is required, it is much simpler to proceed as follows.

Let the quotient be $q(x)$ and the remainder be R.

Then R will be a constant as the division is by a linear function of x.

$$
\begin{array}{r}
x^2 + 2x + 1 \\
(x-1)\overline{\smash{)}x^3 + x^2 - x + 2} \\
\underline{x^3 - x^2} \\
2x^2 - x \\
\underline{2x^2 - 2x} \\
x + 2 \\
\underline{x - 1} \\
3.
\end{array}
$$

Then $x^3 + x^2 - x + 2 \equiv (x - 1)q(x) + R$, an identity and true for all values of x.

By taking $x = 1$, the term involving $q(x)$ becomes zero and we have

$$1 + 1 - 1 + 2 = 0 + R,$$
$$R = 3,$$

the value of $f(x)$ with x put equal to 1, *i.e.* $f(1)$.

General cases. *When a polynomial* $f(x)$ *is divided by* $(x - a)$, *the remainder is equal to* $f(a)$, *the value of* $f(x)$ *with* x *put equal to* a. *More generally, if* $f(x)$ *is divided by* $(ax + b)$, *the remainder is* $f(-b/a)$.

Ex. 2. *When the polynomial* $4x^4 - ax^2 + 5x + 2$ *is divided by* $(2x - 1)$, *the remainder is 4, find the value of* a.

$$\text{Let} \quad f(x) = 4x^4 - ax^2 + 5x + 2.$$

$$\text{Remainder} = f(\tfrac{1}{2}).$$

$$\therefore \ 4 = 4(\tfrac{1}{2})^4 - a(\tfrac{1}{2})^2 + 5(\tfrac{1}{2}) + 2,$$

$$= \tfrac{1}{4} - \tfrac{1}{4}a + \tfrac{5}{2} + 2.$$

$$\tfrac{1}{4}a = \tfrac{3}{4},$$

$$a = 3.$$

Factor theorem. *If* $f(x)$ *is a polynomial such that* $f(a) = 0$, *then* $(x - a)$ *is a factor of* $f(x)$.

This result is a direct consequence of the previous theorem.

Ex. 3. Show that $(x - 2)$ *is a factor of the polynomial* $x^3 - 2x^2 + x - 2$ *and find the other factor.*

$$f(x) = x^3 - 2x^2 + x - 2.$$

$$f(2) = 8 - 8 + 2 - 2 = 0.$$

$$\therefore \ (x - 2) \text{ is a factor of } f(x).$$

Let $$f(x) \equiv (x - 2)g(x).$$

Clearly $g(x)$ is a quadratic function with first term x^2 and constant term 1.

So $$x^3 - 2x^2 + x - 2 \equiv (x - 2)(x^2 + ax + 1).$$

Equating coefficients of x^2, $-2 = a - 2$,

$$a = 0.$$

Hence the second factor is $x^2 + 1$.

Ex. 4. If $(x^2 - x - 6)$ *is a factor of the polynomial* $x^4 + ax^3 - 9x^2 + bx - 6$, *find the values of* a *and* b.

$$x^2 - x - 6 = (x - 3)(x + 2).$$

$\therefore \ (x - 3)$ and $(x + 2)$ are factors of the given polynomial $f(x)$.
By the factor theorem, $f(3) = 0$ and $f(-2) = 0$.

$$\therefore \ 3^4 + a \cdot 3^3 - 9 \cdot 3^2 + b \cdot 3 - 6 = 0,$$

$$27a + 3b - 6 = 0,$$

$$9a + b = 2. \ldots \ldots \ldots \ldots (i)$$

Also $$(-2)^4 + a(-2)^3 - 9(-2)^2 + b(-2) - 6 = 0,$$

$$-8a - 2b - 26 = 0,$$

$$4a + b = -13. \ldots \ldots \ldots (ii)$$

Solving the simultaneous equations (i) and (ii), we have $a = 3$, $b = -25$.

(A) EXAMPLES 23a

1. If $f(x) = 2x^3 - x + 5$, find the values of (i) $f(0)$; (ii) $f(2)$; (iii) $f(-2)$.

2. If $p(x) = 3x^2 + 2x + c$ and $p(1) = 7$, find the value of c.

3. Given polynomials $f(x) = x^3 - x^2 + 1$ and $g(x) = x^3 + ax^2 + 3$ such that $f(-1) = g(1)$, find the value of the constant a.

In ex. 4–11, find in each case, the remainder when the polynomial $f(x)$ is divided by the linear function $l(x)$.

4. $f(x) = x^3 + 1; l(x) = x - 1$.

5. $f(x) = x^2 - 7x + 6; l(x) = x + 1$.

6. $f(x) = 2x^3 - x^2 + x - 3; l(x) = x$.

7. $f(x) = x^4 - 2x^2 + 1; l(x) = x - 2$.

8. $f(x) = 3x^3 + 2x^2 + x - 1; l(x) = x - 3$.

9. $f(x) = x^5 + 3x^3 - 1; l(x) = x - 1$.

10. $f(x) = 4x^3 - 2x^2 + 6x - 3; l(x) = 2x - 1$.

11. $f(x) = 8x^3 - 6x + 5; l(x) = 2x + 3$.

12. Show by the factor theorem that $(x - 3)$ is a factor of $4x^2 - 7x - 15$ and find the other factor.

13. For what value of k does the function $x^2 + kx + 2$ give the same remainder when divided by either $(x - 1)$ or $(x + 1)$?

14. Prove that $(x - 2)$ is a factor of $x^3 - 8$ and find the second factor.

15. When the polynomial $x^5 + ax^3 + ax + 4$ is divided by $(x - 2)$ the remainder is 6. Find the remainder when it is divided by $(x + 2)$.

16. When $x^2 + x - 3$ is divided by $(x - a)$, the remainder is -1. Find the possible values of the constant a.

17. Show that $(x + 3)$ and $(2x - 1)$ are factors of the function $2x^4 + 5x^3 - x^2 + 5x - 3$. Verify that the third factor is $(x^2 + 1)$.

18. The polynomial $2x^3 + ax^2 - 11x + 6$ is divisible by $(x + 2)$, find the value of a. With this value for a, show that the function is also divisible by $(2x - 1)$ and $(x - 3)$.

19. The functions $x^3 + 2x - 1$ and $2x^3 + kx - 5$ leave the same remainder when divided by $(x + 1)$. Find the value of k.

20. Show that $(x^2 - 1)$ is a factor of the polynomial $3x^3 + x^2 - 3x - 1$ and find the remaining factor.

21. Given $f(x) = ax^3 + x^2 - 4x + b$, $f(0) = 1$ and $f(1) = 0$. Find the values of a and b and factorise the resulting expression.

22. Use the factor theorem to show that $(a-b)$ is a factor of a^3-b^3 and a^7-b^7.

23. Verify by the factor theorem that $(x+y)$ is a factor of x^5+y^5.

24. If (x^2-4) is a factor of the polynomial $x^4-2x^3+ax^2+bx+8$, use the factor theorem to find two linear equations connecting a and b and hence find the values of these constants.

25. Given that $(x-1)$ is a factor of the function ax^4+3x^2+x+b and that the remainder when the function is divided by $(x+2)$ is 21, find the values of the constants a and b.

26. Show that the function x^n-y^n is divisible by $(x-y)$ for all positive integral values of n. Under what condition is the function also divisible by $(x+y)$?

Manipulation of surds

A surd is an irrational number of the form $\sqrt{n}$, where n is a positive integer which is not a perfect square.

E.g. $\sqrt{2}, \sqrt{7}, \sqrt{28}$ are examples of surds.

Addition and subtraction

As in the case of fractions, only like surds can be combined by addition and subtraction.

E.g. $3\sqrt{7}-\sqrt{7}=2\sqrt{7}$ and $4\sqrt{11}+2\sqrt{11}=6\sqrt{11}$,

but $\sqrt{7}-\sqrt{5}$ and $3\sqrt{11}+2\sqrt{7}$ cannot be simplified.

Multiplication and division

Surds can be multiplied and divided just like integers.

E.g. $\sqrt{5}\times\sqrt{13}=\sqrt{65}$ and $\sqrt{15}\div\sqrt{3}=\sqrt{5}$.

N.B. *The square of a surd is integral, e.g. $(\sqrt{13})^2=13$.*

Ex. 5. Simplify (i) $4\sqrt{5}-3\sqrt{5}$; (ii) $2\sqrt{5}\times3\sqrt{3}$.

$$\text{(i) } 4\sqrt{5}-3\sqrt{5}=\sqrt{5}.$$
$$\text{(ii) } 2\sqrt{5}\times3\sqrt{3}=6\sqrt{15}.$$

Ex. 6. Expand (i) $(2\sqrt{3}-1)^2$; (ii) $(\sqrt{7}-2)(\sqrt{7}+2)$.

$$\text{(i) } (2\sqrt{3}-1)^2=(2\sqrt{3})^2-2(2\sqrt{3})+1,$$
$$=12-4\sqrt{3}+1=13-4\sqrt{3}.$$
$$\text{(ii) } (\sqrt{7}-2)(\sqrt{7}+2)=(\sqrt{7})^2-2^2$$
$$=7-4=3.$$

Reduction to lowest terms. If the integer n has a factor which is a perfect square, the surd $\sqrt{n}$ can be expressed in terms of a simpler surd,

e.g. $$\sqrt{32} = \sqrt{(16 \times 2)} = \sqrt{16} \times \sqrt{2} = 4\sqrt{2}.$$

Ex. 7. Simplify $\sqrt{27} - \sqrt{12}$.

$$\sqrt{27} = \sqrt{9} \times \sqrt{3} = 3\sqrt{3}; \qquad \sqrt{12} = \sqrt{4} \times \sqrt{3} = 2\sqrt{3}.$$
$$\therefore \ \sqrt{27} - \sqrt{12} = 3\sqrt{3} - 2\sqrt{3} = \sqrt{3}.$$

Fractional forms. If a fraction has a surd in the denominator, the evaluation of the fraction can be much facilitated by *rationalising the denominator* of the fraction as is illustrated in the following example.

Ex. 8. Express the following fractions with rational denominators:— (i) $\dfrac{3}{2\sqrt{5}}$; (ii) $\dfrac{2\sqrt{3}}{2\sqrt{2} - 1}$.

(i) Rationalisation of the denominator in this case is achieved by multiplying numerator and denominator by $\sqrt{5}$.

We have $$\frac{3}{2\sqrt{5}} = \frac{3\sqrt{5}}{2\sqrt{5} \cdot \sqrt{5}} = \frac{3\sqrt{5}}{2 \cdot 5} = \frac{3\sqrt{5}}{10}.$$

(ii) In this case the result $(a+b)(a-b) = a^2 - b^2$ is used to remove the surd in the denominator.

We have $$\frac{2\sqrt{3}}{2\sqrt{2} - 1} = \frac{2\sqrt{3}(2\sqrt{2} + 1)}{(2\sqrt{2} - 1)(2\sqrt{2} + 1)} = \frac{4\sqrt{6} + 2\sqrt{3}}{(2\sqrt{2})^2 - 1^2},$$
$$= \frac{4\sqrt{6} + 2\sqrt{3}}{8 - 1} = \frac{4\sqrt{6} + 2\sqrt{3}}{7}.$$

Ex. 9. Express (i) $\dfrac{1}{(\sqrt{3} + \sqrt{2})^2}$; (ii) $\dfrac{1}{(\sqrt{3} + \sqrt{2})^4}$ in the form $a + b\sqrt{6}$.

(i) Multiplying numerator and denominator by $(\sqrt{3} - \sqrt{2})^2$,

$$\frac{1}{(\sqrt{3} + \sqrt{2})^2} = \frac{(\sqrt{3} - \sqrt{2})^2}{[(\sqrt{3} + \sqrt{2})(\sqrt{3} - \sqrt{2})]^2} = \frac{3 - 2\sqrt{6} + 2}{(3 - 2)^2},$$
$$= 5 - 2\sqrt{6}.$$

(ii) Using the result of (i),

$$\frac{1}{(\sqrt{3} + \sqrt{2})^4} = \left(\frac{1}{(\sqrt{3} + \sqrt{2})^2}\right)^2 = (5 - 2\sqrt{6})^2,$$
$$= 25 - 20\sqrt{6} + 24,$$
$$= 49 - 20\sqrt{6}.$$

(A) EXAMPLES 23b

1. Simplify if possible

(i) $\sqrt{7} + \sqrt{3}$; (ii) $\sqrt{2} \times \sqrt{3}$; (iii) $2\sqrt{5} - \sqrt{5}$; (iv) $\sqrt{14} \div \sqrt{7}$;

(v) $3\sqrt{2}+3\sqrt{2}$;   (vi) $(\sqrt{2})^2$;   (vii) $(2\sqrt{3})^2$;   (viii) $3\sqrt{5}\times2\sqrt{2}$;
(ix) $4\sqrt{3}-2\sqrt{2}$;   (x) $6\sqrt{12}\div3\sqrt{2}$;   (xi) $4\times3\sqrt{5}$;   (xii) $2\sqrt{7}\times4\sqrt{3}$.

2. Expand (i) $3(2-\sqrt{3})$; (ii) $\sqrt{2}(1+\sqrt{3})$; (iii) $(2\sqrt{5})^2$; (iv) $(1+\sqrt{2})^2$;
(v) $(\sqrt{2}+1)(\sqrt{2}-1)$; (vi) $2\sqrt{5}(\sqrt{5}-1)$; (vii) $(2\sqrt{5}-\sqrt{7})(2\sqrt{5}+\sqrt{7})$;
(viii) $(5-3\sqrt{2})^2$; (ix) $(\sqrt{7}-3)(\sqrt{7}+3)$; (x) $(1+\sqrt{2})(2\sqrt{2}-3)$;
(xi) $(3\sqrt{2}-\sqrt{3})(3\sqrt{2}+\sqrt{3})$.

3. Reduce the following surds to their lowest terms:—

   (i) $\sqrt{8}$;   (ii) $\sqrt{18}$;   (iii) $\sqrt{50}$;   (iv) $\sqrt{45}$;   (v) $\sqrt{32}$;

   (vi) $\sqrt{98}$;   (vii) $\sqrt{24}$;   (viii) $\sqrt{72}$;   (ix) $2\sqrt{12}$;   (x) $4\sqrt{54}$.

4. Simplify (i) $\sqrt{18}-\sqrt{8}$; (ii) $3\sqrt{5}-\sqrt{20}$; (iii) $2\sqrt{75}+3\sqrt{12}$.

5. Express with rational denominators (i) $\dfrac{1}{\sqrt{2}}$; (ii) $\dfrac{2}{\sqrt{5}}$; (iii) $\dfrac{1}{2\sqrt{7}}$;

(iv) $\dfrac{\sqrt{2}}{4\sqrt{3}}$; (v) $\dfrac{2\sqrt{5}}{3\sqrt{3}}$; (vi) $\dfrac{\sqrt{6}}{\sqrt{27}}$.

6. Rationalise the denominators of the following fractions:—

   (i) $\dfrac{1}{\sqrt{2}+1}$;   (ii) $\dfrac{1}{\sqrt{2}-1}$;   (iii) $\dfrac{1}{2-\sqrt{3}}$;   (iv) $\dfrac{1}{\sqrt{5}-\sqrt{2}}$;

   (v) $\dfrac{1}{3-2\sqrt{2}}$;   (vi) $\dfrac{2}{3\sqrt{2}+4}$;   (vii) $\dfrac{\sqrt{3}}{4+2\sqrt{3}}$;   (viii) $\dfrac{2\sqrt{2}}{5+3\sqrt{2}}$.

7. Express the following fractions in the form $a+b\sqrt{n}$, where a and b are rational:—

(i) $\dfrac{2-\sqrt{3}}{2+\sqrt{3}}$;   (ii) $\dfrac{\sqrt{5}-1}{\sqrt{5}+2}$;   (iii) $\dfrac{1}{(2\sqrt{3}-3)^2}$;   (iv) $\dfrac{\sqrt{5}}{(2\sqrt{5}-4)^2}$.

8. Find the exact value of $\left(\dfrac{1}{\sqrt{5}+\sqrt{2}}-\dfrac{1}{\sqrt{5}-\sqrt{2}}\right)^2$.

9. Simplify (i) $\dfrac{1}{(\sqrt{5}+\sqrt{2})^2}$; (ii) $\dfrac{8\sqrt{5}}{(\sqrt{5}+2)^2-(\sqrt{5}-2)^2}$.

10. By using the binomial theorem or otherwise, find the exact numerical value of $(2+\sqrt{5})^4+(2-\sqrt{5})^4$.

11. Express $\dfrac{1+\sqrt{7}}{3-\sqrt{7}}$ in the form $p+q\sqrt{7}$, where p and q are integers.

12. Solve the following equations giving the values of x in a form suitable for computation:—

   (i) $x(2\sqrt{3}-3)=4\sqrt{3}$;   (ii) $\sqrt{5}(x-2)=3(1-x)$;   (iii) $2x-1=\sqrt{3}(x+3)$.

13. Verify that both $x=1+\sqrt{3}$ and $x=1-\sqrt{3}$ are solutions of the equation $x^3-x^2-4x-2=0$.

14. If $z = 3 - \sqrt{5}$, express $\dfrac{1}{z} + \dfrac{1}{z^2}$ in the form $a + b\sqrt{5}$, where a and b are rational.

Finite Series

Consider the series, or sequence, of terms

$$u_1, u_2, u_3, u_4, \ldots u_r, \ldots u_n,$$

where u_1 is the first term, u_2 the second term, u_3 the third term and so on, u_r the rth term and u_n the nth term.

The series is completely defined if either (i) u_r is given as a function of r, or (ii) S_n, the sum of the first n terms, is given as a function of n.

Ex. 10. *The rth term of a series is $r(r+1)$. Find (i) the first term; (ii) the second term; (iii) the $(r+1)^{\text{th}}$ term.*

$$u_r = r(r+1).$$

(i) For the 1st term, put $r = 1$;	$u_1 = 1(1+1) = 2.$
(ii) For the 2nd term, put $r = 2$;	$u_2 = 2(2+1) = 6.$
(iii) For the $(r+1)^{\text{th}}$ term, put $r = r+1$;	$u_{r+1} = (r+1)(r+1+1),$
	$= (r+1)(r+2).$

Ex. 11. *The sum of the first n terms of a series is $n^2 + 2n$. Find (i) the first term; (ii) the second term; (iii) the rth term; (iv) the sum of the first $2n$ terms.*

$$S_n = u_1 + u_2 + u_3 + \ldots + u_r + \ldots u_n.$$

(i) Putting $n = 1$, we get s_1, the sum of one term, which is of course the first term u_1.

$$S_1 = u_1 = 1^2 + 2 \cdot 1 = 3.$$

(ii) Putting $n = 2$, we get the sum of the first two terms.

$$S_2 = u_1 + u_2 = 2^2 + 2 \cdot 2 = 8.$$

As $u_1 = 3$, $\qquad\qquad u_2 = 8 - 3 = 5.$

(iii) To find u_r, we get the sum of the first r terms and subtract from it the sum of the first $(r-1)$ terms.

Putting $n = r$, $\qquad S_r = r^2 + 2r.$

Putting $n = r - 1$, $S_{r-1} = (r-1)^2 + 2(r-1).$

$$\therefore\ u_r = S_r - S_{r-1},$$
$$= r^2 + 2r - [(r-1)^2 + 2(r-1)],$$
$$= 2r + 1.$$

(iv) s_{2n} is found by replacing n by $2n$ in the expression for s_n.

$$S_{2n} = (2n)^2 + 2(2n),$$
$$= 4n^2 + 4n.$$

Σ notation. The sum of the series

$$u_1 + u_2 + u_3 + \ldots u_2 + \ldots + u_n,$$

is written $\sum\limits_{r=1}^{r=n} u_r$, or more briefly, $\sum\limits_{1}^{n} u_r$.

Ex. 12. Express the following sums in the $\sum$ notation:—

(i) $1 . 3 + 3 . 5 + 5 . 7 + \ldots n \ terms;$

(ii) $\dfrac{1}{1 . 2} + \dfrac{1}{2 . 3} + \dfrac{1}{3 . 4} + \ldots 2n \ terms.$

(i) In finding u_r, note that $1, 3, 5, \ldots$ is an A.P. whose r^{th} term

is $1 + (r-1)2, \quad i.e. \ 2r - 1.$

So $u_r = (2r-1)(2r+1).$

Sum to n terms, $S_n = \sum\limits_{r=1}^{r=n} (2r-1)(2r+1).$

(ii) In this case, $u_r = \dfrac{1}{r(r+1)}.$

$\therefore$ sum to $2n$ terms, $S_{2n} = \sum\limits_{1}^{2n} \dfrac{1}{r(r+1)}.$

Ex. 13. Evaluate (i) $\sum\limits_{1}^{20} (3r+1);$ (ii) $\sum\limits_{1}^{n} 1:$ (iii) $\sum\limits_{1}^{n} (2+2^r).$

(i) In order to identity the form of the series, find the first three or four terms by giving r the values $1, 2, 3, 4$ successively.
We get $4 + 7 + 10 + \ldots 20 \ terms.$

This series is an A.P. with first term 4 and common difference 3.

$$\therefore \ \text{sum} = \tfrac{20}{2}[8 + (20-1)3] = 650.$$

(ii) In this case, all terms are equal to 1 and the series is

$$1 + 1 + 1 + \ldots n \ \text{terms}.$$

$$\therefore \ \text{sum} = n.$$

(iii) Here $u_1 = 2 + 2^1;$ $u_2 = 2 + 2^2;$ $u_3 = 2 + 2^3 \ldots$ The terms must be kept in this form and not combined.

$$\text{Sum} = (2 + 2^1) + (2 + 2^2) + (2 + 2^3) + \ldots n \ \text{terms},$$

$$= 2 + 2 + 2 \ldots n \ \text{terms} + 2^1 + 2^2 + 2^3 + \ldots n \ \text{terms}.$$

The first n terms have a sum of $2n$ and the second n terms, which form a G.P., have a sum of $2(2^n - 1)$.

$$\therefore \ \text{Sum} = 2n + 2(2^n - 1).$$

(A) EXAMPLES 23c

In each of ex. 1–9 where the rth term of a series is given, find (i) the first term; (ii) the second term; (iii) the $(r+1)^{\text{th}}$ term.

1. $u_r = 2r + 1$.   **2.** $u_r = 2^r$.   **3.** $u_r = 2$.

4. $u_r = r(r+1)$.   **5.** $u_r = r^2(r+1)$.   **6.** $u_r = 3^{r-1}$.

7. $u_r = \dfrac{1}{r(r+2)}$.   **8.** $u_r = \dfrac{1}{2r(r+3)}$.   **9.** $u_r = \dfrac{1}{r(r+1)(r+2)}$.

In each of ex. 10–18 where the sum of the first n terms of a series is given, find (i) the first term; (ii) the second term; (iii) the rth term; (iv) the sum of the first $2n$ terms.

10. $S_n = n$.   **11.** $S_n = 2n + 1$.   **12.** $S_n = n^2$.

13. $S_n = n^2 + n$.   **14.** $S_n = 2n^2 + 3n$.   **15.** $S_n = 2^n - 1$.

16. $S_n = \frac{1}{3}n(n+1)(n+2)$.   **17.** $S_n = \dfrac{n}{n+1}$.   **18.** $S_n = \dfrac{1}{3} - \dfrac{1}{2n+3}$.

Write down (i) the first three terms; (ii) the last term of each of the series in ex. 19–24.

19. $\displaystyle\sum_{r=1}^{r=6} 2r$.   **20.** $\displaystyle\sum_{r=1}^{r=10} r^2$.   **21.** $\displaystyle\sum_{r=1}^{r=20} 2$.

22. $\displaystyle\sum_{r=1}^{r=n} \dfrac{1}{r(r+1)}$.   **23.** $\displaystyle\sum_{r=1}^{r=n+1} (2r^2 + r)$.   **24.** $\displaystyle\sum_{r=1}^{r=2n} r^3$.

For each of the series in ex. 25–30, find the rth term and express the sum in the $\sum$ notation.

25. $1 + 2 + 3 + 4 + \ldots + 30$.

26. $1 + 3 + 5 + 7 + \ldots + 51$.

27. $2 + 4 + 6 + 8 + \ldots + 100$.

28. $1^2 + 3^2 + 5^2 + 7^2 + \ldots n$ terms.

29. $1.2 + 2.3 + 3.4 + 4.5 + \ldots n$ terms.

30. $\dfrac{1}{2.4} + \dfrac{1}{3.5} + \dfrac{1}{4.6} + \dfrac{1}{5.7} + \ldots n$ terms.

31. Evaluate (i) $\displaystyle\sum_{r=1}^{r=20} 4$; (ii) $\displaystyle\sum_{r=1}^{r=16} 2r$; (iii) $\displaystyle\sum_{r=1}^{r=12} (4r - 1)$.

32. Find the sums (i) $\displaystyle\sum_{1}^{10} 2^r$; (ii) $\displaystyle\sum_{1}^{15} (\tfrac{1}{2})^r$; (iii) $\displaystyle\sum_{1}^{20} (3^r + 1)$.

33. Express the sum of the series

$$(1+2)+(2+2^2)+(3+2^3)+\ldots n \text{ terms,}$$

in the $\sum$ notation and evaluate it.

34. Find (i) $\sum\limits_{1}^{n} 2^{r-1}$; (ii) $\sum\limits_{1}^{n} (1+(\tfrac{2}{3})^r)$.

35. The sum to n terms of a series is $\dfrac{2n}{2n+1}$. If u_n is the nth term of the series, find u_n and the sum $\sum\limits_{n+1}^{2n} u_r$.

Mathematical Induction

Mathematical induction is an intuitive process involving *the natural numbers*, the name given to the positive integers $1, 2, 3, 4, \ldots n$. It is a powerful method of proving *stated* theorems or results involving these numbers.

In outline, the method consists basically of two steps,

(i) the verification that the stated result is true for $n = 1$;

(ii) the assumption that the result is true for a particular value of n, say $n = r$, and on this basis, the proof that the result is also true for $n = r+1$.

If both these things can be done, it follows that the stated result is true for all integral values of n.

Many of the applications of mathematical induction deal with the sums of finite series but other results involving the natural numbers can be proved this way. The following worked examples illustrate applications of the method.

Ex. 14. *Prove by induction that the sum of the first n natural numbers is* $\frac{1}{2}n(n+1)$.

We need to prove that $\quad 1+2+3+\ldots+n = \frac{1}{2}n(n+1)$.

Taking $n = 1$, we get $\qquad\qquad\qquad 1 = \frac{1}{2}1(1+1)$

$$= 1.$$

Hence the result is true for $n = 1$.

(ii) Now assume the result is true for $n = r$,

i.e. assume $\qquad 1+2+3+\ldots+r = \frac{1}{2}r(r+1)$.

Then adding the next term of the series, $(r+1)$, to both sides of this equation, it follows that, on the basis of our assumption,

$$1+2+3+\ldots+r+(r+1) = \frac{1}{2}r(r+1)+(r+1),$$
$$= \frac{1}{2}(r^2+3r+2),$$
$$= \frac{1}{2}(r+1)(r+2).$$

As the expression on the R.H.S. is the expression $\frac{1}{2}n(n+1)$ with n equal to $(r+1)$, this is the required result establishing that if the result is true for one value of n, it is also true for the next highest value of n. The result is true for $n = 1$, so it is true for $n = 2, 3$, and so on.

Ex. 15. Prove by induction that $\sum_{1}^{n}(r+1)2^{r-1} = n \cdot 2^n$.

Giving r successive values, the stated result becomes

$$2 \cdot 1 + 3 \cdot 2 + 4 \cdot 2^2 + \ldots + (n+1)2^{n-1} = n \cdot 2^n.$$

(i) Taking $n = 1$, we get $\qquad\qquad 2 \cdot 1 = 1 \cdot 2^1$

$$2 = 2.$$

Hence the result is true for $n = 1$.

(ii) Now assume the result is true for $n = r$,

i.e. assume $\quad 2 \cdot 1 + 3 \cdot 2 + 4 \cdot 2^2 + \ldots + (r+1)2^{r-1} = r \cdot 2^r.$

The $(r+1)^{\text{th}}$ term of the series is $(r+2)2^r$ and adding this to both sides of the above equation, it follows on the basis of our assumption that

$$2 \cdot 1 + 3 \cdot 2 + 4 \cdot 2^2 + \ldots + (r+1)2^{r-1} + (r+2)2^r = r \cdot 2^r + (r+2)2^r,$$
$$= 2^r(2r+2),$$
$$= (r+1)2^{r+1}.$$

This is the required result, establishing that if the result is true for one value of n, it is also true for the next highest value of n.

The result is true for $n = 1$, so it is true for $n = 2, 3$, and so on.

Ex. 16. If n is a positive integer, prove that $f(n) = 7^n(3n+1) - 1$ *is always divisible by* 9.

(i) Taking $n = 1$, $f(1) = 7^1(3+1) - 1 = 27$.

This is divisible by 9, hence the result is true for $n = 1$.

(ii) Now assume the result is true for $n = r$,

i.e. assume

$$f(r) = 7^r(3r+1) - 1 = 9g(r) \text{ where } g(r) \text{ is some function of } r.$$

In this type of example it is usual to write down an expression for $f(r+1) - f(r)$ and establish that this difference is divisible by 9.

We have, $f(r+1) - f(r) = [7^{r+1}(3r+4) - 1] - [7^r(3r+1) - 1],$
$$= [7^r(21r+28) - 1] - [7^r(3r+1) - 1],$$
$$= 7^r(18r+27) = 9 \cdot 7^r(2r+3).$$
$$\therefore f(r+1) = f(r) + 9 \cdot 7^r(2r+3),$$
$$= 9g(r) + 9 \cdot 7^r(2r+3).$$

This is the required result showing that $f(r+1)$ is divisible by 9 and establishing that, if the result is true for one value of n, it is also true for the next highest value of n.

The result is true for $n = 1$, so it is true for $n = 2, 3$, and so on.

(B) EXAMPLES 23d

In ex. 1–10, use the method of mathematical induction to prove the stated results.

1. $1 + 3 + 5 + \ldots + (2n-1) = n^2$.

2. $2 + 4 + 6 + \ldots + 2n = n(n+1)$.

3. $1 + 4 + 7 + \ldots + (3n-2) = \frac{1}{2}n(3n-1)$.

4. $1 + 3 + 3^2 + \ldots + 3^{n-1} = \frac{1}{2}(3^n - 1)$.

5. $1^2 + 2^2 + 3^2 + \ldots + n^2 = \frac{1}{6}n(n+1)(2n+1)$.

6. $1 \cdot 2 + 2 \cdot 3 + 3 \cdot 4 + \ldots + n(n+1) = \frac{1}{3}n(n+1)(n+2)$.

7. $\dfrac{1}{1 \cdot 2} + \dfrac{1}{2 \cdot 3} + \dfrac{1}{3 \cdot 4} + \ldots + \dfrac{1}{n(n+1)} = \dfrac{n}{n+1}$.

8. $1^3 + 2^3 + 3^3 + \ldots + n^3 = \frac{1}{4}n^2(n+1)^2$.

9. $\dfrac{1}{2 \cdot 3} + \dfrac{1}{3 \cdot 4} + \dfrac{1}{4 \cdot 5} + \ldots + \dfrac{1}{(n+1)(n+2)} = \dfrac{n}{2(n+2)}$.

10. $a + ar + ar^2 + \ldots + ar^{n-1} = \dfrac{a(1-r^n)}{1-r}$.

11. Given $f(n) = 5^n - 1$, show that (i) $f(1)$ is divisible by 4; (ii) $f(n+1) - f(n)$ is divisible by 4. Deduce by induction that $5^n - 1$ is divisible by 4 for all positive integral values of n.

12. Prove that $\sum_1^n r(r+2) = \frac{1}{6}n(n+1)(2n+7)$.

13. Use mathematical induction to show that $\sum_1^n (2r-1)^2 = \frac{1}{3}n(4n^2-1)$.

14. Prove the result, $1 \cdot 1! + 2 \cdot 2! + 3 \cdot 3! + \ldots n$ terms $= (n+1)! - 1$.

15. If $f(n) = 3^{2n} + 7$, where n is a positive integer, show that $f(n+1) - f(n)$ is divisible by 8. Hence prove by induction that $3^{2n} + 7$ is divisible by 8.

16. Show that $\sum_1^n \dfrac{r}{(r+1)!} = 1 - \dfrac{1}{(n+1)!}$.

17. If $f(n) = 2^{3n} + 6$, show that $f(n+1) - f(n)$ is divisible by 7 and deduce that $f(n)$ is always divisible by 7, n being a positive integer.

18. Prove that $2.5 + 5.8 + 8.11 + \ldots n$ terms $= n(3n^2 + 6n + 1)$.

19. Prove by induction that $3^{2n} - 5$, where n is a positive integer, is always divisible by 4.

20. Show that $\displaystyle\sum_1^n \frac{1}{(3r-2)(3r+1)} = \frac{n}{3n+1}$.

Algebraic inequalities

Simple linear and quadratic inequalities such as
$$3x + 5 < x - 7,$$
and
$$x^2 - x > 6,$$
are solved by similar methods to those used for the corresponding equalities, or equations.

In dealing with inequalities however, multiplication or division by a negative number or by a term which can be negative should be avoided as such a process leads to a reversal of the inequality sign.

Ex. 17. *For what values of x are the following inequalities true:—*
$$(i)\ 3x + 5 < x - 7;\ (ii)\ x^2 - x > 6?$$

(i) $$3x + 5 < x - 7.$$

Transposing terms as in the case of a linear equation,
$$3x - x < -7 - 5,$$
$$2x < -12,$$
$$x < -6.$$

N.B. division by a negative number was avoided by ensuring that the x-term had a positive coefficient.

(ii) $$x^2 - x > 6.$$

This is a quadratic inequality and we proceed as for a quadratic equation.

Rearranging, $$x^2 - x - 6 > 0.$$

Factorising, $$(x - 3)(x + 2) > 0.$$

The function $(x - 3)(x + 2)$ is zero when $x = -2$ and 3. Take a value of x between these two values, say $x = 0$. For this value of x the function is negative and it will be negative for all values of x between -2 and 3. So the function is positive, as required, for values of x outside the range -2 to 3,
$$\text{i.e. for}\quad x < -2 \quad \text{and} \quad x > 3.$$

Graph of a linear inequality in two unknowns

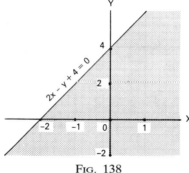

Fig. 138

Consider the linear inequality $2x - y + 4 > 0$. Draw the straight line with equation $2x - y + 4 = 0$, by noting that it cuts the axes at the points $(-2, 0)$ and $(0, 4)$.

Then $2x - y + 4$ is positive on one side of the line and negative on the other.

To find which, make y the subject of the inequality.

We have $y < 2x + 4$.

Now $y = 2x + 4$ for points on the line, $y > 2x + 4$ for points above the line and $y < 2x + 4$ for points below the line.

∴ The given inequality is represented by the shaded area.

Simultaneous linear inequalities in two unknowns

Consider the system of two linear inequalities

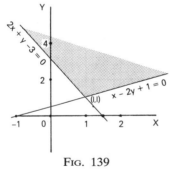

Fig. 139

$$2x + y - 3 > 0,$$
$$x - 2y + 1 < 0.$$

To find the region of the (x, y) plane where both inequalities are satisfied, draw the two straight lines

$$2x + y - 3 = 0,$$
and $\qquad x - 2y + 1 = 0,$

by finding the points of intersection with the axes.

These lines intersect at the point $(1, 1)$.

For the first inequality, $\qquad y > 3 - 2x,$

hence the inequality $2x + y - 3 > 0$ holds for all points above the line $y = 3 - 2x$, i.e. $2x + y - 3 = 0$.

For the second inequality, $\qquad y > \frac{1}{2}(x + 1),$

hence the inequality $x - 2y + 1 < 0$ holds for all points above the line $y = \frac{1}{2}(x + 1)$, i.e. $x - 2y + 1 = 0$.

Consequently both inequalities are simultaneously satisfied in the region shaded in Fig. 139.

Ex. 18. Find the region in the (x, y) plane over which the following inequalities are simultaneously satisfied, $x + y < 3$, $2y > x$ and $x > 1$.

Draw the straight lines

$$x + y - 3 = 0,$$
$$2y - x = 0,$$
and $$x - 1 = 0.$$

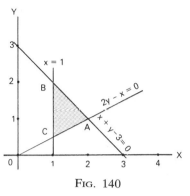

Fig. 140

The points of intersection of the lines are A(2, 1), B(1, 2) and C(1, $\frac{1}{2}$). We find that the inequality $x + y - 3 < 0$ is satisfied by points below the line $x + y - 3 = 0$, and the inequality $2y - x > 0$ is satisfied by points above the line $2y - x = 0$.

Clearly the inequality $x - 1 > 0$ is satisfied by points to the right of the line $x - 1 = 0$.

Consequently the three inequalities are simultaneously satisfied in the area shaded in Fig. 140.

(A) EXAMPLES 23e

In ex. 1–30, find the values of x which satisfy the given inequalities.

1. $2x > 6$.   **2.** $3x < -12$.   **3.** $8 > 2x$.

4. $x + 3 > 7$.   **5.** $2x - 1 < 5$.   **6.** $5x - 6 \geqslant 9$.

7. $6 > -x$.   **8.** $-3x < -18$.   **9.** $2 - x \geqslant 5$.

10. $3x - 1 > 4 - x$.   **11.** $3 - 2x < x + 6$.   **12.** $-2x + 3 > 0$.

13. $2(x - 8) \geqslant -3(x + 5)$.   **14.** $\frac{1}{3}(2x - 1) < 5$.   **15.** $\frac{2}{5}x \leqslant \frac{1}{3}(4 - x)$.

16. $x(x - 2) < 0$.   **17.** $x(3 - x) \geqslant 0$.   **18.** $2x(2x + 5) < 0$.

19. $(x + 1)(2x - 1) \geqslant 0$.   **20.** $(3 - x)(x + 4) \leqslant 0$.   **21.** $x^2 > 9$.

22. $9x^2 \leqslant 16$.   **23.** $x^2 - x - 2 < 0$.   **24.** $x^2 + 3x \geqslant -2$.

25. $2x^2 - 3x > 2$.   **26.** $x^2 < x + 6$.   **27.** $(x - 2)^2 \geqslant 9$.

28. $(x + 3)^2 \leqslant 4$.   **29.** $x^2 + 1 > 0$.   **30.** $(x - 1)^2 \geqslant 0$.

31. If $f(t) = 2t^2 - 3t + 1$ find the values of t for which (i) $f(t) \geqslant 0$; (ii) $f(t) \leqslant 3$.

32. Solve the inequality $(8 - 4m)^2 < 16(3m - 8)$.

33. Show graphically the regions of the (x, y) plane over which each of the following inequalities is satisfied:—

(i) $x + y - 2 > 0$;   (ii) $2x - y - 2 < 0$;   (iii) $x + 2y - 4 > 0$;

(iv) $y - 2x < 0$;   (v) $2y - x > 3$;   (vi) $3y + 5x < 0$.

34. Find graphically the regions of the (x, y) plane over which each of the following sets of simultaneous equations is satisfied:–

(i) $3x + y - 8 > 0$,   (ii) $x + y > 0$,   (iii) $y - x < 1$,
     $x - 2y + 2 < 0$; $y - 2x > 0$; $y + 2x > 4$;

(iv) $x > 0$,   (v) $y > 0$,   (vi) $3x + y - 8 > 0$,
         $y > 0$, $y - x < 0$, $x - 2y + 2 < 0$,
    $x + y < 2$; $x - 3 < 0$; $y - 3 < 0$.

35. Show graphically that there are no values of x and y for which all the inequalities $y - x > 0$, $x + y - 2 > 0$, $x + 2y - 3 < 0$ are simultaneously satisfied.

Asymptotes

Consider the curve with equation $y = \dfrac{2x}{x - 1}$. The value $x = 1$ cannot be substituted as it would make the denominator of the fraction zero.

We can however take values of x nearer and nearer to 1 and see how the graph behaves.

Construct the table of values:

x	1·1	1·01	1·001	1·0001
y	22	202	2002	20002

It is clear that, as x approaches 1 from the right, y becomes increasingly large in the positive direction.

Similarly, as x approaches 1 from the left, y becomes increasingly large in the negative direction. (Fig. 141).

The line $x = 1$ is called an asymptote of the curve; it is parallel to the y-axis.

Rearranging the equation of the curve, we get

$$x = \frac{y}{y - 2},$$

and using a similar argument, it follows that the line $y = 2$ is also an asymptote of the curve; it is parallel to the x-axis.

The curve can be drawn by giving x a series of simple values and its shape and position relative to its asymptotes is shown across.

Ex. 19. *Find the asymptotes parallel to the axes of the curve with equation* $y = \dfrac{2 - x}{2x + 3}$.

Equating to zero the denominator of the fraction, we get

$$2x + 3 = 0; \quad i.e. \quad x = -\tfrac{3}{2}.$$

This is the asymptote parallel to the y-axis.

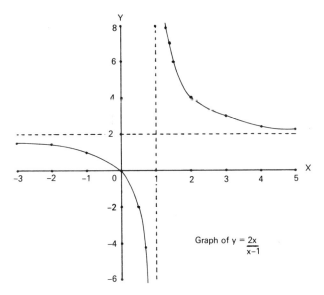

Graph of $y = \dfrac{2x}{x-1}$

FIG. 141

Rearranging the equation to give x in terms of y, we have

$$x = \frac{2 - 3y}{2y + 1}.$$

Equating to zero the denominator of this fraction, we get

$$2y + 1 = 0; \quad i.e. \quad y = -\tfrac{1}{2}.$$

This is the asymptote parallel to the x-axis.

Rational algebraic functions

Functions of the form $\dfrac{P(x)}{Q(x)}$, where $P(x)$ and $Q(x)$ are polynomial functions of x, are called *rational functions*.
Graphs of such functions in the simple cases where $Q(x)$ is a linear function and $P(x)$ a function of degree not higher than the second, will be considered. The essential steps to be followed when drawing such graphs are

  (i) find asymptotes parallel to the axes;
 (ii) determine if, and where, the graph cuts the axes;
(iii) determine if, and where, the graph has turning points;
(iv) obtain additional points by giving x some simple values.

Ex. 20. *Trace the curve with equation* $y = \dfrac{1+x}{1-x}$.

(i) Asymptotes parallel to the axes are $x = 1$ and $y = -1$.

(ii) When $x = 0$, $y = 1$, so the curve cuts the y-axis at the point $(0, 1)$.
When $y = 0$, $x = -1$, so the curve cuts the x-axis at the point $(-1, 0)$.

(iii) Differentiating,
$$\frac{dy}{dx} = \frac{1(1-x)-(-1)(1+x)}{(1-x)^2}$$
$$= \frac{2}{(1-x)^2}.$$

This function is never zero and so the curve has no turning points.

(iv) The following table of values is obtained.

x	-4	-3	-2	-1	0	$\frac{1}{2}$	$\frac{3}{4}$	$1\frac{1}{4}$	$1\frac{1}{2}$	2	3	4	5
y	$-\frac{3}{5}$	$-\frac{1}{2}$	$-\frac{1}{3}$	0	1	3	7	-9	-5	-3	-2	$-1\frac{2}{3}$	$-1\frac{1}{2}$

Using this information the graph can be plotted (Fig. 142)

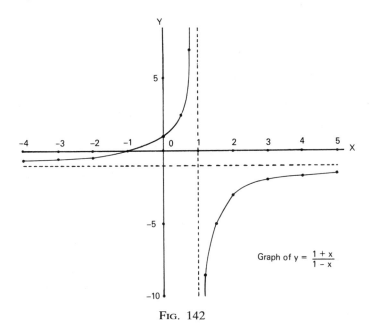

Graph of $y = \dfrac{1+x}{1-x}$

Fig. 142

Ex. 21. *Show that, if x is real, the function* $\dfrac{x^2+2x-5}{2x-3}$ *cannot take* *values between 2 and 3. Find (i) the maximum and minimum points, (ii) the asymptote parallel to an axis, of the graph of* $y = \dfrac{x^2 + 2x - 5}{2x - 3}$ *and sketch the graph.*

$$\text{Let} \quad y = \frac{x^2 + 2x - 5}{2x - 3},$$

then $\qquad\qquad y(2x - 3) = x^2 + 2x - 5,$

and rearranging $\qquad x^2 + x(2 - 2y) + (3y - 5) = 0 \ldots\ldots\ldots\ldots\ldots\ldots$ (i)

As x is real, the roots of this quadratic equation are real,

$$\text{So} \quad b^2 - 4ac \geqslant 0,$$

$$\textit{i.e.} \quad (2 - 2y)^2 - 4(1)(3y - 5) \geqslant 0,$$

$$4y^2 - 20y + 24 \geqslant 0,$$

$$y^2 - 5y + 6 \geqslant 0,$$

Factorising, $\qquad\qquad (y - 2)(y - 3) \geqslant 0, \ldots\ldots\ldots\ldots\ldots$ (ii)

as this inequality is not satisfied for $2 < y < 3$, the given function which is equal to y, cannot take values between 2 and 3.

(i) In cases such as this, the maximum and minimum values of the function can be obtained without use of the calculus.
The solutions of the inequality (ii) are $y \leqslant 2$ and $y \geqslant 3$.

As y is $\leqslant 2$, then $y = 2$ is a maximum value;
as y is $\geqslant 3$, then $y = 3$ is a minimum value.

The value of x corresponding to $y = 2$ is found by substituting for y in equation (i) and solving the equation for x. This gives $x = 1$.

$\therefore$ the point $(1, 2)$ is the maximum point on the graph of $y = \dfrac{x^2 + 2x - 5}{2x - 3}$.

Similarly, we find the minimum point is $(2, 3)$.
(ii) The asymptote parallel to the y-axis is clearly

$$2x - 3 = 0; \qquad \textit{i.e.} \quad x = \tfrac{3}{2}.$$

To find x in terms of y, we solve equation (i) and get

$$x = \frac{-(2 - 2y) \pm \sqrt{(2 - 2y)^2 - 4(3y - 5)}}{2}.$$

As there is no y term in the denominator, there is no asymptote parallel to the x-axis.
Using the above information together with a table of values the graph of the function can be drawn. (Fig. 143).

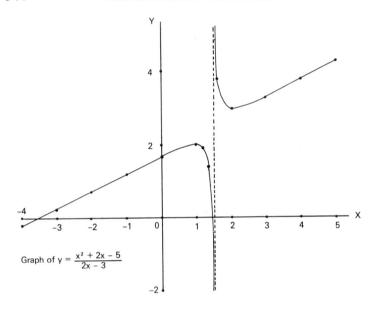

Graph of $y = \dfrac{x^2 + 2x - 5}{2x - 3}$

Fig. 143

(B) EXAMPLES 23f

In each of ex. 1–15, (i) find the asymptotes of the curve whose equation is given; (ii) determine if and where the curve meets the axes; (iii) show that the curve has no turning points; (iv) sketch the curve.

1. $y = \dfrac{x-1}{x-2}$.

2. $y = \dfrac{x}{x-1}$.

3. $y = \dfrac{2-x}{x+2}$.

4. $y = \dfrac{1}{x+1}$

5. $y = \dfrac{2}{2-x}$.

6. $y = \dfrac{1}{x}$.

7. $y = \dfrac{2x-1}{x}$.

8. $y = 1 - \dfrac{2}{x}$.

9. $y = 1 + \dfrac{1-x}{1+x}$.

10. $y = \dfrac{3x}{x+1}$.

11. $y = \dfrac{1-3x}{x}$.

12. $y = \dfrac{2-3x}{x+2}$.

13. $y = 2 - \dfrac{1}{2x-1}$.

14. $y = \dfrac{2x-5}{1-2x}$.

15. $y = \dfrac{4-x}{2x+5}$.

16. Use the method illustrated in worked ex. 21 to show that for each of the following functions there is a range of values which the function cannot take and find this range:—

(i) $\dfrac{x^2+1}{x}$; (ii) $\dfrac{x^2}{x-1}$; (iii) $\dfrac{2x^2}{1+x}$;

(iv) $\dfrac{x^2+4}{2x}$; (v) $\dfrac{3x^2-3}{6x-10}$; (vi) $\dfrac{(x-5)(x-1)}{2x-1}$.

17. For the curve $y=\dfrac{4x^2+8}{2x+1}$, find (i) the range of values which y cannot take; (ii) the asymptote parallel to an axis; (iii) the turning points. Sketch the curve.

18. Show that the function $\dfrac{x^2-1}{x}$ can take all values and sketch its graph.

In ex. 19–24, sketch the graphs of the functions after finding in each case (i) any range of values which y cannot take; (ii) maximum and minimum points; (iii) the asymptote parallel to an axis.

19. $y=\dfrac{x^2+x+4}{x+1}$. **20.** $y=\dfrac{(x-1)^2}{x-2}$. **21.** $y=\dfrac{x^2-3x+9}{x-3}$.

22. $y=x+\dfrac{1}{x-4}$. **23.** $y=x+\dfrac{4}{x}$. **24.** $y=4x+\dfrac{9}{x-1}$.

(B) MISCELLANEOUS EXAMPLES

1. Find the numerical values of p and q which make the function $x^4+px^3+qx-81$ divisible by x^2-2x-3. Find all the factors of the function when p and q have these values.

2. Use the binomial theorem to find the exact value of $(2+\sqrt{5})^5+(2-\sqrt{5})^5$.

3. The sum of the first n terms of a series is $1-(\tfrac{3}{4})^n$. Find an expression for the rth term. Prove that the series is geometric and state the values of the first term and the common ratio.

4. If the roots of the equation $ax^2+bx+c=0$ are real, show that the roots of the equation $a^2y^2-(b^2-2ac)y+c^2=0$ are also real.

5. Sketch the graph of $y=\dfrac{7-3x}{2-x}$ and use the graph to determine the range of values of x for which y is positive and less than 3.

6. When the expression $a(2x-3)^9+b(x-2)^4$ is divided by $(2x-3)$ the remainder is 1 and when it is divided by $(x-2)$ the remainder is 4. Find the values of a and b. With these values for a and b, what is the remainder when the expression is divided by $(x-1)$?

7. Using the result $(\sqrt{x}+\sqrt{y})^2 = x+y+2\sqrt{(xy)}$, determine the square roots of (i) $3+\sqrt{8}$; (ii) $7+\sqrt{48}$, in the forms of surds.

8. Prove by induction that $\sum_1^n (2r-1)^2 = \frac{1}{3}n(4n^2-1)$.

9. Solve the inequalities (i) $(2x-1)^2 > 9$; (ii) $(x+3)^2 < 4(1-x)^2$.

10. Express $\dfrac{1+\sqrt{3}}{(2+\sqrt{3})(3\sqrt{3}+5)}$ in the form $a+b\sqrt{3}$, where a and b are rational.

11. For what value of k is $(x+2)$ a factor of $(x+1)^7+(2x+k)^3$?

12. Show that the curve $y = \dfrac{x}{1+x}$ has no turning points. Find the asymptotes and sketch the curve. Hence find the range of values of x for which $-2 \leqslant \dfrac{x}{1+x} \leqslant 0$.

13. Find the limitations on the values of the constant a in order that the equation $x^2+2ax-a+2=0$ shall have real roots.

14. Evaluate, giving answers correct to two significant figures:—

$$\text{(i) } \sum_1^{20} (1\cdot1)^r; \qquad \text{(ii) } \sum_1^{20} \lg (1\cdot1)^r.$$

15. Sketch the graph of the function $\dfrac{x}{1-x}$ and use the graph to find the range of values of the function $\dfrac{\cos\theta}{1-\cos\theta}$.

16. Prove that the simultaneous equations $(x+a)y = 1$, $(y+b)x = 1$, have no real solutions if $-4 < ab < 0$.

17. Show by using the factor theorem that $(b-c)$, $(c-a)$ and $(a-b)$ are factors of the expression $(b-c)^3+(c-a)^3+(a-b)^3$.
By writing $(b-c)^3+(c-a)^3+(a-b)^3 \equiv A(b-c)(c-a)(a-b)$ and giving a, b, and c any three non-equal values, find A and completely factorise the expression.

18. Prove that the function $\dfrac{x^2-6x+5}{2x-1}$ can take all values except those between -4 and -1. Illustrate by sketching the graph of the function.

19. Prove that $\frac{1}{4}\left[9-\dfrac{1}{9+4\sqrt{5}}\right] = \sqrt{5}$.

20. If $2xy = 1-x$, show graphically that $-1 < x < 1$, provided that y is either positive or less than -1.

21. Use the factor theorem to show that $(b-c)$ is a factor of the expression $b^2c + c^2a + a^2b - bc^2 - ca^2 - ab^2$. Deduce two other factors of the same type and completely factorise the expression.

22. Show graphically that the inequalities $y - x - 2 < 0$, $y - x + 1 > 0$, $3x + y - 6 > 0$ and $3x + y - 8 < 0$ are satisfied by the co-ordinates of points lying within a parallelogram in the (x, y) plane and find the area of this parallelogram.

23. Rationalise the denominator of the fraction $\dfrac{1}{(1+\sqrt{2})+\sqrt{3}}$.

24. Sketch the graphs of (i) $y = \dfrac{x^2+2x+4}{x+2}$; (ii) $y = \dfrac{x^2+2x-4}{x+2}$.

25. Using the same axes, draw the graphs of $y = \frac{1}{5}x^2$ and $y = \dfrac{x-1}{x+2}$ from $x = -6$ to $x = 6$. Use the graphs to find a root of the equation $x^3 + 2x^2 - 5x + 5 = 0$ correct to one decimal place.   (J.M.B.)

REVISION PAPERS

PAPER A(1)

1. Find the equation of the line joining the points $(-1, 1)$, $(3, 2)$ and the length of the perpendicular from the origin to this line.

2. O is the origin and A the point $(3, 0)$. A moving point $P(x, y)$ is such that $AP^2 = 2OP^2$. Find the equation of the locus of P.

3. How many five-digit even numbers can be formed by using the digits $4, 5, 6, 7, 8$ without repetition?

4. Simplify (i) $9^a \times 3^b$; (ii) $\log_{10}25 \div \log_{10}5$.

5. Find the value of the constant a if the remainder when the function $x^3 + ax^2 - 4x + a$ is divided by $(x - 2)$ is 10.

PAPER A(2)

1. Find the equation of the circle which has the points $(-4, 2)$, $(2, 6)$ at the ends of a diameter.

2. A curve has parametric equations $x = 2t^2 - 1$, $y = 4t + 2$. Find the equations of the tangent and normal to the curve at the point with parameter -1.

3. (i) Find x if $\log_2 x = 1 + \log_2 3$; (ii) evaluate $e^{1 \cdot 2}$, taking $e = 2 \cdot 718$.

4. Evaluate (i) $\sum_1^4 r^2$; (ii) $\sum_1^{20} (2r - 1)$.

5. For what values of x are the following inequalities satisfied:–

(i) $8 < -2x$; (ii) $4x - x^2 \geqslant 0$?

PAPER A(3)

1. Find the ratio in which the point $(3, 3)$ divides the line joining the points $(-1, 1)$ and $(5, 4)$.

2. Find the equation of the tangent to the circle $x^2 + y^2 - 2x + 4y - 5 = 0$ at the point $(0, 1)$.

3. A committee of five has to be formed from a group of ten men and eight women. In how many ways can this be done if (i) there are no restrictions; (ii) the committee must consist of three men and two women?

4. (i) Evaluate $x^2 + 4x$ when $x = \sqrt{5} - 2$;
(ii) Find x in the form $a + b\sqrt{3}$, if $x + x\sqrt{3} = 8$.

5. The sum to n terms, S_n, of a series is given by $S_n = 4n^2 - 2n$. Find (i) the first and second terms of the series; (ii) to sum to $(n + 1)$ terms.

PAPER A(4)

1. A straight line has a gradient of 2. A second line, gradient m, makes an angle θ with the first line where $\tan \theta = \frac{3}{4}$. Find the possible values of m.

2. P is the point $(2t + t^{-1}, 2t^{-1})$. Find the (x, y) equation of the locus of P as t varies.

3. Evaluate, using tables if necessary, (i) $\log_4 8$; (ii) $\log_4 10$.

4. Use the factor theorem to find a factor of the polynomial $x^3 + x^2 - 4x - 4$ and complete the factorisation.

5. Show that $\sum_1^{10} (\tfrac{1}{2})^{r-1}$ represents the sum of a geometric series and find (i) the first term; (ii) the common ratio; (iii) the sum of the series.

PAPER A(5)

1. A triangle has vertices A(1, 3), B(4, 2) and the origin O. Find angle AOB and the equation of the line joining O to the point M on AB, where AM : MB = 2 : 3.

2. Find the equation of the circle centre $(-2, 1)$ which touches the line $2y - x + 1 = 0$.

3. How many different arrangements can be made of the letters of the word *root*? In how many of these arrangements will the two o's be separated?

4. Express with rational denominators (i) $\dfrac{3\sqrt{2}}{2\sqrt{3}}$; (ii) $\dfrac{\sqrt{2}}{2\sqrt{2} - \sqrt{5}}$.

5. Use the factor theorem to show that the expression $x^{2n} - 1$, where n is a positive integer, has a factor $x^2 - 1$.

PAPER A(6)

1. Find the equation of the circle, radius 4 units, which is concentric with the circle $x^2 + y^2 - 4x + 6y - 2 = 0$. Show that the second circle lies inside the first.

2. A, B are the points $(m^2, 2m)$, $(3m^2, 6m)$. Find the equation of the locus of the midpoint of AB as m varies.

3. Simplify (i) $(x^{\frac{1}{2}} - x^{-\frac{1}{2}})(x^{\frac{1}{2}} + x^{-\frac{1}{2}})$; (ii) $(x^{\frac{1}{2}} - 1 + x^{-\frac{1}{2}})(x^{\frac{1}{2}} + 1 + x^{-\frac{1}{2}})$.

4. If $\sum_1^n u_r = 2n^2 + 3n$, find (i) u_1; (ii) u_2; (iii) $\sum_1^{2n} u_r$.

5. For what values of x are the following inequalities satisfied:–

         (i) $(x + 1)^2 \geqslant 0$;   (ii) $2x^2 < x + 3$?

PAPER B(1)

1. A, B, C are the points $(3, 4)$, $(9, 1)$, $(7, 7)$ respectively; find the coordinates of the point P which divides AB in the ratio $1 : k$. If CP is perpendicular to AB, find k and hence determine the co-ordinates of the foot of the perpendicular from C to AB.

2. (a) Prove that as t varies, the locus of the point $P(4t + 7, 1 - 3t)$ is a straight line and show that this line is a tangent to the circle $x^2 + y^2 = 25$.

(b) The points O, P, Q are respectively $(0, 0)$, $(h, 0)$, $(3h, 2k)$; R is the midpoint of PQ. Given that h and k vary so that $OR^2 + PQ^2 = 10$, find the equation of the locus of R.   (J.M.B.)

3. (a) A cricket team of eleven men is to be chosen from fourteen men of whom five are bowlers and two others can keep wicket. How many different teams can be chosen if (i) the team is to contain exactly four bowlers and one wicket keeper; (ii) the team is to contain at least four bowlers and at least one wicket keeper?

(b) Prove by induction that $\displaystyle\sum_{r=1}^{r=n} \frac{1}{4r^2 - 1} = \frac{n}{2n + 1}$.

                                                       (J.M.B.)

4. Draw the graph of $y = 3^x$ for values of x from -2 to 2, plotting nine points. Use your graph to estimate (i) the roots of the equation $3^x = 3x + 2$; (ii) the range of values of k for which the equation $3^x = kx$ has no real roots.   (C.)

5. (i) If x, y, z are consecutive terms of a G.P., show that $\log x$, $\log y$, $\log z$ are consecutive terms of an A.P.

(ii) When the expression $ax^3 - 6ax + b$ is divided by $(x - 1)$ the remainder is 1; when the expression is divided by $(x - 2)$ the remainder is 2. Find the remainder when the expression is divided by $(x + 1)$.   (C.)

PAPER B(2)

1. Find the equation of the circle which touches the y-axis at the point A$(0, 3)$ and passes through the point B$(1, 2)$.

The line **OB**, where **O** is the origin, meets the circle again at **C**. Find the equation of the circle which passes through **B**, **C** and the point **D**$(3, 0)$.

(J.M.B.)

2. **A** is the point $(0, 4)$ and **B** the variable point $(2t, 0)$. The perpendicular bisector of **AB** meets **AB** at **M** and the y-axis at **R**. Find, in terms of t, the coordinates of the midpoint **P** of **MR** and find the locus of **P** as t varies.

(C.)

3. (a) Without using tables, evaluate (i) $\log_{\sqrt{2}} 8$; (ii) $\log_a 729 \div \log_a 81$.
(b) Given that $\log_a 2 = m$ and $2^{x-2} = \frac{1}{8}a$, find x in terms of m.
(c) If $p = \log_a \frac{5}{4}$ and $q = \log_a \frac{8}{15}$, show that $2p + q = \log_a \frac{5}{6}$. (J.M.B.)

4. (a) Prove by induction that

$$\frac{1}{1 \cdot 2} + \frac{1}{2 \cdot 3} + \frac{1}{3 \cdot 4} + \ldots + \frac{1}{n(n+1)} = \frac{n}{n+1}.$$

(b) Find the range of values of a for which the roots of the equation $x^2 + 6ax + a^2 + 4x + 2a + 2 = 0$ are not real.

(L.U.)

5. Show that the function $\dfrac{(2x+7)(2x-1)}{x-1}$ has a maximum value of 8 and a minimum value of 32. Draw the graph of the function for values of x from -3 to 4.

PAPER B(3)

1. The ends of one diagonal of a square have coordinates $(2a, 0)$, $(0, a)$. Find the equation of the other diagonal and the equations of the sides of the square.

(J.M.B.)

2. (i) Calculate the distance of the point $(2, 2)$ from the line $3x - 4y = 8$. Prove that this line is a tangent to the circle whose centre is the point $(2, 2)$ and which touches the axes.
(ii) The radius of the circle $x^2 + y^2 - 2x + 3y + k = 0$ is $\frac{5}{2}$, find the value of k. Find also the equation of the diameter of this circle which passes through the point $(5, 2\frac{1}{2})$.

(C.)

3. (a) Ten cards numbered one to ten are placed in a pile. A set of three cards is chosen from the pile, in how many different ways can this be done? How many of these sets of three contain (i) exactly one even-numbered card; (ii) at least one even-numbered card?
(b) If n is a positive integer, prove by induction that $7^n + 5$ is always a multiple of 6.

(J.M.B.)

4. (i) Without using tables, evaluate $\dfrac{1}{(3 - \sqrt{7})^2} + \dfrac{1}{(3 + \sqrt{7})^2}$.

(ii) Show that $x^3 + a^3$ has a factor $(x + a)$ and determine the other factor. Determine whether $x^4 + a^4$ has a factor $(x + a)$. (c.)

5. (i) Solve the equation $2 + \log_{10}x = \log_e x$, given $\log_{10}e = 0 \cdot 4343$.

(ii) The nth term of a G.P. is k times the first term, show that the common ratio r is given by $(n - 1) \log r = \log k$.

PAPER B(4)

1. Find the equation of the circle on the line joining the points $A(7, 10)$ and $B(13, 2)$ as diameter. Verify that the point $(14, 9)$ lies on the circle and find the equation of the tangent to the circle at this point.

Tangents are drawn to the circle from the point $P(0, 1)$. Prove that the gradient of one of the tangents is zero and find the gradient of the other.

(J.M.B.)

2. Find the equation of the locus of the point which moves equidistant from the points $(4, 7)$ and $(6, 3)$.

Find the equation of the locus of the point which moves at a distance of $2\frac{1}{2}$ units from the point $(3, 1\frac{1}{2})$.

Find the coordinates of the points of intersection of the two loci. (c.)

3. (i) Solve the equations (a) $\log_{10}x = 10^{\log_{10}3}$;

(b) $4^y \cdot 20^{2y-2} = 40^y \cdot 2^{2y-1}$.

(ii) Using the substitution $u = 2^x$, solve the equation $2^{2x} - 3 \cdot 2^{x+\frac{1}{2}} + 2^2 = 0$. (c.)

4. Find the sums of the series (i) $\sum\limits_{r=10}^{r=30} (3r - 1)$; (ii) $\sum\limits_{r=1}^{r=10} (1 + 2^r)$;

(iii) $\sum\limits_{1}^{\infty} \sin^n\theta$, where θ is not a multiple of $\frac{1}{2}\pi$.

5. (i) Find the exact value of $\dfrac{\sqrt{5}+1}{\sqrt{5}+2} + \dfrac{\sqrt{5}-1}{\sqrt{5}-2}$.

(ii) When the expression $5x^{23} - 7x^{17} + ax^4$ is divided by $(x + 1)$ the remainder is 5. Find the numerical value of a and the remainder when the expression is divided by $(x - 1)$. (c.)

PAPER B(5)

1. Calculate (i) the angles; (ii) the coordinates of the circumcircle of the triangle formed by the three lines $x + y = 1$, $3x - y = 7$ and $3y = x + 3$.

2. A curve is given parametrically by the equations

$$x = 4 - t^2, \qquad y = 4t.$$

Find the equations of the normals to the curve at the points where it meets the y-axis and show that these normals meet on the x-axis.

(c.)

3. (a) Seven hotels each have one vacancy and accommodation has to be arranged for three men. In how many ways can the men be given accommodation?

(b) Two straight lines intersect at P and six other points are marked on each line. Calculate the number of triangles that can be drawn with three of these points as vertices when (i) P is not used; (ii) P is used.

(J.M.B.)

4. The table shows experimental values of two quantities x and y which are known to be connected by a law of the form $y = kb^x$.

x	1	2	3	4
y	30	75	190	470.

Plot $\log_{10} y$ against x and use your graph to estimate the values of k and b. (c.)

5. (i) If m is a positive integer show that the sum of the series $m + (m+3) + (m+6) + \ldots + 4m$ is five times the sum of the series $1 + 2 + 3 + \ldots + m$.

(ii) Sum to n terms the series $\log_2 x + \log_4 x + \log_{16} x + \log_{256} x + \ldots$, by expressing all the logarithms to base 2.

PAPER B(6)

1. (i) A circle passes through the point $(-2, 3)$ and touches the x-axis. Find the equation of the locus of the centre of the circle and show that the locus passes through the point $(1, 3)$.

(ii) Find the equation of the tangent at the point $(1, 1)$ to the circle $x^2 + y^2 + 6x - 8y = 0$.

2. Find the equation of the line, gradient m, which passes through the point $(3, 1)$. This line meets the x-axis at P and the y-axis at Q. Express the coordinates of P and Q in terms of m and find the equation of the locus of the midpoint of PQ as m varies. (c.)

3. (a) Without using tables, find x and y given that

(i) $\lg x + 2 \lg 5 = 4$;

(ii) $\lg 24 = y \lg 2 + \lg 3$. $\qquad [\lg x \equiv \log_{10} x]$ (J.M.B.)

(b) Use tables to evaluate y correct to 3s.f. given that

$$2^y \cdot 4^{y-1} = 30.$$

4. By eliminating x and y from the equations

$$\frac{1}{x} + \frac{1}{y} = 1, \ x + y = a, \ \frac{y}{x} = m,$$

obtain an equation connecting m and a. Given that a is real, determine the ranges of values of a for which m is real. (J.M.B.)

5. Sketch the graph of $y = \dfrac{9-4x}{2-x}$, showing the asymptotes. Determine the range of values of x for which y is greater than 9. Shade the region in the (x, y) plane defined by the inequalities $0 \leqslant x \leqslant 2$, $0 \leqslant y \leqslant 9$ and $y \leqslant \dfrac{9-4x}{2-x}$.

ANSWERS

(A) Examples 1a. (p. 4)

1. (i) $2\sqrt{2}$; (ii) $\sqrt{10}$; (iii) $3\sqrt{2}$; (iv) $3\sqrt{5}$; (v) $\sqrt{10}$.

2. (i) $AB = \sqrt{13}$; $BC = \sqrt{2}$; $AC = \sqrt{17}$; $\triangle$ not right-angled. (ii) $AB = 2\sqrt{17}$; $BC = 2\sqrt{17}$; $AC = 2\sqrt{34}$; $\triangle$ right-angled at B. (iii) $AB = \sqrt{13}$; $BC = 2\sqrt{5}$; $AC = \sqrt{37}$; $\triangle$ not right-angled.

3. (i) $2\frac{1}{2}$ unit^2; (ii) 34 unit^2; (iii) 8 unit^2.

4. $AB = \sqrt{37}$; $BC = \sqrt{5}$; $CD = \sqrt{17}$; $DA = 5$. Area $= 15$ unit^2.

5. $22\frac{1}{2}$ unit^2.

7. $\sqrt{(\alpha - 2)^2 + (\beta - 3)^2}$. **8.** $6\alpha + 8\beta = 25$.

9. $\alpha = \pm\sqrt{\frac{5}{2}}$. **10.** $3a^2 + 3b^2 + 8a + 16b + 20 = 0$.

11. $A(0, 2\sqrt{5})$; $B(4\sqrt{5}, 0)$; $P(2\sqrt{5}, \sqrt{5})$; $OP = 5$.

13. $D(2, \frac{3}{2})$; $CD = \frac{7}{2}$; $AD = \frac{1}{2}\sqrt{5}$; $AC = \sqrt{17}$; $BC = \sqrt{10}$.

(A) Examples 1b. (p. 6)

1. (i) $(-2, 7)$; (ii) $(\frac{1}{3}, 0)$, $(-1, 0)$; (iii) $(0, -1)$.

2. Ordinate of P is 4; of Q, 1; $PQ = 3\sqrt{2}$. **3.** $3\sqrt{5}$.

4. $8\sqrt{2}$. **5.** $(\frac{-11}{5}, \frac{-2}{5})$; no. **6.** Point of concurrence $(1, 2)$.

7. $(1, 2)$; $(\frac{-1}{3}, \frac{-2}{3})$. **8.** $\sqrt{2}$. **9.** $(1, 2)$; $(0, 0)$; $\sqrt{5}$.

10. $(\frac{3}{5}, \frac{7}{5})$, $(\frac{14}{5}, \frac{-4}{5})$, $(\frac{-8}{5}, \frac{-19}{10})$; lengths of sides $\frac{11}{10}\sqrt{13}$, $\frac{11}{10}\sqrt{17}$, $\frac{11}{5}\sqrt{2}$.

11. (a) $PM = 1$, $QN = 1\cdot21$, $QR = 0\cdot21$, $PR = 0\cdot1$, $\dfrac{QR}{PR} = 2\cdot1$.

(b) $PM = 1$, $QN = 1\cdot0201$, $QR = 0\cdot0201$, $PR = \cdot01$, $\dfrac{QR}{PR} = 2\cdot01$.

12. (i) $\sqrt{7}$; (ii) $\alpha^2 + \beta^2 = 16$; (iii) $x^2 + y^2 = 16$.

13. $y = x = \dfrac{-1 \pm \sqrt{-23}}{2}$. Straight line does not meet the curve in real points.

14. Roots are equal $x = 2$, $y = 4$, so line touches curve.

(A) Examples 2a. (p. 12)

1. (i) 1; (ii) 4; (iii) -6; (iv) $\frac{5}{3}$; (v) -1. **2.** 13. **3.** 2 and -1.

4. Grad. $AB = -1$; grad. $BC = \frac{-5}{4}$; grad. $CA = -2$.

6. $AB = 5$, grad. $AB = \frac{-4}{3}$. **7.** 3; 3. Lines are parallel.

8. $8+2h$; 8.

9. $\dfrac{1}{h^2-1}$.

10. P, $y=a^2-a$; Q, $y=a^2+2ah+h^2-a-h$;
Gradient $=2a+h-1$. As $h \to 0$, gradient $\to 2a-1$.

(A) Examples 2b. (p. 14)

1. (i) $y=2x+1$;   (ii) $y+x=0$;   (iii) $3y=x+11$;
    (iv) $y+ax=a(a-1)$.

2. (i) $y=2x+1$;   (ii) $y=-x-2$;   (iii) $y=x+2$;
    (iv) $11y=6x+11$; (v) $y=-x+a$;   (vi) $y=-x+a+b$.

3. (a) $y=-2x+9$; (b) $3y-4x+3=0$.

4. Equation AC is $2y=3x+4$; equation BD is $4y=-3x+13$; inter-section $x=\frac{5}{9}$, $y=\frac{17}{6}$.

5. Vertices A $(0,0)$; B $(\frac{8}{9}, \frac{16}{9})$; C $(0,4)$; D $(-\frac{8}{9}, \frac{20}{9})$; O $(0,2)$; AC is $x=0$;
BD is $4y+x=8$; AO $=$ OC $=2$; BO $=$ OD $=\frac{2}{9}\sqrt{17}$.

6. (a) Mid-points: PQ, $(\frac{1}{2}, \frac{3}{2})$; QR, $(1, \frac{5}{2})$; RP, $(-\frac{1}{2}, 3)$.
    (b) Median through P, $4y=x+9$; through Q, $5y+4x=13$; through
    R, $y+5x=4$.
    Point of concurrence: $(\frac{1}{3}, \frac{7}{3})$.

7. $y=4x-2$.   **8.** $2y=x$.

9. Equation of PQ is $3y+4x=16$; co-ordinates of Q$(16, -16)$.

10. Co-ordinates of B: $(-2, \pm 3)$.

(A) Examples 2c. (p. 17)

1. (i) Yes; (ii) Yes; (iii) Yes; (iv) Yes; (v) No.

2. (i) No; (ii) Yes; (iii) Yes; (iv) Yes.   **3.** $\frac{4}{3}$; $3y=4x$.

4. $y=7x+17$.   **5.** $-\frac{1}{2}$; $2y+x=0$.   **6.** $y+x=2$.

7. $2y+3x=9$; $(\frac{35}{13}, \frac{6}{13})$.

8. (a) Gradient of AB is 2, of AC is 1; (b) $y=x$; (c) $y=2x-1$;
co-ordinates of D $(1, 1)$.

10. (a) $y=2x+18$; (b) $2y+x+4=0$; co-ordinates of B $(-8, 2)$.

11. $3y+2x=5$.

12. 2nd and 4th lines are parallel; $(-5, -9)$; $(-\frac{5}{12}, \frac{19}{4})$; $(\frac{5}{12}, \frac{9}{4})$; $(0, 1)$.

(A) Examples 2d. (p. 21)

1. (i) $y=1+\dfrac{x}{2}$; (ii) $y=30+\dfrac{3x}{5}$; (iii) $y=60-2x$.

2. (i) $y = -6{\cdot}27 + 8{\cdot}16x$; (ii) $y = 13{\cdot}2 + 16{\cdot}6x$; (iii) $y = 7 - 0{\cdot}12x$.

3. $W = 240 + 20P$.   **4.** $W = -51{\cdot}2 + 12P$.

5. $R = 3{\cdot}7 + 0{\cdot}058W$.   **6.** $P = 90 + 0{\cdot}31W$.

7. $C = 4{\cdot}2 + 0{\cdot}0046N$; 497 passengers.   **8.** $V = 10 + 2{\cdot}72t$.

9. 93, 80, 53.

10. (i) $y = 4 - 2x^2$; (ii) $y = {\cdot}25 + {\cdot}1x^2$; (iii) $y = {\cdot}65 - 2{\cdot}25x^2$.

11. $R = 50 + 1{\cdot}5V^2$.   **12.** $pv = 2000$.

13. $y = 2{\cdot}3 + \dfrac{8{\cdot}2}{x}$.   **14.** $a = 317$, $b = \frac{4}{11}$.

15. $y = 0{\cdot}13 + 0{\cdot}12\sqrt{x}$.

(B) Miscellaneous Examples. (p. 24)

1. $26y + 13x = 19$.

2. Line equation $3y + 4x = 17$, gradient $= \frac{-4}{3}$. Gradient **AC** $= \frac{-4}{3}$.
∴ Line parallel to **AC**; length line $= \frac{5}{2}$; length **AC** $= 5$.

3. $(\frac{7}{2}, \frac{7}{2})$.

4. Equations of perpendiculars:—$y + 2x = 10$, $y - x + 5 = 0$.
Point of intersection $(5, 0)$.

5. (a) $4y + x = 5$; (b) $(4\frac{1}{17}, \frac{4}{17})$; (c) $\dfrac{13}{\sqrt{17}}$. Area $= \frac{13}{2}$.

6. Two of sides are $2\sqrt{5}$ in length. Area $= 6$.

7. (i) $3y + 4x = 0$; (ii) $4y - 3x + 25 = 0$. Area **ABC** $= \frac{25}{44}$.

8. $\dfrac{4}{\sqrt{5}}$.   **9.** $(3, -1)$.

10. Medians are $5y - 9x = -6$; $13y - 15x = 4$; $4y - 3x = 5$; Point of concurrence $(\frac{7}{3}, 3)$.

11. $x + y = 10$, $4y + 3x = 35$; $(5, 5)$.

12. Centre $(-\frac{7}{2}, 3)$, radius $\frac{5}{2}\sqrt{5}$.

13. Centre of circumscribing circle $(1, 1)$.

14. (i) $y = x$; (ii) $4y + x = 14$; (iii) $3y + 2x = 14$;
Orthocentre $(\frac{14}{5}, \frac{14}{5})$.

15. $(6, -2)$.   **16.** $y = x + 1$.

17. (i) $x = (4 + 2\sqrt{3})$, $y = (2 + 2\sqrt{3})$; (ii) $x = (4 - 2\sqrt{3})$, $y = (2 - 2\sqrt{3})$. Perpendicular bisector is $y + x = 6$; meets x-axis at $x = 6$.

18. $(1, 1)$.

19. Curves touches axes at $(-5, 0)$ and $(0, -5)$.

20. $(6, 4)$, $(-6, -4)$.   **21.** $P = 30 + 0{\cdot}2W$.

22. $C = 0{\cdot}227 \tan \theta$.

(A) Examples 3. (p. 33)

1. 4.   **2.** (i) 9; (ii) 4; (iii) 7.

3. $QN = 3 + 4h + h^2$, $QR = 4h + h^2$, $\dfrac{QR}{PR} = 4 + h$. $\dfrac{QR}{PR} \to 4$ as $h \to 0$. Gradient of tangent at $P = 4$.

4. (i) 3; (ii) -1.

5. $PM = x^2 + 2$, $QN = x^2 + 2x\delta x + \delta x^2 + 2$, $PH = \delta x$, $QR = 2x\delta x + \delta x^2$.
$\dfrac{QR}{PR} = 2x + \delta x$. Gradient chord $= 2x + \delta x$.
Gradient curve $= 2x$; 6.

6. $\delta y = 6\delta x - 2x\delta x - \delta x^2$; $\dfrac{\delta y}{\delta x} = 6 - 2x - \delta x$.
Gradient curve $= 6 - 2x$.

7. $-4x$, -4, $+8$.   **8.** 11, $y - 11x + 12 = 0$.

9. (i) 0; (ii) -3; (iii) 2.   **10.** $y + 8x + 9 = 0$.

11.   (i) $\delta y = 6x\delta x + 3\delta x^2$, $\dfrac{\delta y}{\delta x} = 6x + 3\delta x$, $\dfrac{dy}{dx} = 6x$;

     (ii) $\delta y = 2x\delta x + \delta x^2 - \delta x$, $\dfrac{\delta y}{\delta x} = 2x + \delta x - 1$, $\dfrac{dy}{dx} = 2x - 1$;

     (iii) $\delta y = 4\delta x$, $\dfrac{\delta y}{\delta x} = 4$, $\dfrac{dy}{dx} = 4$;

     (iv) $\delta y = 4\delta x - 2x\delta x - \delta x^2$, $\dfrac{\delta y}{\delta x} = 4 - 2x - \delta x$, $\dfrac{dy}{dx} = 4 - 2x$;

     (v) $\delta y = 6x\delta x^2 + 6x^2\delta x + 2\delta x^3$, $\dfrac{\delta y}{\delta x} = 6x\delta x + 6x^2 + 2\delta x^2$, $\dfrac{dy}{dx} = 6x^2$;

     (vi) $\delta y = 6x\delta x + 3\delta x^2 - 2\delta x$, $\dfrac{\delta y}{\delta x} = 6x + 3\delta x - 2$, $\dfrac{dy}{dx} = 6x - 2$;

     (vii) $\delta y = 4x\delta x + 2\delta x^2 + \delta x$, $\dfrac{\delta y}{\delta x} = 4x + 2\delta x + 1$, $\dfrac{dy}{dx} = 4x + 1$.

12. $6t - 1$.   **13.** $6r + 2$.   **14.** 8, -8.

15. (i) -3; (ii) $4 - x$; (iii) $6x^2 - 1$; (iv) $2 + 2x$.

16. $\delta s = 4t\delta t + 2\delta t^2$; $\dfrac{\delta s}{\delta t} = 4t + 2\delta t$; $\dfrac{\delta s}{\delta t} = $ average velocity; $\dfrac{ds}{dt} = 4t = $ velocity at P.

(A) Examples 4a. (p. 40)

1. (i) $\dfrac{dy}{dx} = 9x^2$; (ii) $4x - 7$; (iii) $\dfrac{-2}{x^2}$; (iv) $\dfrac{-6}{x^3}$; (v) $-\dfrac{1}{(1+x)^2}$.

2. (i) $12x^3$; (ii) $\frac{5}{2}x^4$; (iii) $\dfrac{-7}{x^2}$; (iv) $30x^9$; (v) $\dfrac{-6}{x^7}$; (vi) $\dfrac{-3}{4x^4}$; (vii) $3-4x^3$;

(viii) $1+2x$; (ix) $4+8x$; (x) $\dfrac{-1}{x^2}$; (xi) $12x^2+\dfrac{2}{x^3}$; (xii) $77x^{10}+12x^3$;

(xiii) $-\dfrac{2}{x^3}-\dfrac{1}{x^2}$; (xiv) $2x-\dfrac{2}{x^3}$; (xv) $2x-1$; (xvi) $-\dfrac{2}{x^3}+\dfrac{2}{x^2}$.

3. (i) $16-2t$; (ii) $3t^2-4t$; (iii) $u+ft$.   **4.** $50\frac{2}{7}$.   **5.** -1.

6. (i) $\frac{1}{6}$; (ii) $2, -1$; (iii) $\pm\frac{1}{2}$; (iv) $\frac{3}{2}, 0$.

7. 0.   **8.** $-3, 3$.   **9.** $0, 6$.

10. (i) $(1, 0)$; (ii) $(\frac{1}{3}, -\frac{2}{9})$, $(-\frac{1}{3}, \frac{2}{9})$; (iii) $(1, -4)$, $(-\frac{1}{2}, 2\frac{3}{4})$.

11. $(4, -28\frac{2}{3})$, $(-3, 21\frac{1}{2})$.   **12.** $\frac{3}{4}$, $(-2, -\frac{5}{2})$.

13. $\dfrac{dy}{dx}$ is positive when x is greater than $\frac{3}{2}$; $\dfrac{dy}{dx}$ is negative when x is less than $\frac{3}{2}$.

14. Points of intersection; $(1, 3)$, $(-1, 3)$; gradients at $(1, 3)$ are 2 and -2; gradients at $(-1, 3)$ are -2 and 2.

(A) Examples 4b. (p. 44)

1. (i) $y-8x+8=0$; (ii) $y+x=4$; (iii) $y-22x+34=0$;
(iv) $4y+x=4$; (v) $y-12x+15=0$.

2. (i) $4y-x=9$; (ii) $5y+x+21=0$; (iii) $19y+x+210=0$; (iv) $y=x$;
(v) $12y-x=109$.

3. $y-2x=3$; $(-3, -3)$.   **4.** Normal is line $x=0$; $(0, 4)$.

5. $(2, -1)$.   **6.** $(3, 0)$.   **7.** $\frac{1}{3}$, 27

8. $PT=2\sqrt{5}$, $PN=\sqrt{5}$.   **9.** $y+x=1$, $2y-2x+1=0$; $(\frac{3}{4}, \frac{1}{4})$.

10. $5y+x=21$, $OR=\frac{21}{5}$, $OQ=21$, $QR=\frac{21}{5}\sqrt{26}$.

11. $(1, 2)$, $(-1, -2)$; equations of tangents $y=+2$, $y=-2$.

12. $(1, 3)$, $(-\frac{1}{3}, \frac{1}{3})$; tangents $y-6x+3=0$, $3y+6x+1=0$ meet at $(\frac{1}{3}, -1)$.

13. $2y+x=4$; $(0, 2)$ and $(4, 0)$.

15. $(\frac{1}{3}, \frac{1}{9})$, $(-\frac{1}{3}, -\frac{1}{9})$.

16. T is point $(0, -4)$, N is $(0, 6)$, $ST=SP=SN=5$.

Revision Papers I. (p. 46)

Paper A (1)

1. (i) $5\sqrt{2}$; (ii) -1.   **2.** $2y+3x=0$.

3. (i) $16x^3-4x$; (ii) $1+4x-6x^2$.   **4.** $(\frac{1}{6}, \frac{5}{4})$.

5. Tangent $y-4x-3=0$; normal $4y+x+5=0$.

Paper A (2)

1. $y = 6x$.   **2.** $4x = 5y + 22$.   **3.** (i) $4 + 8t$; (ii) $\dfrac{-2}{t^3} + \dfrac{3}{t^4}$.

4. 24; $y - 24x + 54 = 0$.   **5.** $1, 1$.

Paper A (3)

1. 1 sq. unit.   **2.** $\left(\frac{24}{13}, -\frac{1}{26}\right)$.   **3.** 6.

4. $(2, -24)$ and $(-1, 3)$.

5. $(0, 0)$, $(4, 0)$; tangents $y + 4x = 0$, $y - 4x + 16 = 0$; meet at $(2, -8)$.

Paper A (4)

1. Gradient $= \frac{1}{5}$; perpendicular bisector $y + 5x + 6 = 0$.

2. $y = x$.   **3.** (i) $3x^2 + 4x + 3$; (ii) $8x - \dfrac{2}{x^3}$.

4. $\left(\frac{5}{4}, \frac{15}{16}\right)$.   **5.** $\left(24, \frac{31}{2}\right)$.

Paper A (5)

1. $\left(\frac{12}{13}, -\frac{8}{13}\right)$.   **2.** $2x + 4y = 13$.   **3.** (a) $-\frac{5}{2}$; (b) $-\frac{2}{5}$.

4. (i) 8; (ii) -4; (iii) 8.   **5.** $y + 2x = 7$; $\dfrac{7\sqrt{5}}{2}$.

Paper A (6)

1. (i) $y + 2x = 0$; (ii) $2y - x = 5$.

2. AB, -1; BC, $+1$; CD, $-\frac{2}{3}$; DA, ∞. AB and BC are perpendicular.

3. $x = \pm\frac{3}{2}$.   **4.** -1; $y - 3x = 2$; $\left(\frac{1}{9}, \frac{67}{243}\right)$.   **5.** $\frac{2}{5}$, 10.

Paper B (1)

1. $\left(\frac{7}{3}, 3\right)$.   **2.** $OB - \dfrac{x}{x - 3}$.   **3.** (i) $8x + 1$; (ii) $\dfrac{-1}{(x - 2)^2}$.

5. $l = 3 \cdot 98 + \cdot 016W$; Unstretched length $= 3 \cdot 98$ m.

Paper B (2)

1. $AB = 9$, $\dfrac{2}{\sqrt{13}}$; $\frac{25}{12}$.   **2.** $(-1, 2)$.

3. Gradient PQ $= 7 =$ gradient of curve at R.

4. (a) (i) $1 - \dfrac{1}{x^2}$; (ii) $px^{p-1} + \dfrac{p}{x^{p+1}}$; (b) $(1, 3)$, $(4, 12)$.

5. Gradient of curve $= a$; MN $= 2$.

Paper B (3)

1. $(1, 8)$. **2.** $(6\frac{3}{5}, 5\frac{1}{5})$. **3.** $x = \frac{5}{2}$ or $\frac{3}{2}$.

4. Tangent at P is $x + n^2 y - 2n = 0$. **5.** $P = \cdot275W + 4\cdot04$.

Paper B (4)

1. $\frac{49}{40}$. **2.** $(\frac{2}{27}, -\frac{13}{27})$; radius $\frac{5}{27}\sqrt{629}$.

3. L is $\left(\dfrac{2-m}{m}, 2-m\right)$.

4. A $(2, 3)$; B $(-1, -3)$.

5. $9x^2 - 4$; tangent $y - 5x = 1$; tangents at $x = \pm\frac{1}{3}$ are parallel to $3x + y = 0$.

Paper B (5)

1. D $(1, 5)$; E $(-3, -3)$.

2. Parallel AB, $y + x = 6$; perpendicular AB, $y = x$; area $= \frac{15}{2}$.

3. Gradient of tangent $= -\frac{1}{3}$; normal is $y - 3x + 2 = 0$.

4. $(1, 4)$ and $(-\frac{1}{4}, \frac{1}{4})$; tangents $y = 8x - 4$; $4y + 8x + 1 = 0$.

5. $p = \cdot4 + \cdot2y$; $x = \cdot4y + \cdot2y^2$.

Paper B (6)

1. $(6, -2)$; area $= 15$ sq. units. **2.** $(\frac{21}{22}, -\frac{3}{22})$.

3. $x^2 + y^2 = k^2 - a^2$. **4.** (i) $\frac{3}{4}$; (ii) $\frac{9}{5}$; (iii) $\frac{2}{5}\sqrt{106}$.

5. $\dfrac{-2}{(2x+1)^2}$; gradient at any point $= 1 - \dfrac{4}{(2x+1)^2}$; when $x = 0$, gradient $= -3$; $(-1, -3)$.

(A) Examples 5a. (p. 57)

1. $\delta s = 8t\delta t + 4\delta t^2$; $\dfrac{\delta s}{\delta t} = 8t + 4\delta t = $ average velocity; $8t$; 16.

2. $\delta v = 4t\delta t + 2\delta t^2$; $\dfrac{\delta v}{\delta t} = 4t + 2\delta t$; 12.

3. $\delta V = 2\delta t - 2t\delta t - \delta t^2$; $\dfrac{\delta V}{\delta t} = 2 - 2t - \delta t$; $\dfrac{dV}{dt} = $ rate of change of volume; -4.

4. 6. **5.** $\frac{9}{4}$ m s^{-1}; -3 m s^{-2}; 1 sec.

6. $0\cdot2$ m s^{-1}; $-9\cdot6$ m s^{-1}; stone is coming down.

7. $8t - 4$ m s^{-2}; $t = \frac{1}{2}$ s; 0. **8.** 44.

9. (i) $\dfrac{dx}{dt} = 2$; (ii) $\dfrac{dV}{dt} = -3$; (iii) $\dfrac{dy}{dx} = -kx$; (iv) $\dfrac{dp}{dq} = \dfrac{k}{p^2}$.

10. 0.36. **11.** 10.24. **12.** π cm^2.

13. 3.2π cm^3; 1.6π cm^2. **14.** 2.4π cm^3.

15. $\dfrac{dp}{dv} = -\dfrac{500}{v^2}$; $\delta p = \dfrac{-500}{v^2}\,\delta v$; $\delta p = -\frac{1}{5}$ gm cm^{-2}.

(B) Miscellaneous Examples. (p. 58)

1. (i) $-3 + 8x$; (ii) $-4 + 8x$; (iii) $\dfrac{1}{(1-x)^2}$; (iv) $\dfrac{1-2x}{(x^2-x)^2}$.

2. (i) $20t^4 - 12t^3 + 4t$; (ii) $4t + 3t^2 - 10t^4$; (iii) $4t^3 - \dfrac{4}{t^5}$; (iv) $-\dfrac{1}{t^4} + \dfrac{1}{3t^2}$.

3. $6, -2, 3$ at $x = 0, 2, 3$; gradient zero where $x = \dfrac{5 \pm \sqrt{7}}{3}$.

4. Gradient of chord QR $= 6x$.

5. $(3, 15)$. **6.** 3 s, -69 m, 42 m s^{-2}.

7. $.4\%$; $.2\%$. **8.** $75° 58'$; $70° 1'$, to the horizontal.

9. $\dfrac{dy}{dx} = 1 + \dfrac{1}{x^2}$, always positive. **10.** x between $-\frac{1}{2}$ and $+1$.

11. $y + (2a - 2)x = a^2$; $2y(a-1) + 2a(a-2)(a-1) + a = x$.

12. $\left(\dfrac{2t^3}{3t^2-2}\right)$, 0; $(0, -2t^3)$; $\dfrac{2t^3}{3t^2-2}\sqrt{9t^4 - 12t^2 + 5}$.

13. $ty + x = t^3 + 2t$; $\left(\dfrac{-2(t^2+2)}{t}, \dfrac{(t^2+2)^2}{t^2}\right)$.

15. (a) 272 J s^{-1}; (b) 1760 J s^{-1}.

17. $t^2 y + x = 2ct$; area $= 2c^2 = $ constant.

18. $a = 0$ or 2; tangents $y - 3x = 0$, $y + 9x - 12 = 0$.

19. 0 or 4; $y = 0$ and $y - 8x + 16 = 0$.

20. Q$\left(-\dfrac{2}{t}, \dfrac{1}{t^2}\right)$; $\left(t - \dfrac{1}{t}, -1\right)$.

(A) Examples 6a. (p. 63)

1. (i) $24x^5$, $120x^4$, $480x^3$; (ii) $2 - \dfrac{3x^2}{2}$, $-3x$, -3;

(iii) $-\dfrac{2}{x^2}$, $+\dfrac{4}{x^3}$, $-\dfrac{12}{x^4}$; (iv) $2x + \dfrac{2}{x^3}$, $2 - \dfrac{6}{x^4}$, $\dfrac{24}{x^5}$;

(v) $-4x(1 - x^2)$, $-4 + 12x^2$, $24x$.

2. (i) $-18t$; (ii) $\dfrac{2}{t^3}$; (iii) f.

3. −36, 18. **4.** $\frac{3}{2}$, −2. **5.** $\frac{1}{3}$, $\frac{2}{3}$.

6. $x = 3$, $\dfrac{d^2y}{dx^2}$ is positive; $x = -\frac{1}{2}$, $\dfrac{d^2y}{dx^2}$ is negative. **7.** −72.

8. (a) $x > 1$; (b) $x < 1$; (1, −9). **9.** (i) $\frac{2}{3} < x < 0$; (ii) $x > \frac{1}{3}$.

11. $6x + 6$, 6. **12.** $2x - \dfrac{2}{x^3}$, $2 + \dfrac{6}{x^4}$.

(A) Examples 6b. (p. 70)

2. $\frac{3}{2}$; positive, negative; maximum.

3. Positive, positive; $x = 0$; no, no.

4. (i) $(\frac{3}{2}, \frac{9}{4})$, maximum; (ii) $(2, -3)$, minimum; (iii) $(2, 0)$, minimum; (iv) $(\frac{1}{4}, \frac{9}{8})$, maximum; (v) $(1, 2)$, minimum; $(-1, -2)$, maximum; (vi) $(2, -\frac{16}{3})$, minimum; $(-2, \frac{16}{3})$, maximum; (vii) $(1, 9)$, maximum; $(2, 8)$, minimum; (vii) $(2, 12)$, minimum.

5. (i) $(0, 0)$, inflexion; (ii) $(0, 0)$, minimum; (iii) $(0, 1)$, inflexion; (iv) $(0, 0)$, minimum; (v) $(0, 3)$, inflexion.

6. 12. **7.** $-\frac{1}{8}$. **8.** $p = 1$. **9.** 0.

10. (i) maximum 0, minimum $-\frac{4}{27}$; (ii) maximum 48, minimum −77; (iii) maximum $\frac{5}{16}$, minimum $-\frac{13}{27}$; (iv) minimum 2.

11. 150 m. **12.** Minimum when $t = \frac{1}{2}$; minimum value $= -\frac{3}{4}$.

13. $12\frac{1}{2}$. **14.** 18. **15.** Minimum 0, maximum 32.

(A) Examples 6c. (p. 74)

1. (i) Maximum $(-\frac{1}{2}, 83\frac{3}{4})$, minimum $(3, -2)$; (ii) minimum $(\frac{1}{4}, -\frac{1}{256})$, inflexion $(0, 0)$; (iii) inflexion $(1, 1)$; (iv) minimum $(0, 0)$, inflexion $(1, 1)$.

2. Maximum $-(3 + 2\sqrt{2})$, minimum $(2\sqrt{2} - 3)$.

3. Maximum 3456, minimum 0.

4. $A = 100x - \dfrac{x^2}{2}$, 5000 m². **5.** $x = \sqrt[3]{\dfrac{400}{\pi}} = 5\cdot03$ cm.

6. $r = 4$ cm, $h = 2$ cm.

7. $A = \frac{1}{2}l(3 - l)$ m² where $BC = l$ m; $1\frac{1}{8}$ m².

8. Length 9 cm; width $\frac{9}{2}$ cm; height 6 cm.

9. Volume $= 4x(100 - x)(50 - x)$ cm³; $x = 50\left(1 - \dfrac{1}{\sqrt{3}}\right)$ cm.

10. Cost $= £\left(\dfrac{2000}{s} + \dfrac{s^2}{2}\right)$; $s = 10\sqrt[3]{2} = 12\cdot6$ km h⁻¹.

11. $BE = AB = AE = CD = \dfrac{6(6+\sqrt{3})}{11}$ cm;

$BC = DE = \frac{9}{11}(5 - \sqrt{3})$ cm.

12. $v = 4t^3 - 6t^2 - 72t + 2$ m s^{-1}; -160 m s^{-1}.

13. Base $= \sqrt[3]{4000} = 15 \cdot 9$ cm; height $= \dfrac{20}{\sqrt[3]{16}} = 7 \cdot 94$ cm.

14. $a = 2\frac{4}{5}$; $(2\frac{4}{5}, 1\frac{7}{5})$.   **15.** 1 m^3.   **16.** $12\frac{1}{2}$ cm^2.

17. $\dfrac{1000}{3\sqrt{3\pi}} = 108 \cdot 6$ cm^3.   **18.** $x = \frac{1}{3}$.

19. $\dfrac{4a}{3}$.   **20.** $r = R\sqrt{\frac{2}{3}}$; $\dfrac{4\pi R^3}{3\sqrt{3}}$

(A) Examples 7a. (p. 82)

1. $\dfrac{dy}{du} = 4u^3$, $\dfrac{du}{dx} = 2$, $\dfrac{dy}{dx} = 8(2x - 1)^3$.

2. $\dfrac{dV}{du} = 30u^5$, $\dfrac{du}{dt} = -2t$, $\dfrac{dV}{dt} = -60t(1 - t^2)^5$.

3. $\dfrac{dy}{dt} = 4t - 1$, $\dfrac{dx}{dt} = 3$, $\dfrac{dt}{dx} = \frac{1}{3}$, $\dfrac{dy}{dx} = \dfrac{4t - 1}{3}$ or $\dfrac{4x - 7}{9}$.

4. $\dfrac{dy}{du} = -3u^{-4}$, $\dfrac{dx}{du} = 4u^3$, $\dfrac{du}{dx} = \dfrac{1}{4u^3}$, $\dfrac{dy}{dx} = -\frac{3}{4}x^{-\frac{7}{4}}$.

5. (i) $220(4x + 3)^{10}$; (ii) $-80x + 576x^3 - 1152x^5$;
(iii) $64x(2x^2 - 1)^3 - 4x(2x^2 - 1)^{-2}$.

6. (i) $9(3x + 2)^2$; (ii) $-16x(1 - x^2)^7$; (iii) $8(1 - 2x)^{-5}$.

7. (i) $\dfrac{dy}{dt} = 2$, $\dfrac{dx}{dt} = -2$, $\dfrac{dy}{dx} = -1$; (ii) $\dfrac{dy}{dt} = 6$, $\dfrac{dx}{dt} = -3$,

$\dfrac{dy}{dx} = -2$; (iii) $\dfrac{dy}{dt} = 1$, $\dfrac{dx}{dt} = 2$, $\dfrac{dy}{dx} = \frac{1}{2}$.

8. (i) 288; (ii) -128π; (iii) 80.

9. $\frac{3}{4}x^{-\frac{1}{4}}$, $\frac{7}{5}x^{\frac{2}{3}}$, $-\frac{2}{3}x^{-\frac{5}{3}}$, $-1 \cdot 5x^{-2 \cdot 5}$, $3 \cdot 6x^{2 \cdot 6}$, $\frac{4}{3}x^{-\frac{2}{3}}$, $-\frac{1}{4}x^{-\frac{3}{2}}$, $\frac{7}{3}x^{\frac{4}{3}}$, $-\frac{4}{5}x^{-\frac{7}{3}}$.

10. (i) $\dfrac{1}{2\sqrt{x}} - \dfrac{1}{2\sqrt{x^3}}$; (ii) $\dfrac{1}{2\sqrt{1 + x}}$; (iii) $\dfrac{-x}{(x^2 + 1)^{\frac{3}{2}}}$; (iv) $6(3x + 1)^{-\frac{1}{3}}$.

11. $15a^{14}$; $\dfrac{1}{2\sqrt{a}}$; $\frac{3}{4}a^{-\frac{1}{4}}$; $-2a^{-3}$.

(B) Examples 7b. (p. 86)

1. 24π, $\dfrac{dr}{dt} = 2$, 48π cm^2 s^{-1}.

2. 144π, 576π m^3 s^{-1}. **3.** $\dfrac{1}{2\sqrt{\pi}}$.

4. $V = 4\pi h$, $\dfrac{dh}{dt} = \dfrac{3}{2\pi}$ m min^{-1}.

5. $\dfrac{1}{200\pi}$ cm s^{-1}, $\frac{2}{5}$ cm^2 s^{-1}. **6.** 48 unit s^{-1}.

7. $\dfrac{dx}{dt} = 6t$, $\dfrac{dy}{dt} - 3$, $14° 2'$ to the x-axis. **8.** 4 m^3 min^{-1}.

9. $\dfrac{1}{8\pi}$ cm s^{-1}. **10.** $\dfrac{36}{49\pi}$ cm s^{-1}. **11.** $\dfrac{10}{\sqrt[3]{3600\pi}}$ cm s^{-1}.

12. $\sqrt{15}/4 = 0·968$ cm min^{-1}. **13.** 18 cm^3 s^{-1}.

14. 3 m s^{-1}. **15.** 50 gf cm^{-2} s^{-1}.

(A) Examples 8. (p. 91)

1. $y = \dfrac{3x^2}{2} + c$. **2.** $y = x + \dfrac{1}{x} + c$. **3.** $y = \frac{2}{3}x^3 + \frac{1}{3}$.

4. $y = x^4 - 2x^2 - 8$. **5.** $s = 3t - \dfrac{t^3}{3} + 5$; $8\frac{1}{3}$. **6.** $\frac{7}{8}$.

7. $y = x^2 - x - 2$. **8.** $y = \dfrac{2x^3}{3} + \dfrac{3x^2}{2} - x$.

9. $\dfrac{x^5}{5}$, $\dfrac{x^{15}}{15}$, $-\dfrac{1}{2x^2}$, $\dfrac{2}{5}x^{\frac{5}{2}}$, x^4, $-\dfrac{2}{x}$, $\dfrac{3}{5}x^{\frac{5}{3}}$, $\dfrac{5}{6}x^{\frac{6}{5}}$, $x^4 - \dfrac{x^3}{3}$, $2x + \dfrac{1}{2x^2}$, $\dfrac{3}{4}x^{\frac{4}{3}} - \dfrac{3}{2}x^{\frac{2}{3}}$, $-x^3 + 2x^2 - x$. (In each case, the arbitrary constant has been omitted.)

10. $\dfrac{x^8}{8} + c$, $\dfrac{x^3}{3} + \dfrac{x^2}{2} + c$, $\dfrac{t^3}{3} + 2t - \dfrac{1}{t} + c$, $x - x^2 - x^3 + c$, $\dfrac{x^2}{2} + \dfrac{1}{x} + c$.

11. (a) $s = \dfrac{3t^2}{2} - t$; (b) $\dfrac{t^3}{3} + \dfrac{t^4}{4}$; (c) $t - t^2 - t^3$.

12. 12. **13.** $\frac{4}{3}$. **14.** $\dfrac{dy}{dx} = 2x^2$, $y = \frac{2}{3}x^3 + 2$.

15. (a) $\dfrac{t^3}{3}$; (b) $\dfrac{t^2}{2} - \dfrac{t^4}{12}$; (c) $\dfrac{t^3}{3} - \dfrac{t^5}{20}$.

(A) Examples 9a. (p. 99)

1. (i) $\frac{1}{3}$; (ii) $\frac{93}{5}$; (iii) $\frac{14}{3}$; (iv) $\frac{1}{2}$; (v) $-\frac{2}{3}$; (vi) 10; (vii) $\frac{37}{6}$; (viii) $16\frac{1}{2}$; (ix) $-\frac{1}{4}$; (x) $21\frac{1}{3}$.

2. (a) $34\frac{1}{2}$; (b) $27\frac{3}{4}$. **3.** (a) 36; (b) $\frac{7}{3}$.

4. (i) 40; (ii) $\frac{2}{3}$; (iii) $23\frac{1}{3}$; (iv) $2\frac{2}{3}$; (v) $4\sqrt{3} - \frac{4}{3}$.

5. P is $(2, 0)$; area is $\frac{8}{3}$.   **6.** $4\frac{1}{2}$.

7. Between $x = \frac{1}{3}$ and $x = 2$; $2\frac{17}{54}$.

8. (i) $\frac{1}{4}$; $-\frac{1}{4}$. (ii) $-\frac{1}{4}$; $\frac{1}{4}$.

9. $\frac{79}{6}$; $-4\frac{1}{6}$; from 0 to 3 area is $\frac{9}{2}$, from 3 to 5 area is $-\frac{26}{3}$.

10. 0.   **11.** 48.   **12.** $\frac{1}{6}$.

(A) Examples 9b. (p. 103)

1. $\frac{15}{2}\pi$.   **2.** $\frac{32}{3}\pi$.   **3.** 6π.   **4.** $\frac{14}{3}\pi$.   **5.** $\frac{10}{3}\pi$.   **6.** $\frac{17}{6}\pi$.

7. $\frac{781}{5}\pi$.   **8.** $\frac{48}{25}\pi$.   **9.** $\frac{128}{3}\pi$.   **10.** $\frac{50}{3}\pi$.   **11.** 12π.   **12.** 108π.

13. Depth $1 \cdot 6$ cm, volume $12 \cdot 8\pi$ cm^3.   **14.** Volume $= 8 \cdot 1\pi$.

15. OP is $hy = rx$; volume is $\frac{1}{3}\pi r^2 h$.   **16.** $x^2 + y^2 = r^2$; $\frac{2}{3}\pi r^3$.

(B) Examples 9c. (p. 105)

1. $7a^3 + \frac{3}{2}a^2$.   **2.** $\dfrac{c^2(b-a)}{ab}$.   **3.** $5:1$.   **4.** $16\frac{5}{6}$.

5. $(0, 0)$, $(8, 32)$; $42\frac{2}{3}$.   **6.** $4\frac{1}{2}$.   **7.** $2\frac{2}{3}$; $4\frac{4}{15}\pi$.

8. (a) $4\frac{1}{2}$; (b) $\frac{81}{10}\pi$.   **9.** $5:32$.   **10.** $\dfrac{\pi}{1280}$.

11. (a) 36; (b) $5 \cdot 669$ or $\dfrac{9}{\sqrt[3]{4}}$.   **12.** $\frac{2}{3}$.   **13.** $\frac{256}{15}$; $21\frac{1}{3}\pi$.

14. $\frac{2}{3}$.   **15.** $0 \cdot 693$.   **16.** (i) $1 \cdot 19$; (ii) $0 \cdot 91$; (iii) $6 \cdot 08$; (iv) $57 \cdot 3$.

Revision Papers II. (p. 107)

Paper A (1)

1. 1 m s^{-2}; 9 m s^{-1}.   **2.** (a) $8(2x + 1)^3$; (b) -6.

3. (a) $2\frac{1}{4}$; (b) $4\frac{1}{2}$.   **4.** (i) $\dfrac{x^4}{4} - x^2 + x + c$; (ii) $\dfrac{4x^3}{3} - 2x^2 + x + c$; $\frac{5}{6}$, 4.

5. $y = x^3 - x$; $\frac{8}{105}\pi = 0 \cdot 2394$.

Paper A (2)

1. (i) $8x - 4$; (ii) $\dfrac{1}{2\sqrt{(x+2)}}$; (iii) $\dfrac{(3x+2)^3}{9} + c$; (iv) $\frac{2}{5}x^{\frac{5}{2}} + \frac{4}{3}x^{\frac{3}{2}} + c$.

2. $\frac{8}{3}$.   **3.** (a) 14 m s^{-1}; (b) $11 \cdot 014$ m.   **4.** 16 and -16.

5. $\frac{32}{3}$; $\frac{512}{15}\pi$.

Paper A (3)

1. $\dfrac{5\sqrt{5}}{2}$. **2.** 11 m. **3.** −1. **4.** 50π cm^2 min^{-1}; $\cdot 01\pi$.

5. (i) (a) $\dfrac{x^3}{3}+\dfrac{1}{x}+c$; (b) $10\frac{2}{3}$. (ii) $\frac{256}{3}\pi$.

Paper A (4)

1. $6t$, $3t^2-1$, $\dfrac{3t^2-1}{6t}$, $t=\pm\dfrac{1}{\sqrt{3}}$. **2.** −2916.

3. (i) $-\dfrac{4}{(2x-1)^3}$; (ii) $y=x^2+\dfrac{1}{x}-1$. **4.** $0\cdot72\pi$ cm^3.

5. 72; 648π.

Paper A (5)

1. 1. **2.** 11 m s^{-1}; 2 m s^{-2}. **3.** (i) $\dfrac{3}{2(2x+1)^{\frac{1}{2}}}$; (ii) $-\dfrac{9}{4(3x+1)^{\frac{3}{2}}}$.

4. (a) (i) $\dfrac{x^3}{3}-\frac{3}{2}x^2+5x+c$; (ii) $-\dfrac{2}{x}+\dfrac{1}{2x^2}+c$; (b) 1. **5.** $\frac{35}{36}$.

Paper A (6)

1. 8; tangent $y-8x+7=0$; normal $8y+x-9=0$.

2. 108. **3.** $\frac{2}{9}$. **4.** $\frac{1}{6}$. **5.** $\dfrac{\pi}{12}$.

Paper B (1)

1. Greatest velocity 16 m s^{-1}. **2.** Min. 1; max. 2; OT $=\frac{43}{36}$.

3. $\frac{5}{18}$ cm s^{-1}. **4.** 50 cm^2. **5.** (a) $y=x^4-2x^2-7$; (b) $9\frac{1}{3}$.

Paper B (2)

1. (a) (i) $12x^3+\dfrac{6}{x^4}$; (ii) $-\dfrac{3}{(x+1)^4}$. (b) (i) $2\cdot8\pi$ cm^2 s^{-1};

(ii) $3\cdot2\pi$ cm^2 s^{-1}.

3. (b) $x=8$, $y=4$. **4.** (a) $y=12\frac{1}{3}$; (b) $4\frac{1}{2}$ unit^2.

Paper B (3)

1. $y-7x+13=0$; max. $6\frac{5}{27}$; min. 5. **2.** (a) $3x^2$; (b) $0\cdot38$.

3. Length 9 cm, breadth $4\frac{1}{2}$ cm, height 6 cm. **4.** $112\cdot63$ m.

5. (a) $\frac{1}{30}$; (b) $46\cdot95$.

Paper B (4)

1. $\dfrac{1}{72\sqrt{27}}$.

2. (b) $\dfrac{2}{3\pi} = \cdot 212 \text{ m s}^{-1}$.

3. (i) 18 m; (ii) when $t = \dfrac{1}{\sqrt{3}}$.

4. $y - \dfrac{x^4}{4} - \dfrac{2x^3}{3} + \dfrac{x^2}{2} + \dfrac{11}{12}$.

5. 0·646.

Paper B (5)

1. (a) $\dfrac{3}{4(1-x)^{\frac{5}{2}}}$; (b) 315·9.

2. $(-3, 4)$; $(1, 4)$; 4.

3. (a) Min. 0, max. 5.

4. (a) 36; (b) $x = \frac{9}{2}\sqrt[3]{2}$.

5. (a) $11\frac{5}{6}$; (b) $\frac{24}{35}\pi$.

Paper B (6)

1. $(-\frac{3}{4}, \frac{32}{25})$.

2. $\dfrac{4a}{3}$.

3. (a) $\dfrac{2t^3}{3} - \dfrac{7t^2}{2} + 6t$; (b) $4t - 7$. At rest when $s = \frac{27}{8}$ and $\frac{10}{3}$.

4. (a) 0; (b) $\frac{1}{8}$.

5. (i) $8\pi a^3$; (ii) $\frac{32}{5}\pi a^3$.

(A) Examples 10a. (p. 118)

1. (i) 2, $\frac{3}{2}$; (ii) $\frac{1}{3}$, $-\frac{7}{3}$; (iii) 2, 8; (iv) 1, $-\frac{1}{2}$; (v) 2, 1; (vi) $-2p$, $-q$; (vii) $b - a$, b.

2. (i) $x^2 - 6x + 8 = 0$; (ii) $x^2 + 2x + 4 = 0$; (iii) $5x^2 - 3x + 2 = 0$; (iv) $x^2 - 2ax + a^2 = 0$; (v) $x^2 + 3kx + 5k^2 = 0$; (vi) $x^2 - x(2a - b) + a + b = 0$.

3. (i) -11; (ii) 3; (iii) $12\frac{2}{3}$; (iv) $p^2 - 2q$.

4. (i) -17; (ii) 5; (iii) 44; (iv) $-p(p^2 - 3q)$.

5. (i) $\frac{5}{4}$; (ii) $\pm\dfrac{3\sqrt{5}}{4}$; (iii) $\pm\sqrt{5}$; (iv) $18\frac{1}{16}$.

6. (i) Imaginary; (ii) real and different; (iii) imaginary; (iv) real and different; (v) imaginary.

7. $q = \frac{25}{12}$.

8. $\pm 3\sqrt{3}$.

9. $q = 0$.

10. (i) $x^2 - x + 12 = 0$; (ii) $4x^2 + 23x + 36 = 0$; (iii) $6x^2 - x + 2 = 0$; (iv) $8x^2 + 35x + 216 = 0$; (v) $12x^2 - 7x + 49 = 0$; (vi) $2x^2 - 3x + 7 = 0$.

11. Imaginary.

12. $\frac{1}{2}$.

13. (i) $2 < x < 3$; (ii) $\frac{2}{5} < x < \frac{4}{3}$; (iii) $-4 < x < 7$; (iv) $\frac{1}{3} < x < 3$; (v) $-\frac{5}{2} > x > 3$; (vi) $\frac{2}{3} < x < \frac{3}{2}$.

14. (i) Positive; (ii) positive; (iii) negative; (iv) negative; (v) negative.

15. (i) Min. $-\frac{1}{3}$; (ii) max. $2\frac{1}{16}$; (iii) min. 2; (iv) max. $6\frac{1}{4}$.

(B) Examples 10b. (p. 119)

1. 21; 433.   **2.** $a < 3$.   **3.** $2 + \dfrac{r^3}{p^3} + \dfrac{p^3}{r^3}$.

4. $49x^2 - 1150x + 2500 = 0$.   **7.** $q^2x^2 - (p^2 - 2q)x + 1 = 0$.

9. $p = -3$, $q = 19$.   **10.** $p = 2$.   **11.** -3; 19.

12. $-10\frac{1}{8}$.   **14.** $m = -2$, $\frac{14}{9}$.

(A) Examples 11a. (p. 123)

1. $x^2 = 4y$.   **2.** $xy = 9$.   **3.** $x = 1$, $y = 2$; $x = -1$, $y = -2$.

4. $x^2 = y^3$.   **5.** $x = -\frac{3}{2}$, $y = -8$; $x = 4$, $y = 3$.

6. $y^2 = 16(x - 1)$.   **7.** $xy = 16$.   **8.** (i) $x = 5$, $y = 0$; $x = -3$, $y = 4$;
(ii) $x = \frac{1}{6}$, $y = -\frac{1}{3}$; $x = \frac{1}{2}$, $y = 1$.   **9.** $x = 1$, $y = -2$.

10. (i) $x^2 = 16y$; (ii) $y^2 - 4x - 2y + 13 = 0$; (iii) $xy - 2x - y - 14 = 0$;
(iv) $2x - 3y + 11 = 0$.   **11.** (i) $x = 1$, $y = 0$; $x = 2\frac{3}{8}$, $y = 4\frac{1}{8}$;
(ii) $x = 7$, $y = 2$; $x = y = -\frac{1}{2}$.   **12.** $x(a + b) + 1 = 0$.

13. $x = -\frac{1}{3}$, $y = 2$; $x = -\frac{3}{2}$, $y = -\frac{8}{3}$.   **14.** $x^2 - y^2 = 4$.

16. $x = 1$, $y = 1$; $x = -\frac{53}{88}$, $y = -\frac{25}{22}$.

17. $x^2 - 2xy + y^2 - 6x - 3y = 0$.   **18.** $x = -3$, $y = -2$; $x = -2$, $y = -3$.

(A) Examples 11b. (p. 126)

1. (i) A = 2, B = -4; (ii) A = 2, B = 1; (iii) A = 3, B = -2.

2. $3(1 - x)^2 - 2(1 + x)^2$.   **3.** $p = 2$, $q = 1$.

4. (i) $l = 4$, $m = -4$, $n = 1$; (ii) $l = 7$, $m = -5$, $n = 2$; (iii) $l = 0$, $m = \frac{1}{2}$,
$n = -\frac{1}{2}$; (iv) $l = m = \frac{-3}{2}$, $n = \frac{3}{2}$.

5. $\frac{1}{2}(x - 2)(x + 3) + \frac{13}{5}(x + 3)(x - 1) - \frac{11}{10}(x - 1)(x - 2)$.

6. (i) A = $\frac{1}{4}$, B = $\frac{3}{4}$; (ii) A = 2, B = -1; (iii) A = 2, B = -3;
(iv) A = 2, B = -3.   **7.** $4(x - \frac{1}{2})^2 + 8$.   **8.** $a = 2$, $b = -1$.

9. $a = \frac{10}{3}$, $b = \frac{8}{3}$, $c = -6$.   **10.** $a = 1$, $b = 2$.

11. $3x^2 + (x + 2) - 2x(x + 2)$.   **12.** $(x^2 + x + 1)(x^2 - x + 1)$.

13. $p = \frac{3}{2}$, $q = r = \frac{3}{4}$.   **14.** $k = -1$.   **15.** $p = 2$, $q = 3$, $r = 3$.

16. (i) A = -2, B = -1, C = 2; (ii) A = 2, B = -2, C = 1.

(B) Examples 11c. (p. 129)

1. $x = 1$, $y = 0$; $x = -1$, $y = 3$.　　　　**3.** $l = -4$, $m = 4$.

4. $x = \frac{3}{4}a$, $y = \frac{3}{2}a$; $x = \frac{1}{3}a$, $y = -a$.

5. $(1, 2)$, $(-\frac{7}{5}, -\frac{14}{5})$.　　**6.** $4x^2 - 4xy + y^2 - 3x - 3y + 9 = 0$.

7. $a = 1$, $b = -1$, $c = 2$, $d = -3$.　　**8.** $x = 4$, $y = 2$; $x = 2$, $y = 4$.

9. $a = 3$, $b = -2$.

10. $x = 2$, $y = 6$; $x = -2$, $y = -6$ and x, y interchanged.

11. $x = 1 \cdot 36$, $-0 \cdot 39$, -1.　　**12.** (a) $\sqrt[3]{2} = 1 \cdot 26$; (b) $x < 0$ and $x > 1 \cdot 77$.

13. $x = 1 \cdot 28$, $3 \cdot 91$, $-0 \cdot 20$.　　**14.** $x = -0 \cdot 07$, $2 \cdot 72$, $5 \cdot 35$.　　**15.** $x = 1 \cdot 65$.

16. $x = 1 \cdot 23$, $-1 \cdot 15$.　　　　**17.** $x = 2 \cdot 45$, $-2 \cdot 86$.　　　　**18.** $x = 2 \cdot 51$.

(A) Examples 12a. (p. 135)

1. Arithmetical progressions, (i), (ii), (iii), (vi), (vii).

2. (a) (i) 4; (ii) -2; (iii) 1; (vi) -1; (vii) 3.
(b) (i) 31; (ii) -10; (iii) $\frac{15}{2}$; (vi) 13; (vii) 35.
(c) (i) $4n - 1$; (ii) $6 - 2n$; (iii) $n - \frac{1}{2}$; (vi) $21 - n$; (vii) $3n + 11$.

3. (i) 16; (ii) 11, 13; (iii) $19\frac{3}{4}$, $17\frac{1}{2}$, $15\frac{1}{4}$; (iv) $9\frac{4}{5}$, $8\frac{3}{5}$, $7\frac{2}{5}$, $6\frac{1}{5}$.

4. (i) 11; (ii) 13; (iii) 16; (iv) 24.

5. (i) 144; (ii) 288; (iii) 36; (iv) -120; (v) $12a + 66x$.

6. (i) 672; (ii) 80; (iii) 520; (iv) 90.

7. 1875.　　　　**3.** 1950.　　　　**9.** (i) 4949; (ii) 1683; (iii) 3266.

10. (i) 23; (ii) 3; (iii) $4x$.

11. (i) 6, 8, 10; (ii) 16, 21, 26; (iii) $5x$, $8x$, $11x$.

12. 264.　　　**13.** 15.　　　**14.** 2; 500.　　　**15.** 116.

16. (i) 167; (ii) $\frac{14}{83}$; (iii) 11.　　　　**18.** 12 terms.

19. 26.　　　　**20.** 15, 23, 31, 39, 47.

(A) Examples 12b. (p. 140)

1. Geometrical progressions (i), (ii), (v).

2. (i) 3, $2(3)^9$, $2(3)^{n-1}$; (ii) $\frac{1}{2}$, $(\frac{1}{2})^{10}$, $(\frac{1}{2})^n$; (v) $\frac{2}{3}$, $27(\frac{2}{3})^9$, $27(\frac{2}{3})^{n-1}$.

3. (i) 12; (ii) 36, 64; (iii) 150, 90; (iv) 10, 50, 250.

4. (i) 7; (ii) 8; (iii) 5; (iv) $n + 1$.

5. (i) $2^{10} - 1$; (ii) $-\frac{4}{3}(2^8 - 1)$; (iii) $9 - (\frac{1}{3})^5$; (iv) $\dfrac{(3a)^{10} - 1}{3a - 1}$;
(v) $20\{(1 \cdot 05)^{10} - 1\}$.

6. (i) 12; (ii) 27; (iii) x^3.　　　　**7.** (i) 4, 8, 16; (ii) 4, 1, $\frac{1}{4}$.

8. $n-1$.   **9.** 62.   **10.** (i) x^{19}; (ii) $\dfrac{x^{39}-x}{x^2-1}$.

11. (i) $\dfrac{-4}{3^{17}}$; (ii) $27-(\tfrac{1}{3})^7$.   **12.** 2; $\tfrac{3}{11}[2^{10}-1]$.

13. $\tfrac{3}{2}(\tfrac{5}{2})^{11}$; $\tfrac{2}{5}\{(\tfrac{5}{2})^{12}-1\}=(\tfrac{5}{2})^{11}-\tfrac{2}{5}$.   **14.** $\tfrac{3}{2}$.

15. $\tfrac{5}{8}[1-(\tfrac{1}{2})^6]$.   **16.** $3 \cdot 4^6$ or 12,288.

17. $-\tfrac{3}{2}, 4(-\tfrac{3}{2})^{n-1}$.   **18.** $\dfrac{b^3}{a^2}, \dfrac{b^{n-1}}{a^{n-2}}, \dfrac{a^2\left[1-\left(\dfrac{b}{a}\right)^n\right]}{a-b}$.   **20.** $2, \tfrac{2}{3}$.

(A) Examples 12c. (p. 144)

1. (i) 0; (ii) 1; (iii) ∞; (iv) 0; (v) ∞; (vi) 0; (vii) 1; (viii) 2; (ix) 25.
2. (i) $4(1-(\tfrac{1}{2})^n)$—convergent to sum 4;
 (ii) $\tfrac{1}{2}[2^n-1]$—not convergent;
 (iii) $\tfrac{4}{5}[1-(-\tfrac{1}{4})^n]$—convergent to sum $\tfrac{4}{5}$;
 (iv) 2^n-1—not convergent.
3. Convergent:—(ii), (iii), (v), (vi), (vii).
4. (i) $\tfrac{4}{5}$; (ii) $\tfrac{8}{3}$; (iii) $\dfrac{1}{1-x^2}$; (iv) $2+\sqrt{2}=3\cdot414$.
5. $\tfrac{9}{2}$ or $\tfrac{9}{4}$ **6.** $5(\tfrac{1}{2})^9$.   **7.** $3, \tfrac{1}{2}$.   **8.** $\tfrac{16}{3}$.
9. (i) $2+\tfrac{20}{9}=4\tfrac{2}{9}$; (ii) $\dfrac{a}{1-a}+\dfrac{b}{1-b}$.

(B) Examples 12d. (p. 145)

1. $\dfrac{3n}{5}-\dfrac{8}{5}$, $523\tfrac{3}{5}$.   **2.** (i) -45; (ii) 1,287,286.

4. 25, 67.   **5.** $367\tfrac{1}{2}$.   **6.** 1st term $\dfrac{48}{b+1}$; $b=\tfrac{1}{3}$, 2.

7. 786432, 3145728.   **8.** 11 terms.   **9.** 1.   **10.** 31,020.
11. (i) 130, n^2+3n;
 (ii) $65+\dfrac{x(x^{10}-1)}{x-1}$, $\dfrac{n}{2}(n+3)+\dfrac{x(x^n-1)}{x-1}$;
 (iii) $\dfrac{a(a^{10}-1)}{a-1}-\dfrac{b(b^{10}-1)}{b-1}$; $\dfrac{a(a^n-1)}{a-1}-\dfrac{b(b^n-1)}{b-1}$.
12. (i) £235; (ii) 125 m.   **13.** (i) $10(\tfrac{2}{5})^{10}$ m; (ii) $\tfrac{70}{3}$ m.
14. 1900 m.   **15.** £2500$[(1\cdot04)^{15}-1]=$£1997 approx.

16. 9 years.   **17.** 90°.   **18.** £3000[$(1{\cdot}05)^{10} - 1$] = £1887.

19. £401.   **20.** £8800.

21. $S_n = \dfrac{1 - x^n}{(1 - x)^2} - \dfrac{nx^n}{1 - x}$; $S_\infty = \dfrac{1}{(1 - x)^2}$.

22.   (i) $\dfrac{1}{1 - x} + \dfrac{2x}{(1 - x)^2} - \dfrac{2x^n}{(1 - x)^2} - \dfrac{(2n - 1)x^n}{1 - x}$;

    (ii) $\dfrac{1}{1 - x} + \dfrac{4x}{(1 - x)^2} - \dfrac{4x^n}{(1 - x)^2} - \dfrac{(4n - 3)x^n}{1 - x}$;

    (iii) $\dfrac{a}{1 - x} + \dfrac{ax}{(1 - x)^2} - \dfrac{ax^n}{(1 - x)^2} - \dfrac{nax^n}{1 - x}$.

23. $S_n = \dfrac{a}{1 - x} + \dfrac{dx}{(1 - x)^2} - \dfrac{dx^n}{(1 - x)^2} - \dfrac{\{a + \overline{n - 1} \,.\, d\}x^n}{1 - x}$;

    $S_\infty = \dfrac{a}{1 - x} + \dfrac{dx}{(1 - x)^2}$.

(A) Examples 12e. (p. 152)

1. $1 + 4a + 6a^2 + 4a^3 + a^4$; $\ 1 - 5x + 10x^2 - 10x^3 + 5x^4 - x^5$;

    $1 + 12x + 60x^2 + 160x^3 + 240x^4 + 192x^5 + 64x^6$; $\ 1 - x + \dfrac{x^2}{3} - \dfrac{x^3}{27}$;

    $128 + 448x + 672x^2 + 560x^3 + 280x^4 + 84x^5 + 14x^6 + x^7$;

    $243 - 810a + 1080a^2 - 720a^3 + 240a^4 - 32a^5$;

    $a^6 + 6a^5b + 15a^4b^2 + 20a^3b^3 + 15a^2b^4 + 6ab^5 + b^6$.

2. (i) 2016; (ii) $-4032x^5$; (iii) 10.

3. $243x^5 - 1620x^4 + 4320x^3 - 5760x^2 + 3840x - 1024$;

    $16 + 96x + 216x^2 + 216x^3 + 81x^4$.

4. (i) $-3584x^3$; (ii) $-3432 \,.\, (2)^7 \,.\, x^7$; (iii) $-4032x^4y^5$.

5. (i) 5985; (ii) 26,730; (iii) $\frac{455}{27}$; (iv) $126 \,.\, 3^4 \,.\, 2^5$; (v) 60;

    (vi) $\dfrac{n(n - 1)(n - 2)(n - 3)}{4!} a^{n-4}$.

6. (i) $5670x^4$; (ii) $-8064x^5$; (iii) -20; (iv) 8064.   **7.** 100.

8.   (i) $1 + \frac{1}{4}x - \frac{3}{32}x^2 + \frac{7}{128}x^3$; $-1 < x < 1$.

    (ii) $1 - 3x + 6x^2 - 10x^3$; $-1 < x < 1$.

    (iii) $1 - \dfrac{x}{2} + \dfrac{3x^2}{8} - \dfrac{5x^3}{16}$; $-1 < x < 1$.

    (iv) $1 - 4x + 10x^2 - 20x^3$; $-1 < x < 1$.

    (v) $1 - x - x^2 - \dfrac{5x^3}{3}$; $-\frac{1}{3} < x < \frac{1}{3}$.

    (vi) $1 + \dfrac{x}{3} - \dfrac{x^2}{36} + \dfrac{x^3}{162}$; $-2 < x < 2$.

(vii) $1 - 2x + 4x^2 - 8x^3$; $-\frac{1}{2} < x < \frac{1}{2}$.

(viii) $\sqrt[4]{2^3}\left\{1 - \dfrac{3x}{8} - \dfrac{3x^2}{128} - \dfrac{5x^3}{1024}\right\}$; $-2 < x < 2$.

(ix) $\dfrac{1}{27}\left\{1 - 2x + \dfrac{8x^2}{3} - \dfrac{80x^3}{27}\right\}$; $-\frac{3}{2} < x < \frac{3}{2}$.

(x) $1 + 4x + 12x^2 + 32x^3$; $-\frac{1}{2} < x < \frac{1}{2}$.

9. (i) $\frac{189}{128}$; (ii) $\frac{5}{81}$; (iii) $\frac{32}{243}$; (iv) $\dfrac{1}{16a^{\frac{5}{2}}}$.

10. (i) $-\dfrac{135}{16}x^3$; (ii) $\dfrac{63(5)^4}{2^9}x^4$; (iii) $3125x^5$.

11. (i) $1 + \dfrac{x}{2} - \dfrac{5x^2}{8} - \dfrac{3x^3}{16}$; (ii) $1 + 2x + 2x^2 + 2x^3$; (iii) $2 - 3x + 4x^2 - 5x^3$.

12. (i) $1 + 10x + 55x^2 + 210x^3$; (ii) $1 + \dfrac{x}{2} + \dfrac{3}{8}x^2 - \dfrac{3}{16}x^3$.

13. $1 + \dfrac{3x}{2} + \dfrac{43x^2}{8} + \dfrac{279}{16}x^3$.   **14.** $-35x$ and $\dfrac{35}{x}$.   **15.** $\frac{35}{8}$.

16. $2 - x + 5x^2 - 7x^3 + 17x^4$.   **17.** $x + \dfrac{x^2}{2} + \dfrac{5}{5}x^3 + \dfrac{41}{24}x^4$.

18. $1 + 6x + 15x^2 + 20x^3 + 15x^4 + 6x^5 + x^6$; $1 \cdot 06152$.

19. $1 - \dfrac{x}{2} - \dfrac{x^2}{8} - \dfrac{x^3}{16}$; $\cdot 9899(5)$.

20. $1 \cdot 0080$, $1 \cdot 01$, $0 \cdot 9851$, $1 \cdot 1285$, $2 \cdot 0075$, $0 \cdot 2495$.

21. (i) $1 + \dfrac{x}{2}$; (ii) $1 + x$; (iii) $1 + \dfrac{4x}{3}$; (iv) $1 + \dfrac{3x}{5}$; (v) $3^{11}\left[1 - \dfrac{11x}{3}\right]$;

(vi) $-\dfrac{1}{8} - \dfrac{3x}{16}$.

22. $2 + 6x^2 + 18x^4 + 54x^6$.

(B) Examples 12f. (p. 154)

1. $1 + \dfrac{3x}{2} + \dfrac{15x^2}{8} + \dfrac{51x^3}{16} + \dfrac{699}{128}x^4$; $1 \cdot 015$.

2. $1 + \dfrac{19x}{6} + \dfrac{211}{72}x^2$.   **3.** $-\frac{21}{128}$.   **4.** 85.

6. 10, 10.   **7.** $a = 2$, $n = 7$, $b = 2^7$.

8. $a^5 - 5a^4x + 10a^3x^2 - 10a^2x^3 + 5ax^4 - x^5$; $A = 220$, $B = 284$.

9. $-\dfrac{3}{2} - \dfrac{5x}{2} - \dfrac{11x^2}{4} - \dfrac{13x^3}{4}$.   **19.** $1 + \dfrac{11x}{2} + \dfrac{119x^2}{8} + \dfrac{675}{16}x^3$.

11. $-\dfrac{x}{6} + \dfrac{x^2}{36}$.   **12.** $-2 - \frac{47}{12}x$.

13. (i) ·9987; (ii) 1·0228; (iii) 4·0140.   **14.** 1·9698 secs.

15. (i) $1 + 12x + 62x^2 + 180x^3 + 321x^4 + 360x^5 + 248x^6 + 96x^7 + 16x^8$;
(ii) 840.

16. $A = \frac{1}{3}$, $B = \frac{2}{3}$; $1 - x + 3x^2 - 5x^3$; $-\frac{1}{2} < x < \frac{1}{2}$.

17. $A = -3$, $B = 3$; $\dfrac{3}{2} - \dfrac{9x}{4} + \dfrac{21x^2}{8} - \dfrac{45x^3}{16}$.

18. $A = -\frac{13}{25}$, $B = -\frac{13}{25}$, $C = \frac{12}{5}$; $-\dfrac{1}{6} + \dfrac{x}{4} + \dfrac{x^2}{216}$; $-2 < x < 2$.

19. (i) 1;   (ii) $\dfrac{n(n-1)\ldots(n-5)}{6!}$;   (iii) n.

(A) Examples 13a. (p. 164)

1. (i) 2, 2; (ii) 9, 1; (iii) 2, -3; (iv) -3, -5.

3. $\sin 171° = \cdot 1564$, $\sin 311° = -\cdot 7547$, $\sin 452° = \cdot 9994$,
$\sin(-104°) = -\cdot 9703$, $\cos 171° = -\cdot 9877$, $\cos 311° = \cdot 6561$,
$\cos 452° = -\cdot 0349$, $\cos(-104°) = -\cdot 2419$, $\tan 171° = -\cdot 1584$,
$\tan 311° = -1\cdot 1504$, $\tan 452° = -28\cdot 64$, $\tan(-104°) = 4\cdot 0108$,
$\sin 510° = \cdot 5$, $\sin 382° = \cdot 3746$, $\cos 510° = -\cdot 866$,
$\cos 382° = \cdot 9272$, $\tan 510° = -\cdot 5772$, $\tan 382° = \cdot 4040$.

4. 30°, 150°.   **5.** 40°, 320°.   **6.** 56° 19′, 236° 19′.

7. (i) 131° 35′, 228° 25′; (ii) 120° 8′, 300° 8′; (iii) 22° 36′, 157° 24′.

8. 33° 41′, 213° 41′.

9. (i) $19° 31\frac{1}{2}′$, $160° 28\frac{1}{2}′$, $199° 31\frac{1}{2}′$, $340° 28\frac{1}{2}′$; (ii) $100° 39\frac{1}{2}′$, $169° 20\frac{1}{2}′$, $280° 39\frac{1}{2}′$, $349° 20\frac{1}{2}′$; (iii) 31° 39′, 121° 39′, 211° 39′, 301° 39′; (iv) 224° 30′; (v) 268° 18′; (vi) 27° 43′, 92° 17′, 267° 43′, 332° 17′.

10. (i) $\sin(180° - x) = \sin x$, $\sin(180° + x) = -\sin x$;
    (ii) $\cos(180° - x) = -\cos x = \cos(180° + x)$;
    (iii) $\sin(90° - x) = \cos x = \sin(90° + x)$;
    (iv) $\cos(90° - x) = \sin x$, $\cos(90° + x) = -\sin x$.

(A) Examples 13b. (p. 167)

1. (i) $2\sin^2\theta - 1$; (ii) $2 - 4\sin\theta - 2\sin^2\theta$.

2. (i) $3 + 2t^2$; (ii) $1 + 3t^2$.

3. (i) $\sec^2\theta$; (ii) $\tan\theta$.   **4.** $\tan\theta = \pm\sqrt{2}$.

5. $\tan\theta = \pm\dfrac{\sqrt{2}}{2}$, $\cos\theta = \pm\dfrac{\sqrt{6}}{3}$, $\sin\theta = \pm\dfrac{\sqrt{3}}{3}$.   **6.** 45°, 135°.

7. 54° 44′, 125° 16′.   **8.** 90°.   **9.** 45°, 63° 26′.

10. 60°, 180°.   **11.** 60°, 180°.   **12.** 90°.

(A) Examples 14a. (p. 171)

1. (i) $\frac{3}{5}$, (ii) $\frac{4}{3}$, (iii) $\frac{5}{13}$, (iv) $\frac{12}{13}$, (v) $\frac{63}{65}$, (vi) $\frac{33}{65}$, (vii) $\frac{16}{65}$, (viii) $\frac{56}{65}$, (ix) $\frac{63}{16}$, (x) $\frac{33}{56}$.

2. $\sin(x+y) = \frac{56}{65}$, $\cos(x+y) = -\frac{33}{65}$; $\sin(2x+y) = \frac{36}{325}$,
$\cos(2x+y) = -\frac{323}{325}$; $\sin(x+2y) = -\frac{116}{845}$; $\cos(x+2y) = -\frac{837}{845}$.

3. $\sin A = \dfrac{1}{\sqrt{2}}\{\sin(45°+A) - \cos(45°+A)\}$.

5. (i) $\sin 104° = \sin 76°$; (ii) $\sin 4°$; (iii) $\cos 52°$;
(iv) $-\cos 63° = \cos 117°$; (v) $\sin 3x$; (vi) $\cos 2x$.

6. $\sin(180°-A) = \sin A$, $\sin(180°+A) = -\sin A$, $\cos(180°-A) = -\cos A = \cos(180°+A)$, $\sin(90°+A) = \cos A$, $\cos(90°+A) = -\sin A$.

8. $\sin 3x \cos y + \cos 3x \sin y$; $\cos x \cos 2y + \sin x \sin 2y$;
$$\dfrac{\tan 2x + \tan y}{1 - \tan 2x \cdot \tan y}; \quad \dfrac{\tan 3x - \tan y}{1 + \tan 3x \cdot \tan y}; \quad \dfrac{\cot x \cot y - 1}{\cot x + \cot y};$$
$$\dfrac{\cot x \cot y + 1}{\cot y - \cot x}.$$

9. (i) $\dfrac{3\tan x}{2 - \tan^2 x}$; (ii) $\dfrac{3\tan y}{1 - 2\tan^2 y}$.

10. $\cos A \cos B \cos C - \sin A \sin B \cos C - \sin A \sin C \cos B - \sin B \sin C \cos A$.

11. $\dfrac{\tan A + \tan B + \tan C - \tan A \tan B \tan C}{1 - [\tan A \tan B + \tan A \tan C + \tan B \tan C]}$.

13. $1, \dfrac{1}{\sqrt{3}}$. **14.** $\frac{63}{65}$; $\frac{63}{65}$.

(A) Examples 14b. (p. 176)

1. (i) $\dfrac{840}{841}$; (ii) $\dfrac{41}{841}$; (iii) $\dfrac{2}{\sqrt{29}}$; (iv) $\dfrac{5}{\sqrt{29}}$.

2. $\sqrt{2}-1$. **3.** $\dfrac{\sqrt{2-\sqrt{3}}}{2} = \dfrac{\sqrt{6}-\sqrt{2}}{4}$; $\dfrac{\sqrt{2+\sqrt{3}}}{2} = \dfrac{\sqrt{6}+\sqrt{2}}{4}$.

4. $\frac{24}{7}, \frac{1}{3}$. **5.** $-\frac{12}{5}, \frac{12}{13}, -\frac{5}{13}$. **6.** $\dfrac{1+\sqrt{5}}{\sqrt{10+2\sqrt{5}}}$

7. (i) 1·4384; (ii) 0·0367; (iii) 0·3350; (iv) 0·5774.

8. (i) $\dfrac{1+2t-t^2}{1+t^2}$; (ii) $\dfrac{4t^2+6t-4}{1+t^2}$; (iii) $\dfrac{6t+6t^3-t^4+2t^2-1}{1-t^4}$;
(iv) $\dfrac{1+t^2}{2}$; (v) $\dfrac{2t}{3t^2-1}$; (vi) $\dfrac{1+t^2}{2t}$.

9. $\dfrac{\cot^2 A - 1}{2 \cot A}$.   **10.** 2·762.   **11.** $4 \cos^3 x - 3 \cos x$.

12. $\dfrac{3 \tan x - \tan^3 x}{1 - 3 \tan^2 x}$.   **15.** $\dfrac{1 + \cos 4A}{2}$.   **16.** $8 \cos^4 A - 8 \cos^2 A + 1$.

17. (i) $\theta = 0°, 30°, 150°, 180°$;   (ii) $\theta = 30°, 90°, 150°$;
     (iii) $\theta = 0°, 82° \, 49', 180°$;   (iv) $\theta = 0°, 120°$.

(B) Examples 14c. (p. 180)

1. (i) $\sqrt{2} \cos (x + \alpha)$ where $\tan \alpha = 1$; (ii) $5 \cos (x - \alpha)$ where $\tan \alpha = \frac{3}{4}$;
  (iii) $\sqrt{3} \cos (x - \alpha)$ where $\tan \alpha = \dfrac{\sqrt{2}}{2}$; (iv) $2\sqrt{3} \cos (x - \alpha)$ where $\tan \alpha = \dfrac{\sqrt{3}}{3}$.

2. (i) $\sqrt{2} \cos (x + \alpha)$ where $\tan \alpha = 1$; (ii) $5 \cos (x + \alpha)$ where $\tan \alpha = \frac{3}{4}$;
  (iii) $\sqrt{6} \cos (x + \alpha)$ where $\tan \alpha = \dfrac{\sqrt{2}}{2}$; (iv) $13 \cos (x + \alpha)$ where $\tan \alpha = \frac{12}{5}$.

3. (i) $\sqrt{5} \sin (x - \alpha)$ where $\tan \alpha = \frac{1}{2}$; (ii) $\sqrt{19} \sin (x - \alpha)$ where $\tan \alpha = \dfrac{4\sqrt{3}}{3}$; (iii) $5 \sin (x - \alpha)$ where $\tan \alpha = \frac{4}{3}$; (iv) $\sqrt{a^2 + b^2} \sin (x - \alpha)$
  where $\tan \alpha = \dfrac{b}{a}$.

4. (i) $\pm\sqrt{2}, \pm 5, \pm\sqrt{3}, \pm 2\sqrt{3}$; (ii) $\pm\sqrt{2}, \pm 5, \pm\sqrt{6}, \pm 13$; (iii) $\pm\sqrt{5}, \pm\sqrt{19}, \pm 5, \pm\sqrt{a^2 + b^2}$.

5. $x = -59° \, 2', 300° \, 58'$, etc.   **6.** $x = -33° \, 41', 326° \, 19'$, etc.

7. (i) $119° \, 32', 346° \, 42'$; (ii) $19° \, 28', 270°$; (iii) $237° \, 52', 347° \, 21'$; (iv) $0°, 90°, 360°$.

8. $\sqrt{5} \sin (2x + \alpha)$ where $\tan \alpha = \frac{1}{2}$; $x = 0°, 63° \, 26', 180°, 243° \, 26', 360°$.

(A) Examples 14d. (p. 183)

1. (i) $2 \cos 2A \sin A$;   (ii) $2 \cos 2A \cos A$;   (iii) $-2 \sin 2A \sin A$;
  (iv) $2 \sin (A + B) \cos (A - B)$;   (v) $-2 \sin (A + B) \sin (A - B)$;
  (vi) $2 \cos 4A \cos A$; (vii) $2 \sin 5A \cos 2A$; (viii) $2 \sin 4A \sin 3A$;
  (ix) $2 \cos \theta \sin \phi$; (x) $2 \cos \dfrac{3\alpha + 3\beta}{2} \cos \dfrac{\alpha - \beta}{2}$.

2. (i) $\dfrac{\sqrt{6}}{2}$; (ii) $-\dfrac{\sqrt{2}}{2}$.   **5.** $2 \cos \alpha \cos \overline{\beta - \gamma}$.

6. (i) $45°, 90°, 135°$; (ii) $0°, 30°, 90°, 150°, 180°$; (iii) $0°, 45°, 60°, 90°, 120°, 135°, 180°$; (iv) $0°, 45°, 135°, 180°$; (v) $30°, 90°, 150°$; (vi) $0°,$

20°, 30°, 60°, 90°, 100°, 120°, 140°, 150°, 180°; (vii) 0°, 60°, 120°, 180°; (viii) 22° 30′, 60°, 67° 30′, 112° 30′, 120°, 157° 30′.

7. (i) $\tan \dfrac{3x}{2}$;　　(ii) $-\cot \dfrac{3x}{2}$;　　(iii) $-\cot \theta$;　　(iv) $\cot 5\theta$;

(v) $\dfrac{\cos \dfrac{5\theta}{2}}{\cos \dfrac{3\theta}{2}}$;　　(vi) $\dfrac{\sin \theta}{\cos 2\theta}$.

(B) Examples 14e. (p. 183)

1. $\dfrac{41}{841}$, $-\dfrac{840}{41}$, $\dfrac{5\sqrt{29}}{29}$.　　**2.** a.　　**3.** $\frac{1}{4}$.　　**4.** $\frac{1}{4}$.

4. (i) 0°, 99° 36′; (ii) 60°, 70° 32′; (iii) 33° 42′, 45°; (iv) 70° 32′, 120°.

5. (i) 0°, 120°, 240°, 360°; (ii) 60°, 180°, 300°; (iii) 14° 29′, 90°, 165° 31′, 270°; (iv) 0°, 180°, 360°.

6. (i) 45°; (ii) 75°; (iii) 46° 24′, 90°; (iv) 27° 17′, 78° 59′; (v) 106° 16′, 180°; (vi) 109° 28′.

7. −30° or 150°, etc.　　　　**18.** (i) 45°; (ii) 60°; (iii) 108° 29′.

0. (i) 0°, 60°, 90°, 180°; (ii) 0°, 20°, 45°, 90°, 100°, 135°, 140°, 180°; (iii) 45°, 120°, 135°.

(A) Examples 15a. (p. 191)

1. ·3142, 1·257, 2·828, 5·237, 4·713, 1·4835, 1·108, ·3875, 3·338.

2. 60°, 40°, 150°, 420°, 540°, 22° 30′, 67° 30′, 11° 15′, 22° 55′, 37° 14′, 75° 38′.

3. ·5000, ·3827, ·8660, ·6428, 0, ·3894, ·9909.

4. (i) 3·142 cm; (ii) 18·85 cm; (iii) 1·4 cm; (iv) 9·426 cm; (v) 28·28 cm.

5. (i) 9·426 cm²;　(ii) 84·83 cm²;　(iii) 2·8 cm²;　(iv) 70·695 cm²; (v) 141·39 cm².

6. (a) 5·79, 195·29 cm²; (b) 11·98, 189·1 cm²; (c) 22·6, 178·48 cm².

7. 71° 37′.　　**8.** 4·0 cm.　　**9.** 20 cm².

0. (i) 2π rad s⁻¹; (ii) $\dfrac{\pi}{6}$ rad s⁻¹; (iii) $\dfrac{2\pi}{5}$ rad s⁻¹.

1. 28·65 rev min⁻¹.　**12.** 21·15.　　　**13.** 11·71 cm.

4. 1·4285.　**15.** 27·93 cm².　**16.** $\frac{5}{4}$.　　　**17.** 3·142 m.

(A) Examples 15b. (p. 197)

1. (i) $17'$; (ii) $36'$; (iii) $6°\ 18'$; (iv) $6'$; (v) $3'$; (vi) $38'$.

2. $\cdot0087,\ \cdot0190,\ \cdot0029,\ \cdot0381,\ \cdot0035.$ **3.** 153 m.

4. 414 m.   **5.** $21'$.   **6.** $12\ 600$ m.   **7.** $7\cdot20$ m.   **8.** 776 km

(B) Examples 15c. (p. 197)

1. $1 : 10$.   **2.** (a) $16\cdot8$ cm; (b) 34.4 cm^2.   **3.** $33.65 : 1$.

4. $25\cdot200$ cm^3.   **5.** $7\ 333$ cm^3.   **6.** $5\cdot796$ cm^2.

7. (i) $\cdot5075$; (ii) $\cdot8682$; (iii) $\cdot8644$.

8. (i) $1\cdot0059$; (ii) $-1\cdot0188$.   **9.** $0\cdot5$ m.   **10.** $10\ 000$ m.

14. $x = \cdot48$ rad.   **15.** $\theta = 36°\ 23' = \cdot6351$ rad.

16. $\theta = 74°\ 28'$ or $1\cdot2997$ rad.   **17.** $39°\ 30'$.

18. $133°\ 56',\ 332°\ 20'$.   **19.** $71°\ 42'$.   **20.** $149°\ 18'$.

Revision Papers III. (p. 200)

Paper A (1)

1. $1\frac{17}{27}$.

2. (i) $x = \frac{8}{5}$, $y = 10$; $x = -2$, $y = -8$;   (ii) $\mathsf{A} = \frac{1}{2}$, $\mathsf{B} = -\frac{1}{2}$.

3. (i) 1800; (ii) $1\cdot41$.   **4.** $1 - \frac{2}{3}x + \frac{8}{9}x^2 - \frac{112}{81}x^3$.

5. (i) $-0\cdot5736$, $0\cdot8391$, $-0\cdot9397$; (ii) $1\cdot3934$.

Paper A (2)

1. $\frac{3}{4}$.   **2.** $x = -1$, $y = 2$; $x = -\frac{5}{4}$, $y = \frac{17}{8}$.

3. (i) 3780; (ii) $\dfrac{3^{58} - 1}{2}$.   **4.** $-\frac{3}{5}$, $\frac{4}{5}$, $-\frac{3}{4}$.   **5.** $2\cdot58$ cm^2.

Paper A (3)

1. $x^2 - 3x - 15 = 0$.   **2.** $4x + 3y = 12$.

4. (i) $a^{20} - 20a^{19}b + 190a^{18}b^2 - 1140a^{17}b^3$; (ii) $\frac{7}{8}$.

5. (i) $19°\ 29'$; (ii) $20°\ 55',\ 69°\ 5'$.

Paper A (4)

1. $m^2ab = n(a + b)^2$.   **2.** $x = -2$, $y = -3$; $x = -3$, $y = -2$

3. $27 + 29 + \ldots 113$.   **4.** $\sqrt{2}\left(2 - \dfrac{3x}{2} + \dfrac{3x^2}{16} + \dfrac{x^3}{64}\right)$; $|x| < 2$.

5. $0\cdot1305$ m^2.

Paper A (5)

1. $2\frac{1}{4}$.   **2.** $x = 2$, $y = \frac{1}{2}$; $x = \frac{1}{3}$, $y = -2$.

3. (i) 2;   (ii) $(\frac{4}{5})^7$; $\frac{25}{4}\{1 - (\frac{4}{5})^9\}$.   **4.** (i) $1\cdot082$;   (ii) $1\cdot001$.

5. (i) $1\cdot732$;   (ii) $2°$;   (iii) $0\cdot4637$.

Paper A (6)

1. $45x^2 + 98x - 75 = 0$.

2. (i) $\mathsf{A} = 1$, $\mathsf{B} = -2$, $\mathsf{C} = 1$;   (ii) $y^2 - 4x - 2y + 13 = 0$.

3. 1, 7, 13.   **4.** (i) -10;   (ii) $3543750x^4y^4$.

5. (i) $-0\cdot5977$;   (ii) $-1\cdot2817$.

Paper B (1)

1. $k = 8$, $l = 3$.

2. (i) $a = 1$, $b = 7$, $c = 6$, $d = 1$.

3. $a^n + na^{n-1}b + \dfrac{n(n-1)}{2!}a^{n-2}b^2 + \dfrac{n(n-1)(n-2)}{3!}a^{n-3}b^3 + \ldots$;

$a = 2$, $b = \frac{3}{16}$.

4. (a) 31;   (b) $41\cdot24$; $101\cdot1$.

5. (i) $0\cdot1047$;   (ii) 363 000 km, 412 000 km.

Paper B (2)

1. (i) Negative; (ii) $a = \frac{3}{2}$, $b = \frac{11}{4}$; $\frac{11}{4}$.   **2.** 167, $\frac{14}{83}$, 11.   **3.** $x = 2\cdot2$.

4. (i) $1 + \frac{1}{2}x + \frac{3}{8}x^2 + \frac{5}{16}x^3$; $|x| < 1$.   (ii) $1\cdot5106$.

5. $11\cdot9$ cm, $20° 51'$.

Paper B (3)

1. $\mathsf{A} = \frac{3}{16}$, $\mathsf{B} = -\frac{1}{16}$, $\mathsf{C} = -\frac{1}{8}$, $\mathsf{D} = \frac{1}{2}$.   **2.** $k = -2$, $-4\frac{2}{11}$.

3. (i) $\dfrac{1 - x^n}{1 - x}$; $|x| < 1$, $\dfrac{1}{1 - x}$;

(ii) (a) $\dfrac{n}{2}(n + 1)$; (b) $\dfrac{1 - (n+1)x^n + nx^{n+1}}{(1 - x)^2}$.

4. $30°$, $150°$, $228° 35'$, $311° 25'$.   **5.** (a) $0\cdot804$; (b) $\frac{2}{3}$.

Paper B (4)

1. $-\dfrac{b}{a}$, $\dfrac{\sqrt{b^2 - 4ac}}{a}$.

2. (i) $p = 3$, $q = -2$;   (ii) $15x^2 - 30x + 25 - 2k = 0$.

3. $2\cdot75$, $1\cdot15$, $-3\cdot85$.

4. $A = B = 1; \dfrac{3}{2} + \dfrac{3x}{4} + \dfrac{9x^2}{8} + \dfrac{15x^3}{16} + \dfrac{33}{32}x^4 + \ldots$

5. $78 \cdot 8 \text{ cm}^2$.

Paper B (5)

2. (i) $\dfrac{ar + b}{r + 1}, \dfrac{a + rb}{r + 1}$; (ii) £660.

3. $1 + 4x + 2x^2 - 8x^3 - 5x^4 + 8x^5 + 2x^6 - 4x^7 + x^9$; sum of coefficients is 1.

4. (ii) OD, EC = $5 \cdot 6$ cm.

Paper B (6)

1. (i) Positive except for $1\frac{1}{2} < x < 5$; (ii) $k = -\frac{5}{2}$, 4.

2. $x = 1 \cdot 70$ and $4 \cdot 53$ cm.   **3.** $-\frac{1}{3}$.

4. (i) $x^7 + 14x^5 + 84x^3 + 280x + \dfrac{560}{x} + \dfrac{672}{x^3} + \dfrac{448}{x^5} + \dfrac{128}{x^7}$;   (ii) 6.

5. $6 \cdot 65 \text{ m}^2$.

(A) Examples 16a. (p. 219)

1. $2 \cdot 9$.

2. (i) A = $32° 6'$, B = $20° 44'$, C = $127° 10'$;
  (ii) C = $62° 20'$, $b = 23 \cdot 37$, $c = 22 \cdot 62$;
  (iii) $c = 2 \cdot 57$, A = $48° 38'$, C = $91° 22'$.

3. $108° 13'$.   **4.** $26° 21'$.

5. C = $135°$, $a = 5 \cdot 098$, $b = 2 \cdot 395$, area = $4 \cdot 317$; altitude = $3 \cdot 604$.

7. $5 \cdot 083$.   **8.** $4 \cdot 56$ cm, $3 \cdot 8$ cm, $2 \cdot 28$ cm.

9. $11 \cdot 4 \text{ cm}^2$.   **10.** $33 \cdot 13 \text{ unit}^2$.   **11.** $2 \cdot 45$.

12. (i) B = $27° 8'$ or $152° 52'$, C = $132° 52'$ or $7° 8'$, $c = 6 \cdot 427$ or $1 \cdot 088$
  (ii) C = $154° 27'$ or $25° 33'$, A = $10° 33'$ or $139° 27'$, $a = 4 \cdot 244$ or
  $15 \cdot 07$; (iii) B = $48° 35'$ or $131° 25'$, A = $101° 25'$ or $18° 35'$, a
  $15 \cdot 68$ or $5 \cdot 099$; (iv) C = $60°$ or $120°$, B = $90°$ or $30°$, $b = 200$ or
  100.

14. (i) $3 \cdot 651 \text{ unit}^2$; (ii) $14 \cdot 01 \text{ unit}^2$; (iii) $31 \cdot 74 \text{ unit}^2$; (iv) 4330 unit^2.

16. $9 \cdot 143$ m, $7 \cdot 252$ m.   **17.** $15°, 75°$.   **18.** $3 \cdot 361$ cm, $9 \cdot 627$ cm

(B) Examples 16b. (p. 221)

1. (i) A = $42° 36'$, $b = 457 \cdot 7$, $c = 502 \cdot 3$; (ii) C = $31° 16'$, $b = 92 \cdot 98$, a
  $149 \cdot 8$; (iii) B = $76° 29'$, $a = 4 \cdot 855$, $c = 11 \cdot 17$.

2. (i) A = $101° 58'$, B = $61° 16'$, C = $16° 46'$; (ii) A = $94° 34'$, B = $30° 2$
  C = $55°$; (iii) A = $45° 53'$, B = $71° 24'$, C = $62° 45'$.

3. (i) $a = 8 \cdot 894$, $\mathbf{B} = 57° 27'$, $\mathbf{C} = 63° 3'$; (ii) $b = 2363$, $\mathbf{C} = 93° 24'$, $\mathbf{A} = 17° 55'$; (iii) $c = 43 \cdot 06$, $\mathbf{A} = 27° 20'$, $\mathbf{B} = 41° 30'$.

4. Angle $= 116° 15'$, area $= 514 \cdot 7$ cm^2.

5. $\mathbf{A} = 29° 38'$, $\mathbf{B} = 50°$, $\mathbf{C} = 100° 22'$, alt. $= 50 \cdot 08$.

6. $b = 20 \cdot 04$, $\mathbf{A} = 29° 43'$, $\mathbf{C} = 83° 39'$, area $= 107 \cdot 9$.

7. $137 \cdot 6$ cm^2.

8. (i) $14 \cdot 45$ cm; (ii) $58° 44'$; $121° 16'$; (iii) $69° 20'$ or $110° 40'$.

9. Diff. in $a = 25 \cdot 72$.　　　　**10.** $\mathbf{C} = 55° 42'$, $b = 93 \cdot 09$ m.

11. $56° 34'$.　　　**12.** $129 \cdot 8$.　　　**14.** $\mathbf{BD} = 27 \cdot 46$ cm.

15. $71 \cdot 78$ cm^2.　　　**16.** $650 \cdot 6$ m; N. $11° 45'$ E.

17. (i) 10 cm, $7 \cdot 848$ cm; (ii) $38 \cdot 94$ cm^2.　　　**18.** $251 \cdot 3$ m, $140 \cdot 9$ m.

20. $17 \cdot 86$ m.　　　**21.** $10 \cdot 23$ km h^{-1}.　　　**22.** $2 \cdot 782$ km.

24. $c = 26 \cdot 93$ cm, $a = 23 \cdot 0$ cm, $b = 20 \cdot 34$ cm.

(B) Examples 16c. (p. 227)

1. $12\sqrt{3}$ cm; (a) $35° 16'$;　(b) $54° 44'$.

2. $\mathbf{BX} = 14$ cm, $\mathbf{CX} = 14$ cm, $\widehat{\mathbf{BCX}} = 33° 12'$, inclination $25° 22'$.

3. $15° 3'$.　　　**4.** $\mathbf{CD}' = 2 \cdot 56$ m, inclination $69° 28'$.

5. $68° 55'$.　　　**6.** $28° 1'$.

7. (i) $10 \cdot 58$ cm; (ii) $61° 51'$; (iii) $69° 17'$; (iv) $97° 10'$.

8. $95 \cdot 7$ cm, $73° 13'$.　　　　**9.** $24 \cdot 14$ cm, $75°$.

10. $4 \cdot 947$ m, $55° 32'$ and $81° 38'$.　　**11.** $35° 10'$.　　**12.** 560 m.

13. $\mathbf{AB} = 859$ m, N. $75° 39'$ E.　　**14.** $33 \cdot 7$ m.　　**15.** 1069 m.

16. 120 m.　　**17.** 137 m s^{-1}; $23°$ W. of S.　　**18.** $47 \cdot 64$ m.

19. $\mathbf{B}$ is N. $6° 21'$ W. of $\mathbf{A}$, Dist. $= 6485$ m.　　**20.** 2822 m.

21. 119 m.　　**22.** $14 \cdot 9$ m.　　**23.** $\dfrac{d \tan \alpha \tan \beta}{\tan \alpha + \tan \beta}$ m.

(A) Examples 17a. (p. 235)

1. (i) $2 \cos \theta$;　(ii) $-3 \sin \theta$;　(iii) $\dfrac{\sec^2 \theta}{2}$;　(iv) $\cos \theta + \sin \theta$;

(v) $-3 \sin \theta + 4 \cos \theta$;　(vi) $2 \cos \theta - \sec^2 \theta$;　(vii) $3 \sin \theta$;

(viii) $\dfrac{\cos \theta - 4 \sin \theta}{3}$.

2. (i) $-3 \cos x$;　(ii) $\dfrac{\sin x}{2}$;　(iii) $4 \tan x$;　(iv) $-2 \cos x - 3 \sin x$;

(v) $\dfrac{-\cos x + \sin x}{2}$;　(vi) 1; (vii) 4; (viii) 1; (ix) -3; (x) -1.

3. (i) $-3\sin t$;   (ii) $-4\cos t + 3\sin t$;   (iii) $-2\sin t - \cos t$;
  (iv) $2\sin t$.

4. (i) 1;   (ii) -3;   (iii) -2.

5. (i) (a) 1, (b) $\dfrac{1}{\sqrt{2}}$;   (ii) (a) 0, (b) $-\dfrac{1}{\sqrt{2}}$;   (iii) (a) 1, (b) 2;
  (iv) (a) 3, (b) $\sqrt{2}$;   (v) (a) 3; (b) $2(\sqrt{2}-1)$.

6. (i) Tangent $\sqrt{2}y - 1 = x - \dfrac{\pi}{4}$, normal $y + \sqrt{2}x = \dfrac{1}{\sqrt{2}} + \dfrac{\pi}{2\sqrt{2}}$;

  (ii) Tangent $\sqrt{2}y + x = \dfrac{\pi}{4} + 1$, normal $\sqrt{2}y - 2x = 1 - \dfrac{\pi}{2}$;

  (iii) Tangent $y = 2x + 1 - \dfrac{\pi}{2}$, normal $2y + x = 2 + \dfrac{\pi}{4}$;

  (iv) Tangent $y - \sqrt{2}x = 2\sqrt{2} - \dfrac{\pi}{4}\sqrt{2}$, normal $\sqrt{2}y + x = 4 + \dfrac{\pi}{4}$;

  (v) Tangent $y = 2(\sqrt{2}-1)x + 2\sqrt{2} - 1 - 2(\sqrt{2}-1)\dfrac{\pi}{4}$,

   normal $2(\sqrt{2}-1)y + x = 10 - 6\sqrt{2} + \dfrac{\pi}{4}$.

7. (i) $1 - \dfrac{7\sqrt{3}}{2}$;   (ii) $14 + \dfrac{\sqrt{3}}{2}$.

8. (i) $\dfrac{\sin x}{2}$;   (ii) $\dfrac{-\sin x}{2}$;   (iii) $\dfrac{x - \sin x}{2}$;   (iv) $\dfrac{\sin x + x}{2}$.

9. (i) 1;   (ii) 1;   (iii) 1;   (iv) $3 - \sqrt{2}$;   (v) $\frac{5}{2} + \sqrt{3}$.

10. (i) $\dfrac{\sqrt{2}-1}{\sqrt{2}}\pi$;   (ii) $2\sqrt{2}\pi$.

(A) Examples 17b. (p. 240)

1. $8\cos 8x$, $-\frac{1}{2}\sin\dfrac{x}{2}$, $\frac{1}{2}\cos\dfrac{x}{2}$, $3\sec^2 3x$, $\frac{11}{2}\cos\dfrac{11x}{2}$, $-\cdot 3\sin\cdot 3x$,

  $\frac{1}{10}\cos\dfrac{x}{10}$, $20\sec^2 10x$, $-6\sin 3x$, $\cos 2x$, $2\sec^2\dfrac{x}{2}$, $\dfrac{\pi}{180}\cos x°$,

  $-\dfrac{\pi}{180}\sin x°$, $\dfrac{\pi}{180}\sec^2 x°$.

2. $-\dfrac{\cos 3x}{3}$, $\dfrac{\sin 3x}{3}$, $\dfrac{\tan 3x}{3}$, $-2\cos\dfrac{x}{2}$, $3\sin\dfrac{x}{3}$, $4\tan\dfrac{x}{2}$, $\sin 4x$, $-\cos\dfrac{x}{5}$,
  $\frac{1}{4}\sin 2x$, $2\tan 4x$, $\frac{2}{5}\sin\cdot 5x$, $-11\cos\cdot 1x$.

3. (i) $\frac{1}{2}$;   (ii) $\frac{2}{3}$;   (iii) $\frac{1}{2}$;   (iv) π;   (v) $\dfrac{\pi}{2}$;   (vi) -3.

4. (i) $9\cos 3x + 4\sin 2x + 1$; (ii) $-\frac{1}{4}\sin\frac{x}{2} + \frac{3}{2}\cos\frac{x}{2}$;

    (iii) $2\sec^2 x + 3\cos 3x$; (iv) $-\sin 2x$; (v) $\sin 2x$.

5. $2, y - \sqrt{3} = 2\left(x - \frac{\pi}{6}\right)$. **6.** $\dfrac{-7\sqrt{2}}{2}$. **7.** $\dfrac{\pi\sqrt{2}}{40} = \cdot 111$.

8. $0, 1\cdot318, \pi$. **9.** Values of θ between $1\cdot318$ and π.

10. 1 unit. **11.** 4 units. **12.** $\dfrac{2\pi}{3}$.

(A) Examples 17c. (p. 243)

1. $\cos(x+3)$, $-\sin(x-1)$, $\sec^2(x+2)$, $6\cos(3x-1)$, $-\sin(3x+2)$,

    $-2\cos(1-x)$, $-2\sin\left(\dfrac{x}{2}+1\right)$, $-36\sec^2(1-6x)$, $\pi\cos\left(\pi x - \dfrac{\pi}{2}\right)$,

    $-\dfrac{\pi}{2}\sin\left(\dfrac{\pi x}{2} - \dfrac{\pi}{4}\right)$, $-\pi^2\sec^2(1-\pi x)$.

2. $\dfrac{-\cos(2t+1)}{2}$, $\dfrac{\sin(3t-2)}{3}$, $2\tan\left(\dfrac{t}{2}-1\right)$, $2\cos(1-t)$,

    $\dfrac{\sin\left(\pi t + \dfrac{\pi}{2}\right)}{\pi}$, $-2\cos\left(\dfrac{\pi t}{2} - \dfrac{\pi}{4}\right)$.

3. (i) 0; (ii) π.

4. $2\sin x\cos x$, $-2\sin x\cos x$, $-5\cos^4 x\sin x$, $7\tan^6 x\sec^2 x$,

    $12\sin^3 x\cos x$, $-2\cos^3 x\sin x$, $4\sin 2x\cos 2x$, $-4\cos 2x\sin 2x$,

    $4\tan 2x\sec^2 2x$, $2\sin^3\dfrac{x}{4}\cos\dfrac{x}{4}$, $-3\cos^2\dfrac{x}{3}\sin\dfrac{x}{3}$, $\frac{1}{2}\sin^{-\frac{1}{2}} x\cos x$,

    $-\frac{1}{2}\cos^{-\frac{1}{2}} x\sin x, \frac{1}{2}\tan^{-\frac{1}{2}} x\sec^2 x, -2\sin^{-3} x\cos x, 6\cos^{-4} 2x\sin 2x$.

5. 1. **6.** 1 max. 0 min.

7. $y + \sqrt{2}x = \dfrac{1}{\sqrt{2}} + \dfrac{\sqrt{2}\pi}{8}$. **8.** $t = \frac{1}{2}$ or $\dfrac{n\pi+1}{2}$.

9. (i) $\dfrac{\pi}{2}$; (ii) $\dfrac{\pi}{2}$. **10.** $\dfrac{\pi^2}{2}$.

(B) Miscellaneous Examples. (p. 244)

1. (i) $2\cos 2x$; (ii) $-\frac{1}{2}\sin\left(\dfrac{x}{2}\right)$; (iii) $2\sec^2 2x$.

2. (i) $2\cos x\,(1+\sin x)$; (ii) 0; (iii) $\frac{3}{2}\sqrt{\sin x}\cos x$; (iv) $-2\tan x\sec^2 x$.

3. (i) $-n^2 \sin nx$; (ii) $-n^2 \cos nx$.

5. (i) 4; (ii) 2; (iii) $\dfrac{3\pi}{8} + \dfrac{7}{4} - \sqrt{2} = 1 \cdot 514$; (iv) $\dfrac{90}{\pi}$.

6. $-\dfrac{\cos 4x}{8} - \dfrac{\cos 2x}{4} + c$. **7.** (i) $\frac{3}{16}$; (ii) 0. **8.** (i) $\frac{1}{4}$; (ii) $\frac{8}{3}$.

9. (i) Max. $\sqrt{2}$, min. $-\sqrt{2}$; (ii) max. 5, min. -5;

(iii) max. $\dfrac{\pi}{2} - 1$, min. $\dfrac{3\pi}{2} + 1$.

10. Max. point $\left(\dfrac{5\pi}{6}, \dfrac{5\pi}{6} + \dfrac{\sqrt{3}}{2}\right)$, min point $\left(\dfrac{\pi}{6}, \dfrac{\pi}{6} - \dfrac{\sqrt{3}}{2}\right)$.

11. $\dfrac{dy}{dx} = -\dfrac{\sin t}{\cos 2t}$, $t = n\pi$ where n is an integer.

12. $y - x = \pi + 4$. **13.** Subnormal $= \dfrac{2\sqrt{3}}{3}$, subtangent $= \dfrac{\sqrt{3}}{2}$.

14. Increasing. **15.** 4 m s^{-1}, 8 m s^{-2}, $\dfrac{\pi}{4}$ sec.

16. $\dfrac{1}{2} + \dfrac{4}{\pi} = 1 \cdot 77$ unit. **17.** $0 \cdot 37$ unit. **18.** 4 unit^2.

19. $\frac{1}{2}$ unit^2. **20.** $\pi\left(\dfrac{3\pi}{4} - 2\right) = 1 \cdot 12$.

(A) Examples 18a. (p. 251)

1. (i) $12x^3 - 2x$; (ii) $\dfrac{4}{x^3}$; (iii) $\cos x - x \sin x$; (iv) $2x \sin x + x^2 \cos x$;

(v) $\tan x + x \sec^2 x$; (vi) $1 - 3x^2$; (vii) $\dfrac{2}{(1+x)^2}$;

(viii) $\dfrac{-3x^2 + 6x + 1}{(1-x)^2}$; (ix) $\cos^2 x - \sin^2 x = \cos 2x$;

(x) $\dfrac{x \sec^2 x - \tan x}{x^2}$; (xi) $\dfrac{\sin x - x \cos x}{\sin^2 x}$;

(xii) $\dfrac{-(x \sin x + 2 \cos x)}{2x^3}$; (xiii) $\dfrac{2}{(1-x)^3}$;

(xiv) $1 - \sin x - x \cos x$; (xv) $\sin 2x + 2x \cos 2x$;

(xvi) $\dfrac{-2(2x \sin 2x + \cos 2x)}{x^2}$.

2. (i) $2x \cos x - x^2 \sin x - 1$; (ii) $1 - 3x^2 \tan 2x - 2x^3 \sec^2 2x$;

(iii) $\dfrac{-1}{(x+1)^2} + \dfrac{1}{(x+2)^2}$; (iv) $(x+3) \sin x + (2x+1) \cos x$.

3. (i) $\dfrac{-2}{(1+t)^3}$; (ii) $4\cos t - 2t \sin t$; (iii) $\dfrac{4\sin 2t}{t^2} + \dfrac{2\cos 2t}{t^3} - \dfrac{4\cos 2t}{t}$.

4. $-\frac{5}{9}$.   **5.** $\sec^2\theta$, $-\csc^2\theta$.

7. (i) $x = 0, \frac{2}{3}$;   (ii) $x = 2$;   (iii) $x = 1, -\frac{1}{3}$.

9. $\csc\theta + (\pi - \theta)\cot\theta\,\csc\theta$.   **10.** $\frac{8}{9}$.

(B) Examples 18b. (p. 252)

1. (i) $\dfrac{9x\cos 3x - 3\sin 3x}{x^2}$;   (ii) $\dfrac{x-2}{2(x-1)^{\frac{3}{2}}}$;   (iii) $\dfrac{x}{\sqrt{1-x^2}}\{1-3x^2\}$;

(iv) $4\cos 2x\cos 3x - 6\sin 2x\sin 3x$;   (v) $\sin x[\sin x + 2x\cos x]$;

(vi) $\cos^2 x[\cos x - 3(1+x)\sin x]$;   (vii) $\dfrac{1+2x^2}{(1+x)^2(1-2x)^2}$;

(viii) $\dfrac{\cos x + \sin x + 1}{(1+\cos x)^2}$;   (ix) $\dfrac{2x\sec^2 x - 2\tan x}{(\tan x + x)^2}$;

(x) $\dfrac{3\{(1-x)\cos 3x + \sin 3x\}}{(1-x)^4}$;   (xi) $\dfrac{1}{x^2}\sin\left(\dfrac{1}{x}\right)$;

(xii) $3x^2\sin\left(2x + \dfrac{\pi}{2}\right) + 2x^3\cos\left(2x + \dfrac{\pi}{2}\right)$;   (xiii) $1 + \dfrac{x}{\sqrt{x^2+a^2}}$.

2. $\dfrac{4}{x^3}$; $\pm\sqrt{2}$.   **3.** Min. 1, max. $\frac{1}{4}$.

4. Max. $\frac{2}{9}$, min. 2.   **5.** Max. 1, min. 9.

6. Max. 0, min. $-\frac{3456}{3125}$.   **7.** Max. $(2, -1)$, min. $(-2, -\frac{1}{9})$.

8. Tangent $y = 3x + 1$; normal $3y + x + 7 = 0$.

9. (i) $4x^3 + 3x^2 + 2x + 1$;   (ii) $\dfrac{x^2\cos x - x\cos x - \sin x}{(x-1)^2}$;

(iii) $\frac{1}{2}\sin 2x + x\cos 2x$;   (iv) $\dfrac{2x\tan 2x + 2x^2\sec^2 2x}{1+x} - \dfrac{x^2\tan 2x}{(1+x)^2}$.

11. $\dfrac{2t}{t^2-1}$.   **12.** $\dfrac{\sqrt{2}}{2}$.   **14.** $\theta = \cos^{-1}\dfrac{\sqrt{3}}{3} = 54°\,44'$.

15. $42°\,21'$.

(B) Examples 18c. (p. 258)

1. (i) $\dfrac{x}{2y}$;   (ii) $\dfrac{x^2}{y^2}$;   (iii) $-\dfrac{(x+1)}{y}$;   (iv) $\dfrac{2}{y}$;   (v) $\dfrac{1}{y^2}$;

(vi) $-\dfrac{1}{x^2}$ or $-\dfrac{y}{x}$;   (vii) $\dfrac{y-4x}{2y-x+3}$;   (viii) $\dfrac{-y^2}{x^2}$;

(ix) $\dfrac{1-3x^2y^2}{2x^3y}$;   (x) $-\sqrt{\dfrac{y}{x}}$.

2. (i) $\dfrac{r}{2}\tan\theta$; (ii) $\dfrac{r\{\cos\theta-2\cos 2\theta+2\sin 2\theta\}}{\sin 2\theta+\cos 2\theta-\sin\theta-1}$.

3. (i) 1; (ii) 0; (iii) $-\frac{1}{2}$; (iv) $\dfrac{4\sqrt{2}}{5}$.

4. $\dfrac{dy}{dx}=\dfrac{-(x+y)}{x+3y}$.   **5.** -114.

6. Pt. $(0,0)$; gradients ∞ and 0; angle of intn. $=90°$;
   Pt. $(4,2)$; gradients $\frac{1}{4}$ and 1; angle of intn. $=\tan^{-1}\cdot 6=30°\ 58'$.

7. $(1,1)$, $(-1,-1)$; angles of intn. $=\tan^{-1}3=71°\ 30'$.

8. Tangent $2\sqrt{2}\,.\,x-3y=6$, normal $3x+2\sqrt{2}\,.\,y=13\sqrt{2}$.

9. $\dfrac{dy}{dx}=\dfrac{1}{\cos y}=\pm\dfrac{1}{\sqrt{1-x^2}}$ **10.** $\dfrac{dy}{dx}=\pm\dfrac{1}{\sqrt{1-x^2}}$.

11. $\dfrac{dy}{dx}=\pm\dfrac{2}{\sqrt{1-x^2}}$.   **12.** $\dfrac{dv}{dt}=\dfrac{dv}{ds}\times\dfrac{ds}{dt}$; accel. $=-\omega x$.

13. (i) Max. -3, min. $\frac{3}{2}$; (ii) max. 7, min. 1.

14. $\theta=0$, π, 2π, etc.

15. $3\frac{1}{3}$ cm min^{-1}.   **16.** $\dfrac{1}{6}\sqrt{\dfrac{10}{\pi}}=\cdot 297$ cm s^{-1}.

17. $2\sqrt{3}$ cm s^{-1} towards O.   **18.** 8 m s^{-1}.   **19.** $\frac{5}{9}$ cm s^{-1}.

Revision Papers IV. (p. 261)

Paper A (1)

1. $\frac{4}{5}$, $\frac{24}{25}$, $\frac{7}{25}$.   **2.** $36°\ 11'$; $6\cdot 89$ cm.

3. (a) (i) $1-6\cos 2x$, (ii) $3x^2(3\cos x-x\sin x)$;

   (b) (i) $\frac{1}{3}$, (ii) $\sin x+\dfrac{\sin 2x}{2}+c$.

4. Tangent $y-1=\dfrac{\sqrt{3}}{2}\left(x-\dfrac{\pi}{3}\right)$; normal $y-1+2\dfrac{\sqrt{3}}{3}\left(x-\dfrac{\pi}{3}\right)=0$.

5. 8π.

Paper A (2)

1. (i) $\frac{1}{5}$ and -5; (ii) $0°$, $78°\ 28'$, $180°$.

2. (i) A $=116°\ 50'$, C $=43°\ 10'$, $b=2\cdot 3$ cm, $c=4\cdot 6$ cm.
   (ii) A $=23°\ 10'$, C $=136°\ 50'$, $b=5\cdot 22$ cm, $c=10\cdot 44$ cm.

3. $\frac{16}{63}$.   **4.** (a) (i) $6\sec^2 3x$, (ii) $\dfrac{3}{(x+1)^2}$; (b) $\dfrac{11\pi}{6}+\sqrt{3}$.

5. (i) $-2\cos x-\dfrac{x^2}{2}+c$; (ii) 0; (iii) $\pi+2$.

Paper A (3)

1. $\sin(A+B) + \sin(A-B) = 2\sin A \cos B$; $\dfrac{\sqrt{2}}{2}$.   **2.** $80° \ 12'$.

3. (i) $-\dfrac{3}{x^2} + \sec^2 x$; (ii) $\dfrac{3x^2 - 2x - 3}{(3x-1)^2}$; (iii) $2\cos 4x$; (iv) $3\tan^2 x \sec^2 x$.

4. $-0 \cdot 091$.   **5.** (i) $y = x - \cos x + 1$; (ii) 18.

Paper A (4)

2. $60°$.   **3.** $13° \ 38'$.

4. (a) (i) $(6x-1)\sin x + (3x^2 - x + 1)\cos x$; (ii) $9\sin^2 x \cos x$. (b) $\frac{2}{3}$.

5. (i) 2; (ii) $\dfrac{3\pi}{2} + 4$.

Paper A (5)

1. $-\frac{3}{5}, -\frac{4}{5}, \frac{24}{7}, \frac{7}{25}$.   **2.** $31 \cdot 75 \ \text{cm}^2$.   **3.** $2\cos 2x \sin x$.

4. (a) $s = 10$; (b) 5π.   **5.** (i) 2; (ii) π.

Paper A (6)

1. 1; $45°$.   **2.** $BD = 7 \cdot 125 \ \text{cm}$, $AC = 9 \cdot 865 \ \text{cm}$.

3. (i) $-\frac{1}{3}$; (ii) $x^2 \sec^2 x + 2x \tan x$.

4. $\dfrac{dx}{d\theta} = 2 - 2\cos\theta$, $\dfrac{dy}{d\theta} = 2\sin\theta$.   **5.** 1.

Paper B (1)

2. $\hat{A} = 120°$.   **3.** $474 \cdot 8 \ \text{km}$.

4. (i) $x^6(2x-1)^3(22x-7)$; (ii) $4\sin^3 x \cos^3 x \cos 2x$.

Paper B (2)

1. (ii) $30°, 150°$.   **2.** $78° \ 28'$.

4. (i) (a) $\dfrac{x}{(1-x)^3(1+x^2)^{\frac{1}{2}}} + \dfrac{3(1+x^2)^{\frac{1}{2}}}{(1-x)^4}$, (b) $\dfrac{\cos 2x}{\sqrt{\sin 2x}}$; (ii) 0.

5. $\sqrt{5} \ \text{m}$; $\cdot 708 \ \text{m s}^{-2}$.

Paper B (3)

1. $17° \ 10\frac{1}{2}'$, $60° \ 8\frac{1}{2}'$, $197° \ 10\frac{1}{2}'$, $240° \ 8\frac{1}{2}'$.

2. $c = 26 \cdot 59 \ \text{cm}$; $A = 106° \ 47'$; $B = 31° \ 41'$.

3. (i) $\dfrac{a - ky}{kx}$; (ii) $9° \ 28'$.

4. (i) $-\dfrac{4x+11}{(x-1)^4}$; $2(x^2+1)\cos 2x+2x\sin 2x$. (ii) $0{\cdot}16$ mm s^{-1}.

5. $4{\cdot}82$.

Paper B (4)

1. $68°\,42'$, $164°\,26'$.

2. BC $= 4{\cdot}25$ m, AD $= 3{\cdot}94$ m.

4. (i) $(a)\ \dfrac{1-6x^2-7x^3}{2\sqrt{1+x}}$, $(b)\ \sec x(\sec^2 x+\tan^2 x)$; (ii) max. 25, min. 1.

5. $0{\cdot}342$.

Paper B (5)

1. 802 m.   **2.** $59{\cdot}7$ m.   **3.** $\dfrac{\pi}{2},\dfrac{\pi}{6}$; max. $\tfrac34$; min. $\tfrac12$.

4. (i) $(a)\ -\dfrac{1+5x^3}{2\sqrt{x^3}}$, $(b)\ 3\{x^2\cos 6x-2(x^3-1)\sin 6x\}$; (ii) $\dfrac{m(2-m^3)}{1-2m^3}$.

5. $\sin\pi t-\cos\pi t+1$; $-\pi^2$.

Paper B (6)

1. (ii) $x=15°,\ 75°,\ 90°,\ 135°$.   **3.** 8378 m.

4. (i) $(a)\ 3x^2\sin(2x+1)+2x^3\cos(2x+1)$; $(b)\ \dfrac{1-2x-2x^2}{(x+1)^2(x+2)^2}$.

5. (i) $\tfrac13 x^3+2x-\dfrac{1}{x}+c$; (ii) $\tfrac32 x+2\sin x+\tfrac14\sin 2x+c$.

  $16y=36+12x-3x^2$; $(-2,0)$, $(6,0)$; 16.

(A) Examples 19a. (p. 272)

1. $(-\tfrac38,\tfrac{17}{8})$.   **2.** 2 units.   **3.** 3.

5. $116°\,34'$.   **6.** (i) $(2,3\tfrac12)$; (ii) $(1,3)$.

7. C$(3,-4)$, D$(1,1)$.   **9.** opposite sides.

10. (i) $18°\,26'$; (ii) $8°\,8'$; (iii) $8°\,8'$; (iv) $8°\,8'$.

11. $(-\tfrac74,\tfrac74)$; $(-13,-2)$. **12.** (i) $(1,3)$; (ii) $\dfrac{24}{\sqrt{34}}$; (iii) $75°\,58'$.

13. $(5\tfrac12,3)$.   **14.** $61°\,56'$.   **15.** $\sqrt5$ units.

16. 1 and $\tfrac17$.   **17.** $7y=16x$.   **18.** $\dfrac{1}{\sqrt5},\dfrac{3}{\sqrt5};\dfrac{2}{\sqrt5}$.

20. $\tfrac{39}{20}$.   **21.** 2 units   **22.** $y+3x=8$, $3y-x+6=0$.

(A) Examples 19b. (p. 276)

1. $x^2 + y^2 - 2x - 2y - 2 = 0.$ 2. $x^2 + y^2 - 4y + 3 = 0.$
3. $x^2 + y^2 - 9 = 0.$ 4. $x^2 + y^2 + 2x - 6y + 7 = 0.$
5. $x^2 + y^2 + 4y - 16 = 0.$ 6. $x^2 + y^2 + 6x - 8y = 0.$
7. $x^2 + y^2 - 8x = 0.$ 8. $x^2 + y^2 + 6x + 8y = 0.$
9. $x^2 + y^2 + 2x - 4y - 21 = 0.$ 10. $x^2 + y^2 + 6x - 2y - 15 = 0.$
11. (i) $x^2 + y^2 - 6x - 2y + 8 = 0;$ (ii) $x^2 + y^2 - 4x + 4y - 2 = 0;$
     (iii) $x^2 + y^2 + 2x - 2y - 8 = 0;$ (iv) $x^2 + y^2 - 3x + 3y = 0.$
12. (i) 4 units;   (ii) $x^2 + y^2 - 4x - 8y + 4 = 0.$
13. $x^2 + y^2 + 2x - 4y + 4 = 0.$
14. (i) $(1, -2)$, 4 units;   (ii) $(-2, 3)$, 1 unit;   (iii) $(0, 2)$, $\frac{3}{2}$ units;
     (iv) $(-\frac{3}{2}, 0)$, $\sqrt{5}$ units;   (v) $(1, 2)$, $\sqrt{5}$ units; (vi) $(4, 0)$, 3 units;
     (vii) $(3, -1)$, 4 units;   (viii) $(\frac{1}{2}, -\frac{1}{2})$, $\frac{1}{2}\sqrt{2}$ units;   (ix) $(-\frac{3}{2}, -1)$, $\frac{3}{2}$ units;
     (x) $(2, -\frac{3}{2})$, $\frac{1}{2}\sqrt{15}$ units;   (xi) $(\frac{5}{4}, -\frac{3}{4})$, $\frac{1}{4}\sqrt{42}$ units;   (xii) $(-\frac{4}{3}, \frac{1}{3})$, $\frac{1}{3}\sqrt{35}$ units.
16. $x^2 + y^2 + 2x - 4y - 20 = 0.$ 17. $x^2 + y^2 - 8x - 2y = 0.$
18. (i) $x^2 + y^2 - 4x = 0;$ (ii) $x^2 + y^2 - 1 = 0;$
     (iii) $5(x^2 + y^2) - 20x + 10y + 16 = 0;$ (iv) $x^2 + y^2 + 4x - 5 = 0.$
19. $y - x + 3 = 0.$ 22. $x^2 + y^2 - 8x - 8y + 16 = 0.$
23. $(0, 0), (-3, 3); x^2 + y^2 + 3x - 3y = 0.$
24. $x^2 + y^2 + 2x - 12y + 12 = 0, x^2 + y^2 - 10x + 4y + 4 = 0.$
26. $x^2 + y^2 - 3x - 4y = 0.$

(A) Examples 19c. (p. 280)

1. (i) $\sqrt{21}$;   (ii) 4;   (iii) $\sqrt{11}$;   (iv) 1;   (v) $\sqrt{7}$;   (vi) 3;   (vii) 4;
     (viii) $\sqrt{2}$ units.
2. (i) $x + y - 2 = 0;$ (ii) $3x - 4y + 25 = 0;$ (iii) $x + 2y - 2 = 0;$
     (iv) $4x - 3y - 13 = 0;$ (v) $5x - 4y = 0;$ (vi) $y + 2 = 0;$
     (vii) $3x - y - 1 = 0;$ (viii) $11x + 4y - 33 = 0.$
3. (i) $x - y = 0;$ (ii) $4x + 3y = 0;$ (iii) $2x - y - 4 = 0;$
     (iv) $3x + 4y + 9 = 0;$ (v) $4x + 5y = 0;$ (vi) $x + 1 = 0;$
     (vii) $x + 3y + 3 = 0;$ (viii) $4x - 11y - 12 = 0.$
5. $3x - 4y - 19 = 0, 3x - 4y + 31 = 0.$ 6. $x + y + 3 = 0, x - y - 1 = 0.$
7. $x + y - 4 = 0; x + y + 4 = 0.$ 8. $5x - 8y + 21 = 0, 5x - 8y - 68 = 0.$
9. 3 units.   10. $(2, 2), x + y - 4 = 0.$ 11. (i) $\frac{3}{4}; -\frac{2}{11}.$
12. $\pm\frac{3}{4}.$

(B) Miscellaneous Examples. (p. 281)

1. $\dfrac{5}{\sqrt{13}}, \dfrac{12}{\sqrt{13}}$, opposite.   **2.** $1:5$.   **3.** $3x - y + 2 = 0$, $x + 3y + 4 = 0$.

4. $(3, 1)$.   **6.** $(-1, 5)$.   **7.** $3x - 4y + 7 = 0$, $(11, 10)$.

8. $\tan \theta = \frac{1}{2}$, $\tan \varphi = \frac{4}{3}$.

9. $y - 2 = m(x - 1)$; $P\left(\dfrac{2m - 9}{2m - 3}\right)$, $Q\left(\dfrac{m + 10}{m + 1}\right)$; $m = \frac{1}{2}$, $P(4, \frac{7}{2})$, $Q(7, 5)$.

10. $(-1, 4)$, $x^2 + y^2 + 2x - 8y + 4 = 0$.

11. $5(x^2 + y^2) - 20x + 10y + 12 = 0$.

12. $x^2 + y^2 - 6x - 2y + 9 = 0$, $x - 4 = 0$.

13. $(13, 4)$.   **14.** $\pm\frac{4}{5}$, $77°\ 20'$.   **15.** $x^2 + y^2 - 2x - 4y - 5 = 0$.

16. Centre $(\frac{2}{5}, \frac{11}{5})$, radius 1.   **17.** $(\frac{19}{5}, \frac{7}{5})$.

18. $x - 7y + 81 = 0$, $x + 7y + 11 = 0$; $16°\ 16'$.   **19.** Radii 9, 7; $\sqrt{130}$.

20. $x^2 + y^2 - 10x - 8y + 16 = 0$; $(8, 8)$.   **21.** $x = 0$, $y = 0$.

22. $x^2 + y^2 - 2x - 2y + 1 = 0$, $x^2 + y^2 - 10x - 10y + 25 = 0$; $x + y - 3 = 0$.

23. $x^2 + y^2 - 5x - 4y + 4 = 0$.   **24.** $21y + 20x = 64$.

25. $x^2 + y^2 - 2x - 6y = 0$, $x^2 + y^2 - 12x - 16y + 80 = 0$; $x + y - 8 = 0$.

26. $25(x^2 + y^2) - 50y + 9 = 0$, $25(x^2 + y^2) - 150x - 100y + 244 = 0$.

28. $x^2 + y^2 - 28x - 10y + 196 = 0$.

(A) Examples 20a. (p. 286)

1. $x^2 + y^2 = 4$.   **2.** $y = \pm 3$.   **3.** $y = x$.   **4.** $x + y = 0$.

5. $x^2 + y^2 - 2x - 2y + 1 = 0$.   **6.** $5x + 3y - 10 = 0$.   **7.** $x = \pm 2y$.

8. $y^2 = 4(x + 1)$.   **9.** $x^2 + y^2 - 2x = 0$.   **10.** $3(x^2 + y^2) - 2y - 22 = 0$.

11. $y = 2x$.   **12.** $2x + y - 2 = 0$.   **13.** $7x + 4y - 25 = 0$.

14. $2y^2 - x - 4y + 2 = 0$.   **15.** $x = 1$.   **16.** $xy = 16$.   **17.** $y = x^2 + 2x$.

18. $xy - y - 2 = 0$.   **19.** $y^3 = x^2$.   **20.** $xy + x + y - 1 = 0$.

21. $3(x^2 + y^2) - 32x + 64 = 0$.   **22.** $5x^2 = 32y$.

(A) Examples 20b. (p. 288)

1. $(15, 4)$.   **2.** $(-2, 0)$.   **3.** $(0, 2)$.   **4.** $(27, 9)$.

5. $(0, 10)$.   **6.** $(\frac{3}{2}\sqrt{2}, \frac{1}{2}\sqrt{2})$.   **7.** $t = 2, (7, 0)$.   **8.** $(0, \frac{3}{4})$.

9. $(-6, 2), (0, 2)$. **10.** 4 units.   **11.** $(-17, 6)$.   **13.** 2.

14. -1.   **15.** $\frac{1}{2}$.   **16.** $-\frac{1}{2}$.   **17.** 0.   **18.** 1.   **19.** $y - mx + 2m^2 = 0$.

20. $py - p^3x + 2(p^4 - 1) = 0$.   **21.** $qx - y = q^4$.

(B) Examples 20c. (p. 290)

1. $y^2 = 4ax$.   **2.** $21(x^2 + y^2) + 174x + 189 = 0$.

3. $3x^2 - y^2 + 4x - 4 = 0$.   **4.** $x^2 + y^2 + 2x - 4y + 3 = 0$.

5. $2(x^2 + y^2) - 2x + 3y - 4 = 0$ where $y > 0$.

6. $x^2 - 2xy + y^2 - 2x - 2y = 0$.   **7.** $x^2 + 2xy + y^2 - x - 2y = 0$.

8. $x^2 - y^2 = 4$.   **9.** $x + y - 1 = 0$.

10. $4x^2 + 9y^2 = 36$.   **11.** $x^2 + 2y - 2 = 0$.

12. $py - x = ap^2$; $qy - x = aq^2$ **13.** $\frac{1}{3}\sqrt{3}$

14. $x - 2y - 6 = 0$; $t = 5$.   **15.** $x^2 - 4x - 2y + 5 = 0$.

16. $2x^2 - 3x + y = 0$.   **17.** $x^2 + y^2 = 14$; $\sqrt{14}$.

18. $\left(3 + \dfrac{1}{2t}, \, 2t\right)$; $xy - 3y - 1 = 0$.   **21.** $(\frac{1}{2}ap^2, \frac{3}{2}ap)$; $2y^2 = 9ax$.

22. $(p^2 + pq + q^2, \, pq(p + q))$; $pq = -1$.   **24.** $x + 2y = 1$.

25. $(y - 1) = m(x - 1)$.   **26.** $y + px = 4p + 2p^3$; $y^2 = 2x - 12$.

27. $y = 2a$.   **28.** $t^2 y + x = 2t$.

(A) Examples 21a. (p. 294)

1. 12.   **2.** 20.   **3.** 168.   **4.** 27.

5. 3; 6.   **6.** 12.   **7.** 30.   **8.** 180.

9. 30.   **10.** (i) 20; (ii) 25.   **11.** 6.   **12.** 1540.

13. (i) 380; (ii) 400.   **14.** 15; 30.   **15.** 24.   **16.** 6.   **17.** 9.

(A) Examples 21b. (p. 297)

1. 6.   **2.** 20.   **3.** 6.   **4.** (i) 120;

(ii) 30; (iii) 120; (iv) 24.   **5.** 720.   **6.** 24.

7. 40320.   **8.** 1680.   **9.** (i) 60; (ii) 120; (iii) 120.

10. 720.   **11.** 60.   **12.** 90.   **13.** 10080.

14. (i) 24; (ii) 120; 96.   **15.** 4896.   **16.** 40320; 1680.

17. 1296.   **18.** 40320; 576.   **19.** 4320.

20. 720; 720; 518400.   **21.** 120; 24.

(A) Examples 21c. (p. 299)

1. (i) 20; (ii) 15; (iii) 70; (iv) 1680.   **2.** 12.

3. 20.   **4.** 1260.   **5.** 90.   **6.** 1680.

7. 70.   **8.** 1260.   **9.** 180.   **10.** 210.

11. 90.   **12.** 9979200.   **13.** 5040.   **14.** 3003.

15. 11.   **16.** 129729600.

(A) Examples 21d. (p. 301)

1. 10. **2.** 10. **3.** 6. **4.** 35.
5. 210. **6.** 56. **7.** 56. **8.** 495.
9. 5005. **10.** 1287. **11.** (i) 3; (ii) 10; (iii) 35; (iv) 1.
12. 10. **13.** 495. **14.** 840. **15.** 56.
16. 1585584. **17.** (i) 184756; (ii) 45045; (iii) 6300.
18. 6435; 56; 360360.

(B) Examples 21e. (p. 303)

1. (i) 8; (ii) 6. **2.** (i) 1440; (ii) 240. **3.** 36.
4. 9856. **5.** (i) 48; (ii) 72. **6.** (i) 4; (ii) 6; (iii) 4; 14.
7. 12, 2. **8.** 144. **9.** 166320. **10.** 120, 96.
11. 4084080. **12.** 192. **13.** 55. **14.** (i) 20; (ii) 70.
15. (i) 360; (ii) 900. **16.** (i) 24; (ii) 52. **17.** 90000, 18000.
18. (i) 720; (ii) 360; (iii) 192. **19.** 3360; 200.
20. (i) 24; (ii) 48. **21.** 1812. **22.** 4464.

(A) Examples 21f. (p. 305)

1. $\frac{1}{2}$. **2.** $\frac{1}{3}$. **3.** $\frac{1}{10}$. **4.** $\frac{1}{13}$. **5.** $\frac{3}{13}$.
6. $\frac{2}{5}$. **7.** $\frac{1}{15}$. **8.** $\frac{1}{6}$. **9.** $\frac{1}{4}$. **10.** $\frac{2}{7}$.
11. $\frac{5}{13}$. **12.** $\frac{1}{28}$. **13.** $\frac{2}{5}$. **14.** $\frac{1}{80}$.
15. (i) $\frac{1}{36}$; (ii) $\frac{1}{18}$. **16.** $\frac{3}{20}$.

(B) Examples 21g. (p. 307)

1. $\frac{1}{8}$. **2.** (i) $\frac{1}{11}$; (ii) $\frac{1}{55}$. **3.** $\frac{1}{221}$. **4.** $\frac{2}{5}$.
5. $\frac{1}{12}$. **6.** $\frac{3}{95}$. **7.** $\frac{5}{48}$. **8.** $\frac{27}{512}$; $\frac{9}{56}$.
9. $\frac{9}{64}$. **10.** $\frac{1}{3}$. **11.** $\frac{1}{8}$; $\frac{3}{8}$. **12.** $\frac{2}{5}$.
13. $\frac{857}{1105}$. **14.** $\frac{13}{34}$. **15.** $\frac{1}{15}$. **16.** (i) $\frac{16}{81}$; (ii) $\frac{32}{81}$.

(A) Examples 22a. (p. 310)

1. (i) 1; (ii) $\frac{1}{6}$; (iii) 16; (iv) $\frac{1}{27}$; (v) 9.
2. (i) a; (ii) $a^{\frac{19}{12}}$; (iii) $\sqrt{a}$; (iv) $a^{\frac{2}{3}}b^{\frac{1}{3}}$.
3. (i) 2^{11}; (ii) 2^{2a-b}; (iii) 2^{13}; (iv) 2^{x+3y}; (v) 2^{4a-3b}; (vi) $2^{\frac{17}{12}}$.

4. (i) $x = -4$; (ii) $x = 0$; (iii) $x = -4$; (iv) $x = \frac{3}{2}$; (v) $x = \frac{5}{3}$.

5. (i) $(2^x)^2$, (ii) $(2^x)^2$; (iii) $2 \cdot 2^x$; (iv) $\sqrt{2} \cdot (2^x)^2$.

6. (i) 2; (ii) 3; (iii) 2.   **7.** (i) $a^{\frac{3}{4}} + 2a^{\frac{2}{3}}b^{\frac{2}{3}} + b^{\frac{3}{4}}$; (ii) $a^{\frac{3}{4}} - b^{\frac{3}{4}}$.

8. (i) $a^{\frac{2}{3}}b^{\frac{1}{6}}$; (ii) $a^{\frac{3}{4}}$; (iii) $a^{2p}b^{p-1}$.

9. (i) 3; (ii) 5; (iii) 6; (iv) -1; (v) -4; (vi) $\frac{1}{2}$; (vii) $\frac{2}{3}$; (viii) $\frac{3}{2}$.

10. (i) 4; (ii) -1; (iii) $\frac{1}{2}$; (iv) $1 \cdot 5$.

11. (i) 4; (ii) 4; (iii) -1; (iv) $\frac{1}{2}$; (v) 0; (vi) -3; (vii) $\frac{3}{2}$.

12. (i) $n = 16$; (ii) $n = 25$; (iii) $n = \frac{1}{27}$; (iv) $n = \frac{1}{2}$; (v) $n = 1$.

13. $x = 4$.   **14.** (i) $1 \cdot 39794$; (ii) $0 \cdot 30103$; (iii) $1 \cdot 30103$.

15. (i) $1 \cdot 54407$; (ii) $1 \cdot 14613$; (iii) $\bar{1} \cdot 44716$ or $-0 \cdot 55284$.

16. 4; $\frac{1}{4}$.   **17.** (i) $\log_{10} 12$; (ii) $\log_3 6$; (iii) $\log_5 \frac{16}{27}$; (iv) $\log_3 8$; (v) $\log_5 225$; (vi) $\log_2 \frac{3}{343}$; (vii) $\log_2 10$; (viii) $\log_4 8$; (ix) $\log_{10} 200$.

18. (i) $x = 4$; (ii) $x = \frac{25}{4}$; (iii) $x = \frac{3}{2}$; (iv) $x = 5$.

19. (i) $2 \log a + 3 \log b + \log c$; (ii) $-\log a - \log b - \log c$; (iii) $\log a + \frac{1}{2} \log b + \frac{3}{2} \log c$.

20. (i) $1 \cdot 079181$; (ii) $1 \cdot 924270$; (iii) $-0 \cdot 892790$; (iv) $0 \cdot 158408$.

21. a, $\log_2 3$; 3.   **22.** (i) 2; (ii) 4; (iii) b.

(B) Examples 22b. (p. 312)

1. $\frac{3}{2}, \frac{2}{3}$.   **2.** $x^{3q-2p-r}$.   **3.** (i) $a + b$; (ii) $1\frac{1}{2}$.

4. (i) $1 \cdot 26$; (ii) $3 \cdot 32$; (iii) $2 \cdot 20$.   **5.** (i) 3; (ii) $\frac{1}{6}$; (iii) $3 \log_{10} 3$.

6. $\dfrac{2}{t}$.   **7.** $\frac{1}{2}(\log_2 x + \log_2 y)$.   **8.** (i) $1 + n$; (ii) $n - 1$; (iii) $\dfrac{1}{n-1}$.

10. $x = 9, \frac{1}{9}$.   **11.** (i) 9^u; (ii) $u + \frac{1}{2}$; (iii) $\dfrac{2}{u}$.   **12.** $x = 5$.

13. $x = 3$, $y = 1$.   **14.** $a = 4$, $b = 2$.   **15.** $x = 5$, $y = 2$.

16. $x = 125$, $\frac{1}{125}$.   **17.** $3\log_8 3$, $\frac{3}{2}\log_8 3$; 3^6.

18. $x = 16$, $y = 2^{\frac{3}{2}}$; $x = 4$, $y = 8$.

19. $x = \frac{1}{2}$.   **20.** $a = 3 \cdot 6, n = 1 \cdot 15$.   **21.** $a = 180, n = -\frac{2}{3}$.

22. $a = 4 \cdot 1, n = 2 \cdot 45$.   **23.** $a = 5 \cdot 7, n = 2 \cdot 7$.   **24.** $a = 1 \cdot 5, n = 1 \cdot 5$.

(A) Examples 22c. (p. 318)

1. (i) $4 \cdot 482$; (ii) $1 \cdot 649$; (iii) $0 \cdot 6703$; (iv) $20 \cdot 08$; (v) $1 \cdot 284$.

2. (i) 1; (ii) 0; (iii) $1 \cdot 54$; (iv) $0 \cdot 521$.

3. (i) $0 \cdot 6931$; (ii) $1 \cdot 504$; (iii) $3 \cdot 219$; (iv) $-0 \cdot 9163$.

5. (i) 2; (ii) $\frac{1}{2}$; (iii) -1; (iv) $\frac{3}{2}$.

6. (i) $x = \log 2$; (ii) $x = 0$; (iii) $x = \frac{1}{5} \log 2$; (iv) $x = \frac{1}{10} \log 2$.

7. $\log 2$; 2.   **8.** (i) 3; (ii) x; (iii) 4; (iv) $\sqrt{x}$; (v) $\dfrac{1}{x}$.

9. (i) $\log \dfrac{x^2(x+1)}{(1-x)^3}$; (ii) $\log(x+1)$.

10. (i) $x = e^3$; (ii) $x = e^{-2}$; (iii) $x = 2e$; (iv) $x = e$.

11. (i) $x = \frac{1}{2}$; (ii) $x = -1$; (iii) $x = 1$.

12. (i) $1\cdot58(5)$; (ii) $2\cdot20$; (iii) $0\cdot646$; (iv) $1\cdot63$.

(B) Examples 22d. (p. 318)

1. $t = -9\cdot16$.   **2.** $t = 0\cdot811$.   **3.** $t = 0\cdot520$.

4. (i) n_0; (ii) $t = 0\cdot139$.   **5.** $\frac{1}{14}(m+2n)$.

6. (i) $x = 8\cdot63$; (ii) $x = 0\cdot642$; $x = 0\cdot557$.   **7.** $x = 1, 1\cdot585$.

9. $1\cdot25$ g.   **10.** $x = -1, 2$.   **12.** (i) $x = 4, y = 3$; (ii) $x = 5, y = 2$.

13. $x = 1\cdot763$.   **14.** $x = 1\cdot818$.   **16.** $x = 2\cdot39, y = 2\cdot06$.

(A) Examples 22e. (p. 322)

1. $\dfrac{1}{x}$.   **2.** $\dfrac{1}{x+2}$.   **3.** $\dfrac{3}{3x+1}$.

4. $\dfrac{2}{2x-5}$.   **5.** $-\dfrac{1}{1-x}$.   **6.** $-\dfrac{3}{4-3x}$.

7. $\dfrac{2x}{x^2+1}$.   **8.** $\dfrac{2x-1}{x^2-x+1}$.   **9.** $-\dfrac{2x}{3-x^2}$.

10. $-\dfrac{1}{x}$.   **11.** $\dfrac{3x^2}{x^3+2}$.   **12.** $-\dfrac{1}{x+1}$.

13. $3e^x$ **14.** $3e^{3x}$.   **15.** $-3e^{-3x}$.

16. $4e^{2x}$.   **17.** $-4e^{-x}$.   **18.** e^{x+1}.

19. $4e^{(4x-1)}$.   **20.** $-6e^{(1-2x)}$.   **21.** $4e^{-4x}$.

22. $-2e^{-2x}$.   **23.** $-e^{(1-x)}$.   **24.** $-\frac{9}{2}e^{-3x}$.

25. (i) $\frac{1}{2}\log x + c$;   (ii) $\frac{4}{3}\log x + c$;   (iii) $\log(x-1)+c$;

(iv) $\log(x+4)+c$;   (v) $\log(x-7)+c$;   (vi) $\frac{1}{2}\log(2x+1)+c$;

(vii) $2\log(3+x)+c$; (viii) $-\log(4-x)+c$;   (ix) $\frac{1}{2}\log(x+1)+c$;

(x) $-\log(1-3x)+c$; (xi) $\frac{1}{4}\log(4x+3)+c$; (xii) $\frac{1}{3}\log(2x-5)+c$.

26. (i) $2e^x+c$;   (ii) $\frac{1}{2}e^x+c$;   (iii) $\frac{1}{2}e^{2x}+c$;

(iv) $-2e^{-x}+c$;   (v) $e^{4x}+c$;   (vi) $-\frac{1}{2}e^{-2x}+c$;

(vii) $\frac{1}{2}e^{6x}+c$;   (viii) $e^{(x-2)}+c$;   (ix) $-e^{(1-x)}+c$;

(x) $\frac{1}{2}e^{(2x+3)}+c$;   (xi) $-\frac{1}{2}e^{-2x}+c$; (xii) $e^{(x-1)}+c$.

27. $\log 4$. **28.** $\log 4$. **29.** $\frac{1}{2}\log\frac{5}{3}$.

30. $\frac{1}{2}(e^2-1)$. **31.** e^2-e. **32.** $\dfrac{1}{2}\left(1-\dfrac{1}{e^4}\right)$.

33. $\log\frac{3}{2}$. **34.** $\log 2$. **35.** $\frac{2}{3}\log\frac{7}{4}$.

36. $\frac{1}{2}(e^3-1)$. **37.** $\dfrac{1}{2}\left(e-\dfrac{1}{e}\right)$. **38.** $\dfrac{1}{4}\left(e+\dfrac{1}{e}-2\right)$.

(B) Examples 22f. (p. 323)

1. $x^2(3\log x+1)$. **2.** $e^{3x}(1+3x)$. **3.** $-\tan x$.

4. $e^x(\sin x+\cos x)$. **5.** $6e^{-2x}(x-x^2)$. **6.** $e^{\sin x}\cos x$.

7. $e^{-x}(1-x)$. **8.** $\dfrac{2}{x^2+1}-\dfrac{\log(x^2+1)}{x^2}$.

9. $-\dfrac{e^x}{(e^x+1)^2}$. **10.** $\dfrac{1}{x(x+1)}$. **11.** $e^{-x^2}(1-2x^2)$.

12. $\dfrac{x}{x^2+1}$. **13.** 1. **15.** $\log 3$.

16. Minimum $(e^{-\frac{1}{2}}, -\frac{1}{2}e^{-1})$. **17.** 4. **19.** $-e^{-1}$.

20. $\log 3$. **21.** $4-5\log 3$. **22.** (i) $1-\log 2$; (ii) $\log 2-\frac{1}{2}$.

23. $\frac{1}{2}\pi(e^2-1)$. **24.** (i) $\frac{1}{2}e^2+2e-\frac{3}{2}$; (ii) $e^2-e^{-2}+4$.

(A) Examples 23a. (p. 327)

1. 5; (ii) 19; (iii) -9. **2.** $c=2$. **3.** -5.

4. 2. **5.** 14. **6.** -3. **7.** 9.

8. 101. **9.** 3. **10.** 0. **11.** -13.

12. $4x+5$. **13.** $k=0$. **14.** x^2+2x+4. **15.** 2.

16. 1, -2. **18.** -3. **19.** -3. **20.** $3x+1$.

21. $a=2$, $b=1$; $(x-1)(2x^2+3x-1)$. **24.** $a=-6$, $b=8$.

25. $a=1$, $b=-5$. **26.** n even.

(A) Examples 23b. (p. 329)

1. (i) no simpler form; (ii) $\sqrt{6}$; (iii) $\sqrt{5}$; (iv) $\sqrt{2}$; (v) $6\sqrt{2}$; (vi) 2; (vii) 12; (viii) $6\sqrt{10}$; (ix) no simpler form; (x) $2\sqrt{6}$; (xi) $12\sqrt{5}$; (xii) $8\sqrt{21}$.

2. (i) $6-3\sqrt{3}$; (ii) $\sqrt{2}+\sqrt{6}$; (iii) 20; (iv) $3+2\sqrt{2}$; (v) 1; (vi) $10-2\sqrt{5}$; (vii) 13; (viii) $43-30\sqrt{2}$; (ix) -2; (x) $1-\sqrt{2}$; (xi) 15.

3. (i) $2\sqrt{2}$; (ii) $3\sqrt{2}$; (iii) $5\sqrt{2}$; (iv) $3\sqrt{5}$; (v) $4\sqrt{2}$; (vi) $7\sqrt{2}$;
(vii) $2\sqrt{6}$; (viii) $6\sqrt{2}$; (ix) $4\sqrt{3}$; (x) $12\sqrt{6}$.

4. (i) $\sqrt{2}$; (ii) $\sqrt{5}$; (iii) $16\sqrt{3}$.

5. (i) $\dfrac{\sqrt{2}}{2}$; (ii) $\dfrac{2\sqrt{5}}{5}$; (iii) $\dfrac{\sqrt{7}}{14}$; (iv) $\dfrac{\sqrt{6}}{12}$; (v) $\dfrac{2\sqrt{15}}{9}$; (vi) $\dfrac{\sqrt{2}}{3}$.

6. (i) $\sqrt{2}-1$; (ii) $\sqrt{2}+1$; (iii) $2+\sqrt{3}$; (iv) $\dfrac{\sqrt{5}+\sqrt{2}}{3}$; (v) $3+2\sqrt{2}$;
(vi) $3\sqrt{2}-4$; (vii) $\frac{1}{2}\sqrt{3}(2-\sqrt{3})$; (viii) $\frac{2}{7}\sqrt{2}(5-3\sqrt{2})$.

7. (i) $7-4\sqrt{3}$; (ii) $7-3\sqrt{5}$; (iii) $\frac{7}{3}+\frac{4}{3}\sqrt{3}$; (iv) $5+\frac{9}{4}\sqrt{5}$.

8. $\frac{8}{9}$.   **9.** (i) $\frac{7}{9}-\frac{2}{9}\sqrt{10}$; (ii) 1.   **10.** 322.   **11.** $5+2\sqrt{7}$.

12. (i) $8+4\sqrt{3}$; (ii) $\frac{1}{4}(3\sqrt{5}-1)$; (iii) $11+7\sqrt{3}$.

14. $\frac{13}{8}+\frac{5}{8}\sqrt{5}$.

(A) Examples 23c. (p. 333)

1. (i) 3; (ii) 5; (iii) $2r+3$.   **2.** (i) 2; (ii) 4; (iii) 2^{r+1}.

3. (i) 2; (ii) 2; (iii) 2.   **4.** (i) 2; (ii) 6; (iii) $(r+1)(r+2)$.

5. (i) 2; (ii) 12; (iii) $(r+1)^2(r+2)$.   **6.** (i) 1; (ii) 3; (iii) 3^r.

7. (i) $\frac{1}{3}$; (ii) $\frac{1}{8}$; (iii) $\dfrac{1}{(r+1)(r+3)}$.   **8.** (i) $\frac{1}{8}$; (ii) $\frac{1}{20}$; (iii) $\dfrac{1}{2(r+1)(r+4)}$.

9. (i) $\frac{1}{6}$; (ii) $\frac{1}{24}$; (iii) $\dfrac{1}{(r+1)(r+2)(r+3)}$.

10. (i) 1; (ii) 1; (iii) 1; (iv) $2n$.   **11.** (i) 3; (ii) 2; (iii) 2; (iv) $4n+1$.

12. (i) 1; (ii) 3; (iii) $2r-1$; (iv) $4n^2$.

13. (i) 2; (ii) 4; (iii) $2r$; (iv) $4n^2+2n$.

14. (i) 5; (ii) 9; (iii) $4r+1$; (iv) $8n^2+6n$.

15. (i) 1; (ii) 2; (iii) 2^{r-1}; (iv) $2^{2n}-1$.

16. (i) 2; (ii) 6; (iii) $r(r+1)$; (iv) $\frac{2}{3}n(2n+1)(2n+2)$.

17. (i) $\frac{1}{2}$; (ii) $\frac{1}{6}$; (iii) $\dfrac{1}{r(r+1)}$; (iv) $\dfrac{2n}{2n+1}$.

18. (i) $\frac{2}{15}$; (ii) $\frac{2}{35}$; (iii) $\dfrac{2}{(2r+1)(2r+3)}$; (iv) $\dfrac{1}{3}-\dfrac{1}{4n+3}$.

19. (i) $2+4+6$; (ii) 12.   **20.** $1+4+9$; (ii) 100.

21. (i) $2+2+2$; 2.   **22.** (i) $\frac{1}{2}+\frac{1}{6}+\frac{1}{12}$; (ii) $\dfrac{1}{n(n+1)}$.

23. (i) $3+10+21$; $2(n+1)^2+n+1$.

24. (i) $1+8+27$; (ii) $8n^3$.   **25.** r; $\displaystyle\sum_1^{30} r$.

26. $(2r-1)$; $\displaystyle\sum_1^{26} (2r-1)$.   **27.** $2r$; $\displaystyle\sum_1^{50} 2r$.

28. $(2r-1)^2$; $\displaystyle\sum_1^n (2r-1)^2$.   **29.** $r(r+1)$; $\displaystyle\sum_1^n r(r+1)$.

30. $\dfrac{1}{(r+1)(r+3)}$; $\displaystyle\sum_1^n \dfrac{1}{(r+1)(r+3)}$.   **31.** (i) 80;   (ii) 272;   (iii) 300.

32. (i) 2046;   (ii) $1-(\tfrac{1}{2})^{15}$;   (iii) $20+\tfrac{3}{2}(3^{20}-1)$.

33. $\displaystyle\sum_1^n (r+2^r)$; $\tfrac{1}{2}n(n+1)+2(2^n-1)$.   **34.** (i) 2^n-1;   (ii) $n+2(1-(\tfrac{2}{3})^n)$.

35. $\dfrac{2}{(2n-1)(2n+1)}$; $\dfrac{2n}{(2n+1)(4n+1)}$.

(A) Examples 23e. (p. 339)

1. $x>3$.   **2.** $x<-4$.   **3.** $x<4$.   **4.** $x>4$.

5. $x<3$.   **6.** $x\geqslant 3$.   **7.** $x>-6$.   **8.** $x>6$.

9. $x\leqslant -3$.   **10.** $x>\tfrac{5}{4}$.   **11.** $x>-1$.   **12.** $x<\tfrac{3}{2}$.

13. $x\geqslant \tfrac{1}{5}$.   **14.** $x<8$.   **15.** $x\leqslant \tfrac{20}{11}$.   **16.** $0<x<2$.

17. $0\leqslant x\leqslant 3$.   **18.** $-\tfrac{5}{2}<x<0$.   **19.** $x\leqslant -1$, $x\geqslant \tfrac{1}{2}$.

20. $x\leqslant -4$, $x\geqslant 3$.   **21.** $x<-3$, $x>3$.   **22.** $-\tfrac{4}{3}\leqslant x\leqslant \tfrac{4}{3}$.

23. $-1<x<2$.   **24.** $x\leqslant -2$, $x\geqslant -1$.   **25.** $x<-\tfrac{1}{2}$, $x>2$.

26. $-2<x<3$.   **27.** $x\leqslant -1$, $x\geqslant 5$.   **28.** $-5\leqslant x\leqslant -1$.

29. All real values of x.   **30.** All real values of x.

31. (i) $t\leqslant \tfrac{1}{2}$, $t\geqslant 1$;   (ii) $-\tfrac{1}{2}\leqslant t\leqslant 2$.   **32.** $3<m<4$.

(B) Examples 23f. (p. 344)

1. (i) $x=2$, $y=1$;   (ii) $(1,0)$, $(0,\tfrac{1}{2})$.   **2.** (i) $x=1$, $y=1$;   (ii) $(0,0)$.

3. (i) $x=-2$, $y=-1$;   (ii) $(2,0)$, $(0,1)$.   **4.** (i) $x=-1$, $y=0$;   (ii) $(0,1)$.

5. (i) $x=2$, $y=0$;   (ii) $(0,1)$.   **6.** (i) $x=0$, $y=0$;   (ii) none.

7. (i) $x=0$, $y=2$;   (ii) $(\tfrac{1}{2},0)$.   **8.** (i) $x=0$, $y=1$;   (ii) $(2,0)$.

9. (i) $x=-1$, $y=0$;   (ii) $(0,2)$.   **10.** (i) $x=-1$, $y=3$;   (ii) $(0,0)$.

11. (i) $x=0$, $y=-3$;   (ii) $(\tfrac{1}{3},0)$.

12. (i) $x=-2$, $y=-3$;   (ii) $(\tfrac{2}{3},0)$, $(0,1)$.

13. (i) $x=\tfrac{1}{2}$, $y=2$;   (ii) $(\tfrac{3}{4},0)$, $(0,3)$.

14. (i) $x=\tfrac{1}{2}$, $y=-1$;   (ii) $(\tfrac{5}{2},0)$, $(0,-5)$.

15. $x=-\tfrac{5}{2}$, $y=-\tfrac{1}{2}$;   (ii) $(4,0)$, $(0,\tfrac{4}{5})$.

16. (i) -2 to $+2$;   (ii) 0 to 4;   (iii) -8 to 0;   (iv) -2 to $+2$;   (v) $\tfrac{1}{3}$ to 3;   (vi) -4 to -1.

17. (i) -8 to $+4$; (ii) $x = -\frac{1}{2}$; (iii) $(-2, -8)$ Max., $(1, 4)$ Min..

19. (i) -5 to $+3$; (ii) $(-3, -5)$ Max., $(1, 3)$ Min.; (iii) $x = -1$.

20. (i) 0 to 4; (ii) $(1, 0)$ Max., $(3, 4)$ Min.; (iii) $x = 2$.

21. (i) -3 to $+9$; (ii) $(0, -3)$ Max., $(6, 9)$ Min.; (iii) $x = 3$.

22. (i) 2 to 6; (ii) $(3, 2)$ Max., $(5, 6)$ Min.; (iii) $x = 4$.

23. (i) -4 to $+4$; (ii) $(-2, -4)$ Max., $(2, 4)$ Min.; (iii) $x = 0$.

24. (i) -8 to $+16$; (ii) $(-\frac{1}{2}, -8)$ Max., $(\frac{5}{2}, 16)$ Min.; (iii) $x = 1.$

(B) Miscellaneous Examples (p. 345)

1. $p = 10$, $q = -90$; $(x + 1)(x - 3)(x + 3)(x + 9)$.   **2.** 1364.

3. $\frac{1}{4}(\frac{3}{4})^{r-1}$; $\frac{1}{4}, \frac{3}{4}$.   **5.** $x > \frac{7}{3}$.   **6.** $a = 4$, $b = -16$; 12.

7. (i) $\pm(1 + \sqrt{2})$; (ii) $\pm(2 + \sqrt{3})$.   **9.** (i) $x > 2$, $x < -1$; (ii) $x < -\frac{1}{3}$, $x > 5$.

10. $7 - 4\sqrt{3}$.   **11.** $k = 5$.   **12.** $x = -1$, $y = 1$; $-\frac{2}{3} \leqslant x \leqslant 0$.

14. (i) 63; (ii) $8 \cdot 7$.   **15.** $-\frac{1}{2}$ to $+\infty$.   **17.** $A = 3$.

21. $-(b - c)(c - a)(a - b)$.   **22.** $\frac{3}{2}$ unit^2.   **23.** $\frac{1}{4}(2 + \sqrt{2} - \sqrt{6})$.

25. $x = -3 \cdot 7$.

Revision Papers V. (p. 348)

Paper A (1)

1. $x - 4y + 5 = 0$; $\dfrac{5}{\sqrt{17}}$.   **2.** $x^2 + y^2 + 6x - 9 = 0$.   **3.** 72.

4. (i) 3^{2a+b}; (ii) 2.   **5.** $a = 2$.

Paper A (2)

1. $x^2 + y^2 + 2x - 8y + 4 = 0$.

2. tangent $y + x + 1 = 0$; normal $y - x + 3 = 0$.

3. (i) $x = 6$; (ii) $3 \cdot 320$.   **4.** (i) 30; (ii) 400.

5. (i) $x < -4$; (ii) $0 \leqslant x \leqslant 4$.

Paper A (3)

1. $2 : 1$.   **2.** $x - 3y + 3 = 0$.   **3.** (i) 8568; (ii) 3360.

4. (i) 1; (ii) $-4 + 4\sqrt{3}$.   **5.** (i) $2, 12$; $4n^2 + 6n + 2$.

Paper A (4)

1. $\frac{1}{2}, -\frac{11}{2}$.   **2.** $y^2 - 2xy + 8 = 0$.   **3.** (i) $\frac{3}{2}$; (ii) $1 \cdot 661$.

4. $(x + 1)(x - 2)(x + 2)$.   **5.** (i) 1; (ii) $\frac{1}{2}$; $2(1 - (\frac{1}{2})^{10})$.

Paper A (5)

1. $45°$; $11y - 13x = 0$.   **2.** $x^2 + y^2 + 4x - 2y = 0$.

3. 12; 6.   **4.** (i) $\frac{1}{2}\sqrt{6}$; (ii) $\frac{1}{3}(4 + \sqrt{10})$.

Paper A (6)

1. $x^2 + y^2 - 4x + 6y - 3 = 0$.   **2.** $y^2 - 8x - 0$.

3. (i) $x - x^{-1}$; (ii) $x + 1 + x^{-1}$.   **4.** (i) 5; (ii) 9; (iii) $8n^2 + 6n$.

5. (i) all values; (ii) $-1 < x < \frac{3}{2}$.

Paper B (1)

1. $\left(\dfrac{3k+9}{1+k}, \dfrac{4k+1}{1+k}\right)$; $k = 2$; $(5, 3)$.

2. (a) $3x + 4y = 25$; (b) $2x^2 + 5y^2 = 10$.

3. (a) (i) 70; (ii) 252.   **4.** (i) $x = -0·47, 1·82$; (ii) $0 \leqslant k \leqslant 2·9$.

5. (ii) 11.

Paper B (2)

1. $x^2 + y^2 - 2x - 6y + 9 = 0$; $x^2 + y^2 - 6x - 4y + 9 = 0$.

2. $(\frac{1}{2}t, 2 - \frac{1}{4}t^2)$; $y = 2 - x^2$.   **3.** (a) (i) 6; (ii) $\frac{3}{2}$; (b) $x = \dfrac{1}{m} - 1$.

4. (b) $-1 < a < -\frac{1}{4}$.

Paper B (3)

1. $2y - 4x + 3a = 0$; $3y - x - 3a = 0$, $3y - x + 2a = 0$, $y + 3x - a = 0$, $y + 3x - 6a = 0$.

2. (i) 2 units; (ii) $k = -3$; $2y - 2x + 5 = 0$.

3. (a) 120; (i) 50; (ii) 110.

4. (i) 8; (ii) $x^2 - ax + a^2$; no.   **5.** (i) 34·3.

Paper B (4)

1. $x^2 + y^2 - 20x - 12y + 111 - 0$; $4x + 3y - 83 = 0$; $\frac{4}{3}$.

2. $x - 2y + 5 = 0$; $x^2 + y^2 - 6x - 3y + 5 = 0$; $(1, 3)$, $(3, 4)$.

3. (i) (a) $x = 1000$; (b) $y = 2·301$; (ii) $x = \frac{1}{2}, \frac{3}{2}$.

4. (i) 1239; (ii) $8 + 2^{11}$; (iii) $\dfrac{\sin \theta}{1 - \sin \theta}$.   **5.** (i) 6; (ii) $a = 3$; 1.

Paper B (5)

1. (i) $63° 26'$, $53° 6'$, $63° 26'$; (ii) $(2, 1)$.

2. $x - y + 8 = 0$, $x + y + 8 = 0$.

3. (a) 210; (b) (i) 180; (ii) 36. **4.** $k = 12$, $b = 2·5$.

5. (ii) $2(1 - (\frac{1}{2})^n) \log_2 x$.

Paper B (6)

1. (i) $x^2 + 4x - 6y + 13 = 0$; (ii) $4x - 3y - 1 = 0$.

2. $y - 1 = m(x - 3)$; $2xy - x - 3y = 0$.

3. (a) (i) $x = 400$; (ii) $y = 3$; (b) $y = 2·30$.

4. $m^2 + m(2 - a) + 1 = 0$; $a \le 0$, $a \ge 4$.

5. $1\frac{4}{5} < x < 2$.

INDEX